Alexandre Grothendieck:
A Mathematical Portrait

Grothendieck

Alexandre Grothendieck: A Mathematical Portrait

edited by
Leila Schneps
Institut de Mathématiques de Jussieu
Université Pierre et Marie Curie, Paris

International Press
www.intlpress.com

Alexandre Grothendieck: A Mathematical Portrait

Editor: Leila Schneps (Institut de Mathématiques de Jussieu, Université Pierre et Marie Curie, Paris)

ISBN: 978-1-57146-282-4

Printed in the United States of America.

Fifth printing, 2022

Front cover image: "Portrait of Daniel-Henry Kahnweiler",

The frontispiece depicts Grothendieck lecturing in the mid-1960s.
Illustration by Coralie Colmez.

Contents

Foreword

The present book is, at least in part, the outcome of a conference organized in the tumbledown little stone village of Peyresq in the French Alps, entitled *Alexandre Grothendieck: Biography, Mathematics, Philosophy*[1]. While there, a group discussion amongst the participants was devoted to the subject of how one could best write a book that would explain, to the generations of young people who didn't even begin to study mathematics until long after Grothendieck had totally disappeared from the scene, what the special and extraordinary nature of his contribution to mathematics really was.

Together, we conceived the idea of a collection of articles, each devoted to one or another aspect of Grothendieck's work. Far from the usual sort of math publication, however, these articles were not to concentrate on the actual mathematical content of Grothendieck's contribution—after all, thousands of pages already exist devoted to all aspects of his work, not least those written by his own hand—but on the features which constitute in some sense his personal mathematical signature. These identifying traits are very familiar to those who know Grothendieck's work: the search for maximum generality, the focus on the harmonious aspects of structure, the lack of interest in special cases, the transfer of attention from objects themselves to morphisms between them, and—perhaps most appealingly—Grothendieck's unique approach to difficulties that consisted in turning them, somehow, upside down, and making them into the actual central point and object of study, an attitude which has the power to subtly change them from annoying obstacles into valuable tools that actually help solve problems and prove theorems.

In order to really comprehend what Grothendieck brought to each subject that he worked in, it helps to have a relatively clear idea of the state of affairs before he came. Placed against their proper background, the fundamental simplicity and the extraordinary power of many of his ideas stand out clearly—unexpected shifts of focus, generalizations to precisely the situations that were classically avoided, new use of an approach or a technique familiar from another domain. Of course, Grothendieck also possessed tremendous technical prowess, not even to mention a capacity for work that led him to concentrate on mathematics for upward of sixteen hours a day in his prime, but those are not the elements that characterize the magic in his style. Rather, it was the absolute simplicity (in his own words, "nobody before me had stooped so low") and the total freshness and

[1]August 24-29, 2008, organized by myself together with Pierre Lochak and Winfried Scharlau

fearlessness of his vision, seemingly unaffected by long-established views and vantage points, that made Grothendieck who he was.

In the years following the Peyresq conference, some of the originally planned contributions to the book fell by the wayside, while some new people—Joe Diestel, David Mumford, Frans Oort, Yuri Manin—came on board to fill lacunae in the mathematical biography. While the end result is far from complete (how could it ever be?), the articles do cover the major aspects of Grothendieck's work from the topological vector spaces of his graduate student days in the early 1950s through the work on K-theory, schemes, fundamental groups and cohomology of his heyday, to the motives that were still no more than an elusive and enchanting concept governed by a "yoga" based on his powerful intuition in 1970, when, at the age of 42, he brutally left his job at the IHES south of Paris and ruptured with his mathematical companions. Written by people who knew him personally, some by mere acquaintance and others extremely well, some as students and others as colleagues or latecomers to the Grothendieckian scene, the articles in this book contain a wealth of personal memories, explanations and anecdotes about the effect that Grothendieck's personality, ideas and mathematics had on those who came near him.

Like a cubist work of art, the mathematical portrait of Grothendieck painted in this book is made up of a multiplicity of different planes and different angles. I hope that, taken as a whole, it will make it possible for those who never knew him to discern something of the inimitable character and style of the original.

Leila Schneps
Paris
September 2013

Grothendieck and Banach space theory

Joe Diestel

It's been over a half century since Grothendieck burst upon the functional analysis scene solving basic questions about Fréchet spaces and their inductive limits. He spent far less than a decade in the area, but changed forever the landscape of the subject. This note is devoted to just how he made such a profound difference. We restrict ourselves in the main to his contributions in and around Banach space theory, the aspect of functional analysis nearest to his mathematical heart-of-hearts.

A general comment seems à propos: while Grothendieck was a theory-builder par excellence, along the way to building structures he often solved problems, sometimes in startling fashion.

To be sure, this discussion will focus on five of his papers:

Sur les applications linéaires faiblement compactes d'espaces du type $C(K)$, Canad. J. Math. **5** (1953), 129-173.

Produits tensoriels topologiques et espaces nucléaires, Memoirs AMS **16** (1955).

Résumé de la théorie métrique des produits tensoriels topologiques, Bol. Soc. Math. São Paulo **8** (1953), 1-79 (1956).

Sur certaines classes de suites dans les espaces de Banach et le théorème de Dvoretzky-Rogers, Bol. Soc. Math. São Paulo **8** (1953), 81-110 (1956).

Une caractérisation vectorielle-métrique des espaces L^1, Canad. J. Math. **7** (1955), 552-561.

In his '53 Canadian Journal paper, he gives an early indication of his natural homological bent of mind. He defines classes of Banach spaces by properties of operators acting on the spaces to other Banach spaces - this is a first. For instance, he says a Banach space X has the **Dunford-Pettis**

Emeritus Professor, Department of Mathematical Sciences, Kent State University, Kent, OH 44242, USA. j_diestel@hotmail.com.

property if every weakly compact linear operator from X to any Banach space Y (one that takes bounded sets into weakly compact sets) is completely continuous (takes weakly convergent sequences to norm convergent ones); a classical theorem of Dunford and Pettis says that all L-spaces (throughout we will call any space that's isometrically isomorphic to a space of absolutely integrable functions with respect to a countably additive measure an L-space) have the property and Grothendieck goes on to show all $C(K)$-spaces do, too. In sum, he isolates three properties (the **Dunford-Pettis** property, the **reciprocal Dunford-Pettis** property and the **Dieudonné** property), each relating how operators act on the space; he is particularly interested in weakly compact operators. He provides deep insights into weakly compact subsets of measures and along the way makes a first installment to the solution of a problem of **Banach** and **Mazur**.

The problem of **Banach** and **Mazur** originated in the thirties as a result of their study of universal spaces in the category of Banach spaces. They showed that every Banach space is (isomorphic to) a closed linear subspace of a $C(K)$-space and a quotient of an L-space. They wondered what could be said about Banach spaces that are isomorphic to a subspace of an L-space and a quotient of a $C(K)$-space. Grothendieck notes in his '53 Canadian Journal paper that any such space must be reflexive. He relies on a classical result of **Steinhaus** to the effect that L-spaces are weakly sequentially complete and his own discovery that every operator from a $C(K)$-space to a weakly sequentially complete space is weakly compact; since a weakly compact quotient map can only have a reflexive range, *the only quotients of $C(K)$-spaces that reside as subspaces of an L-space must be reflexive.*

This groundbreaking paper continues to be a source of inspiration; a list of mathematicians who've improved on its results reads like a 'Who's Who' of modern functional analysis. Names like **Ando, Pełczyński, Rosenthal, Bourgain, Kisliakov** and **Talagrand** are among the luminaries who've looked to this paper for inspiration. To state just one result that highlights the prominence of the *ideas* found in the paper, we state a beautiful structure theorem of **Elton** and **Odell**: *every infinite dimensional Banach space contains a closed linear subspace that is either isomorphic to c_0, l^1, or fails the Dunford-Pettis property.*

Note: As mentioned earlier, Grothendieck's work served to solve problems already extant as well as create theories. We have started following his thoughts on the problem of Banach and Mazur, and will continue with such in our discussion of the *Résumé*, right through to his complete and surprising solution. Sometimes, he solved problems that were 'in the wind', so to speak, that were known but had not made it into print. We take special note of a result he published in the Canadian Journal in 1954 in which he showed that if E is a closed linear subspace of $L^p(m)$ ($p \geq 1$ and m a finite measure) that consists entirely of m-essentially bounded functions, then E is finite dimensional. This was apparently a problem being bandied about in harmonic analysis circles at the time, and asked of Grothendieck. It was an easy consequence of his '53 Canadian Journal paper. This is an unusual case; it was almost as though the mathematics was produced on the spot, just so the right question could be asked!

Undoubtedly, Grothendieck's most famous work in functional analysis is his *Memoir*. For many, the *Memoir* is known as the birthplace of *nuclear spaces*, a class of locally convex spaces that includes many of the most important examples of linaer spaces found in analysis that are not Banach spaces – it is a basic feature of this theory that any space that is simultaneously nuclear and normed is finite dimensional. However important the class of nuclear spaces might be, the *Memoir* contains much more of importance, tensor products of locally convex spaces is a main theme of the *Memoir* and along for the ride comes the beginnings of Grothendieck's work on tensor products of Banach spaces.

John von Neumann and several of his coworkers had broached the subject of tensor products, with special attention paid to the situation of Hilbert spaces. A roadblock to progress was encountered when trying to compute the dual of the so-called *injective* tensor product; here Grothendieck gives a startling and elegant proof identifying the dual, connecting the problem with measure theory. The idea is so simple (once conceived) that we feel we must present it – it is also the cornerstone of Grothendieck's later work, his *Résumé*.

Let X and Y be Banach spaces, and consider their tensor product, $X \otimes Y$. For a typical vector $u = x_1 \otimes y_1 + \cdots + x_m \otimes y_m$ in $X \otimes Y$, define the *injective norm* $||u||_\vee$ of u by

$$||u||_\vee = \sup\{|\sum_{k=1}^{m} x^*(x_k)y^*(y_k)|\},$$

where the supremum is taken over all x^*, y^* in the closed unit balls of X^* and Y^* respectively. With this norm, $X \otimes Y$ is a normed linear space; unless either X or Y is finite dimensional, it's unlikely to be complete, so complete it – the result is the *injective tensor product of* X *and* Y, $X \overset{\vee}{\otimes} Y$. The question is to identify the dual space of $X \overset{\vee}{\otimes} Y$. Keep in mind that the basic raison d'être for tensor products is that the dual of a tensor product is a space of bilinear functionals and so it's natural to look to the continuous bilinear functionals for members of the dual. Look carefully at the very definition of the norm $||u||_\vee$ for $u \in X \otimes Y$: realize that the norm of u is the supremum of $|(x^* \otimes y^*)(u)|$ as x^* ranges over the dual ball of X^* and y^* ranges over the dual ball of Y^*; recall that these dual balls have a natural topology, called the weak* topology, and in this topology these balls are compact;

a compact Hausdorff space makes its presence known, namely the product K of the balls in their weak* topologies; by all that's holy, $(X \otimes Y, || \cdot ||_\vee)$ embeds in a linear isometric fashion into $C(K)$. The dual of $(X \otimes Y, || \cdot ||_\vee)$ can, therefore, be identified with a family of regular Borel measures on K, thanks to the **F. Riesz** et al representation theorem. It was natural therefore for Grothendieck to call the associated bilinear functionals *integral*; indeed, a bilinear functional ϕ on the product of X and Y is integral precisely when there is a (regular Borel) measure m defined on K (as above), so that for any $x \in X$ and $y \in Y$, we have

$$\phi(x, y) = \int_K x^*(x)y^*(y)dm(x^*, y^*),$$

the norm of ϕ as a member of the dual of the injective tensor product of X and Y is the total variation of the measure m.

Now, what must be understood here is that Grothendieck used tensor products to study the operators between Banach spaces that arise from (continuous) bilinear functionals. So too it was with the *integral linear operators*: the operator u from the Banach space X to the Banach space Y is *integral* if the bilinear functional it induces on the product of X and Y^* is integral. It is a sign of the genius of the man that he recognized the central role to be played by these operators in organizing the structure theory of Banach spaces. As a hint at what could be done (in the right hands) with the integral operators, Grothendieck studied the *approximation property* and the *metric approximation property* for a Banach space: X has the *approximation property* if any operator u with domain X can be approximated uniformly on compact sets of X by finite rank operators; if one can further approximate by finite rank operators with norms no greater than that of u then we say X has the *metric approximation property.* In his *Memoir*, Grothendieck establishes a list of equivalences of this property, many of which were apparently known to the mathematicians of the Polish school gathered around **Banach**; he further recorded examples of classical Banach spaces which enjoyed these properties. It was in this study that he provided a few great surprises, surprises that resulted from his deep understanding of integral operators.

Here are a few such:

Integral operators with reflexive domains are nuclear (absolute sums of 1 dimensional operators).

If the domain of the integral operator has a separable dual then the operator is nuclear.

Consequently, a Banach space that is either reflexive or a separable dual and has the approximation property has the metric approximation property.

Note: There is an intermediate property to the metric approximation and the approximation properties, called the *bounded approximation property,* wherein one asks that there be a uniform bound to the finite rank approximants of the identity operator on the space. Grothendieck's result above can be marginally improved to say that a dual space with the approximation property and the *Radon-Nikodym property* has the bounded approximation property; **it is unknown to this day whether any dual space with the approximation property has the bounded approximation property**.

Another result from Part I of the *Memoir* that is well-known is Grothendieck's identification of the *projective tensor product* of any L-space (with respect to a measure m) and X with the space of Bochner m-integrable, X-valued functions. If X and Y are Banach spaces, then the *projective tensor norm* of a vector $u \in X \otimes Y$ is given by

$$||u||_\wedge = \inf\{\sum_{k=1}^{m} ||x_k|| \, ||y|| : u = x_1 \otimes y_1 + \cdots + x_m \otimes y_m\};$$

again, unless either X or Y is finite dimensional, it is unlikely that this norm on $X \otimes Y$ is complete, so we complete it and denote by $X \overset{\wedge}{\otimes} Y$ the resulting Banach space and call this space *the projective tensor product of X and Y*. The dual of the projective product is the linear space of all continuous bilinear functionals on the product of X and Y. The identification of $X \overset{\wedge}{\otimes} Y$ is unknown, untidy or unwieldy. (We will return to this topic later, where this initial example in which the projective product is known is highlighted.)

The *Memoir* actually has two parts. It seems that Part II is less studied than Part I. More's the shame. Part II is *mostly* about locally convex spaces, establishing as it does the basic theory of *nuclear spaces*; we make no comments about the importance of which work, and refer instead to the fascinating and informative book "*History of Banach Spaces and Linear Operators*" by **Albrecht Pietsch**, especially the discussion in and around pages 391-392, for an insightful description of what was done in the *Memoir* in this direction. Part II also has some basic contributions to Banach space theory, especially in the early pages. Here Grothendieck introduces the p-nuclear operators, proves some remarkable formulas for the composition of such operators, makes headway in the relationships between traces and determinants (which curiously went unnoticed for more than a decade) and proved substantial generalizations of some old results of **Hermann Weyl** on eigenvalue distributions of operators. A few words here about this last subject.

Early in Part II, Grothendieck introduces operators he calls *applications de puissance p-ème sommable*; these operators are now known (in Banach space circles) as the *p-nuclear operators*. Though he never said much about this, it seems clear that Grothendieck knew the relationship between these operators and the **Schatten-von Neumann** classes $\mathcal{S}_p$ of operators in case the operators were acting on a Hilbert space. In any case, he showed (assuming that $0 < p \leq 1$) that while Weyl's theorem (for a p-nuclear operator acting on a Hilbert space, the sequence of eigenvalues belongs to l^p) doesn't generalize to Banach spaces, all is not lost. Indeed, Grothendieck shows that for any (complex) Banach space X, if $u : X \to X$ is p-nuclear, then its sequence of eigenvalues belongs to l^r, where

$$\frac{1}{r} = \frac{1}{p} - \frac{1}{2};$$

moreover, in case $p = 1$, this is best possible.

Many abstract analysts are surprised to find that Grothendieck paid such careful attention to the eigenvalue distribution of compact operators, but a close look at his work in functional analysis should give broad hints that he cared for the most important mathematical matters regardless of their possible technical complexity.

We come now to Grothendieck's most lasting contribution to Banach space theory, his *Résumé*. In this remarkable document, Grothendieck lays out his view of Banach space structure theory. He deals only with Banach spaces, saying on page 1: "*En effet, la lecture de [4]* [the *Memoir*] *montrera que presque toutes les questions de la théorie générale, y comprise la théorie des espaces nucléaires, se ramènent en réalité à des questions sur les espaces de Banach.*" He uses tensor products as a foundation on which he builds classes of operators between Banach spaces; he views three classes of spaces as the basic building blocks of this theory: Hilbert spaces, $C(K)$-spaces and L-spaces. Reading the *Résumé* is a challenge.

The attentive reader needs patience and perseverence. The opening section is very dry, arid in fact. In it, he carefully outlines the fundamental features of *tensor norms*, broaching the natural duality between a tensor norm α and the α'-forms, those bilinear functionals that act continuously on the α-tensor product of two *finite dimensional Banach spaces*. The fact that he builds his theory around the behavior of tensor products of finite dimensional spaces is critical to the eventual success of this theory. Indeed, Grothendieck understood that many of the most critical structural properties of a Banach space rely on the positioning of its finite dimensional subspaces; the *local theory* of Banach spaces is born. What's more, he sees that the glue to piece together the finite dimensional spaces according to the dictates of a given structural question is the ideal of the operators of type α' corresponding to the α'-forms. *A telling point here is that he rushes to observe that a bilinear form ϕ on the product of two Banach spaces X and Y is an α'-form (with α'-functional norm no more than k) precisely when for any pair of finite dimensional subspaces E and F of X and Y respectively belonging to some cofinite family of finite dimensional subspaces, the restriction of ϕ to the product of E and F is an α'-form of α'-functional norm no more than k.*

The serious reader now needs courage.

In section 2 of the *Résumé*, the special roles of $C(K)$- and L-spaces start to be clarified. Here various *hulls* of tensor norms are defined: *left injective, right injective, injective, left projective, right projective* and *projective hulls* are defined. Indeed, the injective and projective norms are tensor norms with the injective norm the least and the projective norm the greatest of all tensor norms; it is so that the pointwise supremum and infimum of any family of tensor norms is also a tensor norm, and from this it's plain sailing to show the existence of the various hulls mentioned.

Along the way, the operators generated by these operations on tensor norms are characterized, and the magic of Grothendieck's genius starts making itself felt. He establishes *factorization schemes* for operators; while it *may* have occurred to earlier mathematicians to factor operators through more familiar spaces, Grothendieck showed how it could be done and, more to the point, how this meant something important about the structure of the Banach spaces under study. Section 2 gives promise that the rather arid landscape drawn in the first section can be reclaimed and is actually quite fertile, given the proper tender love and care.

In section 3, tensor norms related to Hilbertian affairs are studied, and among other delicacies the tensor norm h is defined with the unmentioned purpose that operators of type h will factor through Hilbert spaces. Grothendieck's goal of using Hilbert spaces, $C(K)$-spaces and L-spaces as building blocks for the structure theory of Banach spaces starts becoming viable.

For instance, in section 2 he derives two special *ideals* of operators from his tensor norms, operators that factor through a subspace of some L-space and those that factor through a quotient of some $C(K)$-space. He then takes note that if E is n-dimensional Euclidean space with inner product $(\cdot, \cdot)$ and one considers the rotation invariant probability on the unit sphere of E, then the map that takes a vector $x \in E$ into the function $(\cdot, x)$ in the space of integrable functions on the sphere is a linear isometry of E into this L-space. It follows that E, being finite dimensional, is isomorphic to its dual, a quotient of the dual of an L-space; but it's well-known that the dual of an L-space is a $C(K)$-space. Now for most, this would lead to the fact that E is isomorphic to a subspace of an L-space and a quotient of a $C(K)$-space, but Grothendieck had better eyes than most! His conclusion: **every** *Hilbert space is isomorphic to a subspace of an L-space and a quotient of a $C(K)$-space.*

When first reading the *Résumé*, this conclusion certainly left me baffled. For a while I thought maybe 'proof by example' was okay if Grothendieck did it; soon I realized that this was not at all the case. Rather, he had in mind the fact that the identity operator on any Hilbert space must belong to the aforementioned ideals, since the restriction of this identity to any finite dimensional subspace belongs to each, as proved. But the inclusion of the identity operator in each of these ideals of operators is tantamount to the

Hilbert space itself being both a subspace of an L-space and a quotient of a $C(K)$-space.

Yes, Grothendieck's hands were faster than the eyes could follow. Magic was in the air.

It's in section 3 that Grothendieck starts showing how his grand view of how Banach space theory should develop can also tell us new things about old friends. A very special class of operators on Hilbert spaces, the *Hilbert-Schmidt* class, is given special Banach space treatment. Using factorization techniques he shows, for example, that if u is a linear operator from $L^2(m)$ to itself, with m a finite measure, and if all u's values are essentially bounded, then u is a Hilbert-Schmidt operator (it's not particularly obvious that u is even compact)! Similarly, if u is a linear operator from l^2 to itself and all its values are actually in l^1, then u is Hilbert-Schmidt. Proving new results about old friends is a known mathematical technique for making new mathematical friends.

The final section of the *Résumé* contains probably the most profound Banach space result that Grothendieck discovered, his famous *fundamental inequality of the metric theory of tensor products.* This inequality bounds the *preintegrale* norm of the identity operator on any Hilbert space; the *preintegrale tensor norm* is the largest *injective* tensor norm. He proves that there is a constant K such that the preintegral tensor norm of the identity *on any Hilbert space* is no more than K. This constant has come to be known as **Grothendieck's constant**, and to date its exact value is unknown; it depends on the underlying scalar field and it is known that the constant in the complex case is strictly less than the real case and in each case exceeds 1.

Knowing such a constant exists, Grothendieck was able to discover a myriad of truths. Ever the great storyteller, he expressed some of his consequences in beguiling manner. Here's one of my favorites, the 'Six Theorem' (which was taught me by Dan Lewis): *Let u be a bounded linear operator between two Banach spaces and suppose u factorizes through* **6** *spaces U, V, W, X, Y and Z where each of these spaces is one of type $C(K)$-, L- or Hilbert-space, with successive factors never the same type and each type used twice; then u is nuclear.* Remarkable, possibly useless, but certainly entertaining. **Note:** If I were completely honest, I would mention that the wonderful 'Six Theorem' doesn't appear as such in the *Résumé*; rather, it was a way for Dan Lewis to keep my attention. It does follow from arguments like those in section 4.b of chapter 4 of the *Résumé.*

On a more serious note, he also solved the Banach-Mazur problem mentioned earlier. His conclusion: **if X is a Banach space which is simultaneously a subspace of an L-space and a quotient of a $C(K)$-space, then X is an isomorph of a Hilbert space.**

The *Résumé* is chock full of delicacies, too numerous to list here. To be sure, it remained more than a bit of a mystery for better than a decade, when **Joram Lindenstrauss** and **Aleksander Pełczyński** wrote their fundamental call to arms "*Absolutely summing operators in L^p-spaces and their applications*". Among other things, they give an alternative formulation of Grothendieck's fundamental theorem, one that's palatable to ordinary research mathematicians. They admit "*the theory of tensor products constructed in Grothendieck's paper has its intrinsic beauty", however warn that the paper is "quite hard to read and its results are not generally known even to experts in Banach space theory*". The citations for the Lindenstrauss-Pełczyński paper include papers by most of the leading abstract analysts of the past forty years. It was a monumental contribution if only because it led many to return to the *Résumé* for another (or perhaps a first) look at what all the excitement was about.

In a paper that appeared in the same issue of the Bulletin of the São Paulo Mathematical Society, immediately following the *Résumé*, Grothendieck gave his view of the Dvoretzky-Rogers theorem. Recall that in the *Scottish Book*, problem **122**, the question was asked whether or not in any infinite dimensional Banach space one could find a series that was unconditionally summable but not absolutely summable. In 1947, **Macphail** showed that such a series could be shown to exist in the space l^1 of all absolutely summable sequences of real numbers; incidentally, Macphail's proof involves an insightful use of the Rademacher functions and Khinchine's inequalities – it is Hilbertian in character. A couple of years later, **A. Dvoretzky** and **C.A. Rogers** solved the problem by showing that indeed, in every infinite dimensional Banach space, there is an unconditionally convergent series which is not absolutely convergent. Grothendieck held the Dvoretzky-Rogers work in the highest regard and was intensely interested in this theorem (and its proof). In his *Memoir*, he gave an operator-theoretic proof that revolved around what's now known as *absolutely summing operators*; now he turns his attention to a profound geometric lemma found in the Dvoretzky-Rogers paper, and extends their theorem in many directions. Building on the *Résumé*, he shows that if p is a real number > 1, then one can find in any infinite dimensional Banach space a series that is weakly p-summable but not absolutely p-summable.

In the last section of this paper, he expresses the belief that the Dvoretzky-Rogers lemma hints at a much deeper truth: maybe if X is any infinite dimensional Banach space then for any n and any $\epsilon > 0$, one can locate inside X an n-dimensional subspace that's an ϵ-perturbation of n-dimensional Euclidean space.

It's a wonderful tradition in functional analysis, started by the Polish school, to pose problems with an eye to challenging future generations.

Grothendieck continued this tradition. In his *Memoir*, Grothendieck wondered whether or not every Banach space has the approximation property (and even has therein the amazing "Proposition" 37). Here we must say that the approximation problem goes back at least as long ago as the Polish school around **Banach** and **Mazur**; indeed, problem **153** of the *Scottish Book*, a problem posed by **Mazur**, was well-known to be equivalent to the approximation problem.

In the *Résumé*, Grothendieck lists six classes of open problems; in this sequel to the *Résumé*, he asks about the possibility of improving the Dvoretzky-Rogers lemma (in quite a dramatic fashion). The progress of a discipline can be measured by how well succeeding generations answer the questions of their predecessors.

Remarkably enough, it was the question at the end of the sequel that was answered first; in 1960, **Dvoretzky** (who was, incidentally, the reviewer for Mathematical Reviews of both the Bol. Soc. Mat. São Paulo papers) showed that if X is any infinite dimensional Banach space, n is any positive integer and $\epsilon > 0$, then one can find in X a subspace of dimension n that is an ϵ-perturbation of a Euclidean space of dimension n. This stunning result is known as **Dvoretzky's Spherical Sections Theorem**, and remains one of the deepest results in all of functional analysis.

The last paper we want to discuss, albeit very briefly, is Grothendieck's '55 Canadian Journal paper. As mentioned earlier, he had identified the projective tensor product of any L-space with any Banach space X with the space of Bochner integrable X-valued functions. A consequence is that if X is a closed linear subspace of the Banach space Y, then for any L-space L, the projective tensor product $L \overset{\wedge}{\otimes} X$ is a closed linear subspace of the projective tensor product $L \overset{\wedge}{\otimes} Y$. **That's it:** If a Banach space Z has the property that $Z \overset{\wedge}{\otimes} X$ is a closed linear subspace of Y, the Z is an L-space. Along the way, he shows that any $C(K)$-space that is a dual space has (isometrically) a unique predual, and guesses that his proof can be modified to cover the non-commutative case, that is, if X is a von Neumann algebra, then need X have a unique predual?

Again, this paper had much in it that was unsaid. In particular, its main result, characterizing as it did L-spaces in a tensorial manner, also hinted that for the classical reflexive Lebesgue spaces, their vector-valued cousins did **not** behave in a tensor sensitive manner. This opens the very real possibility that many classically important operators (such as the *Riesz projection*, the *Hilbert transform*, the *Fourier transform*) might not act as one might naively believe in the vector-valued setting; this was indeed to be the case, and the understanding of the phenomena involved was among the greatest accomplishments in abstract analysis of the 70's and 80's.

Among the many contributors, we find such as **Bourgain, Burkholder, Kisliakov, König, Maurey** and **Pisier**.

What's the status of the problems posed by Grothendieck?

Rather than belabor the point, here are the names of those who have contributed to their solution (anyone who wants to see what came of challenging a latter day generation can take a longer look at some of the best that's been done): **Enflo** (showed there are Banach spaces without the approximation property) and **Szankowski** (who demonstrated that there is a classical space, the space of bounded linear operators on an infinite dimensional Hilbert space, that fails the property); **Gordon** and **Lewis** proved that the tensor norm $\mathbf{L}$ is not dominated by $\mathbf{L}'$ by showing that not every absolutely summing operator factors through an L-space, and **Pełczyński** showed that the absolutely summing natural inclusion of the disk algebra into the Hardy space H^1 does not so factor; to date, the exact values of Grothendieck's constants (real/complex) remain unknown, with best estimates for the real case due to **Krivine** and in the complex case to **Haagerup**; **Pisier** (again) and **Haagerup** extended Grothendieck's fundamental inequality to a C^*-algebra setting; **Pisier** (once more), **Kisliakov** and **Bourgain** showed that there are non-L-spaces X for which every operator from X to Hilbert space is absolutely summing; **Pisier** constructed a Banach space P such that the injective and projective tensor product of P with itself are the same; **Sakai** proved that every von Neumann algebra has a unique predual. To be sure, the list of those who've contributed mightily to the problems set by Grothendieck extends considerably beyond what's just been mentioned. Again, an excellent place to read about these developments and beyond is **Pietsch**'s *History*.

As is usual in such affairs, the solution of suitably formulated problems leads to new problems; so too is it with the questions asked by Grothendieck, and the references listed below contain more than enough to keep the coming generations challenged. The Grothendieck legacy in functional analysis is alive and well.

Note: In his own listing of when he wrote his papers, Grothendieck lists his '55 Canadian Journal paper as having been written before his *Résumé*. This is a curious happenstance. The result of this '55 paper is just the kind of result that he asks about in the problems of this remarkable document, yet he makes no mention of the result. To be sure, he does state many results from his *Memoir*, so calling on earlier work was not anathema to him.

Bibliography

R. Daniel Mauldin, editor, *The Scottish Book: Mathematics from the Scottish Cafe*, Birkhäuser (Boston, Basel, Suttgart), 1981.

S. Banach, S. Mazur, Zur Theorie der linearen Dimension, *Studia Math.* **4** (1933), 100-112.

J. Lindenstrauss, A. Pełczyński, Absolutely summing operators in L^p-spaces and their applications, *Studia Math.* **29** (1968), 275-326.

M.S. Macphail, Absolute and unconditional convergence, *Bull. AMS* **53** (1947), 121-123.

A. Dvoretzky, C.A. Rogers, Absolute and unconditional convergence in normed linear spaces, *Proc. Nat. Acad. Sci. (USA)* **38** (1950), 192-197.

A. Dvoretzky, Some results on convex bodies and Banach spaces, in *Proceedings of the International Symposium on Linear Spaces*, Jerusalem, 123-160.

P. Enflo, A counterexample to the approximation problem for Banach spaces, *Acta Math.* **158** (1973), 309-317.

A. Szankowski, $B(H)$ does not have the approximation property, *Israel J. Math* **30** (1981), 89-108.

Y. Gordon, D.R. Lewis, Absolutely summing operators and local unconditional structures, *Acta Math.* **133** (1974), 27-48.

A. Pełczyński, Sur certaines propriétés isomorphiques nouvelles des espaces de Banach de fonctions holomorphes A et H^∞, *C.R. Acad. Sci. Paris* **279:A**, 9-12.

J.L. Krivine, Constantes de Grothendieck et fonctions de type positif sur les sphères, *C.R. Acad. Sci. Paris* **284:A**, 377-379, 445-446.

U. Haagerup, A new upper bound for the complex Grothendieck constant, *Israel J. Math.* **60** (1987), 199-224.

G. Pisier, Grothendieck's theorem for non-commutative C^*-algebras, *J. Funct. Anal.* **29** (1978), 397-415.

G. Pisier, Une nouvelle classe d'espaces de Banach vérifiant le théorème de Grothendieck, *Ann. Inst. Fourier* **28** (1978), 69-90.

S.V. Kisliakov, Spaces with small annihilators, *J. Soviet Math.* **16** (1981), 1181-1184.

J. Bourgain, New Banach space properties of the disc algebra and H^∞, *Acta Math.* **152** (1984), 1-48.

J. Bourgain, W.J. Davis, Martingale transforms and complex uniform convexity, *Trans. AMS* **294** (1986), 501-515.

G. Pisier, Counterexamples to a conjecture of Grothendieck, *Acta Math.* **151** (1983), 181-203.

L'influence d'Alexandre Grothendieck en K-théorie

Max Karoubi

Dans sa démonstration du théorème de Riemann-Roch en géométrie algébrique en 1957, Grothendieck avait introduit un groupe mystérieux à l'époque, noté $K(X)$. Ici X est une variété algébrique quasi-projective non singulière disons sur le corps des complexes. Ce groupe $K(X)$ est le quotient du groupe libre engendré par les classes d'isomorphie $[E]$ de fibrés algébriques par le sous-groupe engendré par les relations $[E] = [E'] + [E"]$, chaque fois qu'on a une suite exacte

$$0 \longrightarrow E' \longrightarrow E \longrightarrow E'' \longrightarrow 0$$

On peut considérer les éléments de $K(X)$ comme des "classes" formelles de fibrés vectoriels sur X, d'où la terminologie "K-théorie", la lettre K suggérant le mot "Klassen" en allemand, la langue maternelle de Grothendieck. Une définition équivalente en termes de faisceaux algébriques cohérents est possible ; cf. [**BS**] théorème 2 et la remarque au bas de la page 108.

Le groupe $K(X)$ est en fait un foncteur contravariant de X (pour l'image réciproque des fibrés). C'est aussi un foncteur covariant pour les morphismes propres $f : X \longrightarrow Y$. En effet, Grothendieck leur associe des "homomorphismes de Gysin" $f_*^K : K(X) \longrightarrow K(Y)$ qui jouissent de nombreuses propriétés formelles qu'on ne développera pas ici. Cependant, il est important de mentionner le cas où Y est réduit à un point et X projective. Alors $f_*^K(E)$ est la somme alternée suivante (bien connue par Serre et Hirzebruch [**H**], 1954):

$$\sum (-1)^k dim(H^k(X; E))$$

en convenant d'identifier E au faisceau de ses sections algébriques.

Professeur émérite à l'UFR de Mathématiques de l'Université Paris Diderot, 13 rue Albert Einstein, 75205 Paris Cedex 13, France. max.karoubi@gmail.com.

Si E est un fibré algébrique, on peut le voir comme un fibré topologique sur l'espace topologique obtenu en munissant X de la topologie transcendante ; il a donc des classes caractéristiques de Chern $c_i(E)$ appartenant à $H^{2i}(X;Z)$. En suivant Hirzebruch, écrivons la classe totale de Chern

$$c(E) = 1 + c_1(E) + c_2(E) + \ldots$$

comme un produit formel

$$c(E) = \prod (1 + x_i)$$

où les x_i sont de degré 2. La "classe de Todd" du fibré E est alors définie comme le produit formel

$$Todd(E) = \prod \frac{x_i}{1 - e^{-x_i}}$$

qu'on exprime en chaque degré comme un polynôme en les fonctions symétriques élémentaires des x_i, donc des $c_i = c_i(E)$. D'après Hirzebruch [H] 1954, on a plus précisément les formules suivantes en bas degrés (dans la cohomologie rationelle)

$$\begin{array}{lcl} Todd_0(E) & = & 1 \\ Todd_1(E) & = & c_1/2 \\ Todd_2(E) & = & (c_2 + c_1^2)/12 \\ Todd_3(E) & = & c_2c_1/24 \\ Todd_4(E) & = & (-c_4 + c_3c_1 + 3c_2^2 + 4c_2c_1^3 - c_1^4)/720 \\ \ldots & & \end{array}$$

Par sa définition même, la classe de Todd est "multiplicative", c'est-à-dire vérifie l'identité

$$Todd(E \oplus F) = Todd(E).Todd(F)$$

On définit par le même formalisme un "caractère de Chern", qui est un homomorphisme d'anneaux de $K(X)$ vers la cohomologie rationelle en degrés pairs $H^{pair}(X)$, par la formule suivante (où n est le rang de E)

$$Ch(E) = \sum_{i=1}^{n} exp(x_i)$$

Parallèlement à f_*^K, un homomorphisme de Gysin est classiquement défini en cohomologie

$$f_*^H : H^*(X) \longrightarrow H^*(Y)$$

Cependant, le diagramme évident (qui représente bien la "vision fonctorielle" de Grothendieck)

$$\begin{array}{ccc} K(X) & \xrightarrow{f_*^K} & K(Y) \\ {\scriptstyle Ch}\downarrow & & \downarrow{\scriptstyle Ch} \\ H^*(X) & \xrightarrow{f_*^H} & H^*(Y) \end{array}$$

n'est PAS commutatif. La déviation de commutativité est donnée par les classes de Todd des fibrés tangents TX et TY à X et Y respectivement. Plus précisément, pour tout fibré E sur X, le théorème de Riemann-Roch-Grothendieck (RRG) s'écrit ainsi :

$$f_*^H(Ch(E).Todd(TX)) = Ch(f_*^K(E)).Todd(TY)$$

En raison du caractère linéaire des deux membres par rapport à E, cette formule est équivalente à la suivante (pour tout élément x de $K(X)$)

$$f_*^H(Ch(x).Todd(TX)) = Ch(f_*^K(x)).Todd(TY)$$

En multipliant par $(Todd(TY))^{-1}$, elle s'écrit également

$$f_*^H(Ch.Todd(Tf)) = Ch(f_*^K(x))$$

où $Todd(Tf) = Todd(TX).f^*(Todd(TY)^{-1})$ est par définition la classe de Todd du "fibré tangent le long des fibres" de f, soit formellement $[TX] - [f^*TY]$.

Si Y est un point, cette formule se réduit à celle de Riemann-Roch-Hirzebruch (cf. [**H**] 1954) :

$$f_*^H(Ch(E).Todd(TX)) = \sum (-1)^k dim(H^k(X;E))$$

qui a été démontrée avant Grothendieck par des méthodes tout à fait différentes. En particulier, le premier membre (à priori rationnel) est un nombre entier, ce qui n'est nullement évident à priori.

Il est important de noter que dans la formule de RRG, on peut remplacer la cohomologie usuelle par une cohomologie plus "algébrique" qui est l'anneau gradué $A(X)$ des classes de cycles algébriques sur X pour l'équivalence linéaire, la graduation étant déterminée par la codimension des cycles. Cet anneau est un invariant plus fin que la cohomologie usuelle qui ne dépend que de la topologie de X. Dans ce cadre plus algébrique, Grothendieck a construit une théorie des classes de Chern et du caractère de Chern tout à fait analogue à la théorie classique [G1]. Le caractère de Chern induit alors un isomorphisme

$$K(X) \otimes_{\mathbb{Z}} \mathbb{Q} \xrightarrow{\cong} A(X) \otimes_{\mathbb{Z}} \mathbb{Q}$$

La formule de RRG énoncée plus haut est loin d'être la plus générale. Dans SGA6 [**G2**], Grothendieck et ses collaborateurs se débarrassent de plusieurs hypothèses gênantes (nécessité d'un corps de base k, régularité de X et Y, hypothèses trop restrictives de quasi-projectivité pour X et Y, etc.). Pour réaliser ce programme, il faut redéfinir le groupe $K(X)$ pour un schéma X le plus général possible. Dans ce cadre, la "bonne" catégorie n'est pas celle des fibrés vectoriels mais celle des complexes parfaits. Par définition, un tel complexe est localement quasi-isomorphe à un complexe borné de fibrés vectoriels. La K-théorie est alors décrite à partir de triangles de complexes plutôt que de suites exactes. Ce type de théorie a trouvé des applications importantes dans de nombreux travaux ultérieurs en géométrie et en topologie : Waldhausen [**W**], Thomason et Trobaugh [**TT**][1], Schlichting [**Sc**], etc. Nous y reviendrons un peu plus loin.

Revenons cependant à la K-théorie "traditionnelle". Quelques années après Grothendieck, Atiyah et Hirzeburch [**A**] ont exploité cette idée pour construire une "K-théorie topologique", $K^{top}(X)$, autrement dit une théorie construite avec des fibrés vectoriels topologiques (disons à fibre un espace vectoriel complexe pour commencer). Ils ont aussi démontré un théorème de Riemann-Roch différentiable sous la forme suivante [**AH**] : soit $f : X \longrightarrow Y$ une application différentiable propre entre deux variétés C^∞ telle que le fibré tangent le long des fibres $Tf = [TX] - [f^*TY]$ soit muni d'une structure complexe stable (ou même seulement ^{c}spinorielle, cf. [**K1**] p.289). On a alors une formule de Riemann-Roch dans ce contexte différentiable qui s'écrit ainsi

$$f_*^H(Ch(x).Todd(Tf)) = Ch(f_*^K(x))$$

Comme en K-théorie algébrique, celle-ci mesure la non commutativité du diagramme

$$\begin{array}{ccc} K^{top}(X) & \xrightarrow{f_*^K} & K^{top}(Y) \\ \downarrow {\scriptstyle Ch} & & \downarrow {\scriptstyle Ch} \\ H^*(X) & \xrightarrow{f_*^H} & H^*(Y) \end{array}$$

Ici f_*^K est un homomorphisme de Gysin en K-théorie topologique dont la définition est sensiblement différente de celle de Grothendieck. Pour un plongement par exemple, elle utilise essentiellement "l'isomorphisme de Thom"

$$K^{top}(X) \to K_c^{top}(\nu)$$

(ν étant le fibré normal à X dans Y et K_c^{top} désignant la K-théorie à supports compacts). Celui-ci se déduit des théorèmes de périodicité de Bott qui expriment que la K-théorie topologique des sphères S^n est périodique par rapport à n (période 2 dans le cas complexe, 8 dans le cas réel).

En fait, le théorème de Riemann-Roch-Grothendieck dans le cas des fibrés algébriques complexes et pour la cohomologie ordinaire résulte du théorème d'Atiyah-Hirzebruch mentionné ci-dessus et d'un théorème de

[1]En fait, Thomason et Trobaugh supposent toujours quelques conditions du type X quasi-compact et quasi-séparé.

Baum-Fulton-Mac Pherson [**BFP**] qui exprime la commutativité du diagramme suivant

$$\begin{array}{ccc} K(X) & \xrightarrow{f_*^K} & K(Y) \\ \downarrow & & \downarrow \\ K^{top}(X) & \xrightarrow{f_*^K} & K^{top}(Y) \end{array}$$

Les deux f_*^K sont les homomorphismes de Gysin en K-théorie algébrique ou topologique respectivement.

Une différence essentielle entre les K-théories algébrique et topologique est que cette dernière est plus calculable. Par exemple le caractère de Chern induit un isomorphisme

$$K^{top}(X) \otimes_{\mathbb{Z}} \mathbb{Q} \cong H^{pair}(X; \mathbb{Q})$$

Par contre, les deux K-théories ont beaucoup de points en commun, notamment l'existence d'opérations non additives en général, définies à l'aide des puissances extérieures de fibrés vectoriels, soit

$$\lambda^k : K(X) \to K(X)$$

qui font de $K(X)$ ce que Grothendieck appelle un λ-anneau. D'autres opérations ψ^k en K-théorie furent introduites ultérieurement par Adams [**K1**]. Elles s'expriment comme fonctions polynomiales des λ^i, soit $\psi^k = Q_k(\lambda^1, \ldots, \lambda^k)$, où Q_k est le polynôme de Newton. Ces opérations λ^k et ψ^k sont beaucoup plus manipulables que leurs analogues cohomologiques. Pour s'en convaincre, il suffit de parcourir les ouvrages classiques de topologie algébrique et constater la technicité délicate requise pour définir avec soin les opérations de Steenrod...

La K-théorie topologique a connu des applications spectaculaires dans les années 60 et a été une des inspirations d'Atiyah et Singer pour leur célèbre théorème de l'indice [**AS**]. La manière la plus élégante d'énoncer ce théorème est de définir deux indices "analytique" et "topologique" pour une variété C^∞ compacte X

$$i_a : K_c^{top}(T^*X) \to \mathbb{Z} \ et \ i_t : K_c^{top}(T^*X) \to \mathbb{Z}$$

L'homomorphisme i_a est défini grâce aux opérateurs pseudo-différentiels sur X alors que i_t est défini en utilisant essentiellement l'homomorphisme de Gysin en K-théorie topologique. L'égalité de ces deux indices est une généralisation du théorème de Riemann-Roch-Hirzebruch (RRH). En considérant des familles d'opérateurs, Atiyah et Singer obtiennent même une généralisation de RRG dans un cadre différentiable mais il est clair que les idées fondamentales de Grothendieck ont inspiré une grande partie de leur formalisme.

L'influence de Grothendieck s'est étendue bien au delà de ces dernières considérations, grâce à un mélange intime de géométrie algébrique et de topologie. Si nous revenons à la géométrie algébrique classique, il paraissait clair dans les années 60 que le problème d'une bonne définition des "foncteurs dérivés" de la K-théorie se posait, $K(X)$ n'étant que le premier groupe $K_0(X)$ de cette série de foncteurs. Des définitions de K_1 et de K_2 avaient déjà été proposées par Bass et Milnor [**Ba**],[**M1**], motivées par des applications arithmétiques intéressantes. Le déclic est venu de Grothendieck lui-même qui avait eu le premier l'idée d'associer un ensemble simplicial à une catégorie arbitraire (appelé aussi nerf de cette catégorie), bien qu'il ne l'ait jamais publiée. Ainsi les simplexes de dimension 0 sont les objets, ceux de dimension 1 les flèches, ceux de dimension 2 les flèches composables, etc. La réalisation géométrique de cet ensemble simplicial est un objet intéressant en soi. On retrouve par exemple l'espace classifiant d'un groupe discret G, en considérant une catégorie avec un seul objet, les morphismes étant les éléments du groupe.

Pour définir les groupes $K_n(X)$ (avec $K_0(X) = K(X)$), Quillen [**Q**] associe à X une catégorie astucieuse $Q(X)$, dont les objets sont les fibrés vectoriels E sur X, un morphisme de E vers F étant donné par un sous-fibré F' de F et une surjection de F' sur E

$$\begin{array}{ccc} F' & \longrightarrow & F \\ \downarrow & & \\ E & & \end{array}$$

La loi de composition dans cette catégorie s'exprime simplement par une juxtaposition de diagrammes. Quillen définit alors les groupes $K_n(X)$ comme les groupes d'homotopie de l'ensemble simplicial associé à $Q(X)$ (avec un décalage de degrés). On les notera simplement $\pi_{n+1}(Q(X))$. Il n'est pas difficile de voir que $\pi_1(Q(X))$ est bien le groupe $K(X)$ de Grothendieck.

Quillen a explicité de nombreuses applications de sa définition. Par exemple, si $X = Spec(A)$, le calcul (rationnel) de $K_*(X)$ est équivalent à celui de l'homologie rationnelle du groupe linéaire discret infini $GL(A) = colim\ \ GL_n(A)$. Plus précisément $H_*(GL(A);\mathbb{Q})$ est une algèbre extérieure graduée sur les groupes $K_*(X) \otimes_{\mathbb{Z}} \mathbb{Q}$. Le cas où A n'est pas commutatif peut être traité de manière analogue en remplaçant les fibrés vectoriels sur X par les modules projectifs de type fini sur A.

Par exemple, si A est l'anneau des entiers algébriques d'un corps de nombres avec r_1 places complexes et r_2 places réelles, on déduit des calculs de Borel [**Bo**] et du fait que les $K_n(A)$ sont des groupes abéliens de type fini [**Q**] les résultats suivants (pour $j > 0$):

$$\begin{array}{rcl} K_{2j}(A) & & est\ un\ groupe\ fini \\ K_{4j-1}(A) & = & Z^{r_1} \oplus\ un\ groupe\ fini \\ K_{4j+1}(A) & = & Z^{r_1+r_2} \oplus\ un\ groupe\ fini \end{array}$$

(1)

Comme illustrations de ces techniques, voici deux applications spectaculaires de la K-théorie algébrique "supérieure" que nous venons de définir.

Théorème de Matsumoto [**M1**] *Si F est un corps, $K_2(F)$ est isomorphe naturellement au quotient de $F^* \otimes_{\mathbb{Z}} F^*$ par le sous-groupe engendré par les relations $u \otimes (1-u)$, où $u \in F - \{0, 1\}$.*

Théorème de Merkurjev Suslin [**Sr**]. *Si F est un corps, on a un isomorphisme naturel*

$$K_2(F)/nK_2(F) \cong H^2(G; (\mu_n)^{\otimes 2})$$

où μ_n désigne le groupe des racines de l'unité dans une clôture algébrique de F, de groupe de Galois G. Si on suppose en outre $\mu_n \subset F$, ce groupe $K_2(F)/nK_2(F)$ est isomorphe à la n-torsion du groupe de Brauer de F.

Exemple : si $n = 2$, on montre ainsi que tout élément de la 2-torsion du groupe de Brauer d'un corps quelconque est la classe du produit tensoriel d'algèbres de quaternions.

Un lien plus direct avec la géométrie algébrique découvert par Quillen [**Q**] est le suivant : soit $\mathcal{K}_p$ le faisceau sur X associé au préfaisceau

$$U \to K_p(U)$$

Le groupe de cohomologie $H^p(X; \mathcal{K}_p)$ est alors isomorphe au groupe de Chow $A^p(X)$. Nous renvoyons le lecteur au livre de Srinivas [**Sr**] pour de nombreuses autres applications de la théorie de Quillen.

Une idée maîtresse dans l'œuvre de Grothendieck a été l'introduction des catégories dérivées. Ce formalisme a permis à Waldhausen [**W**] puis à Thomason-Trobaugh [**TT**] de définir les groupes $K_n(X)$ avec de bonnes propriétés formelles pour des schémas quasiment arbitraires, ce que nous avions déjà évoqué un peu plus haut. En effet, dans SGA6, Grothendieck et ses collaborateurs [**G2**] définissent le groupe K_0 d'une catégorie dérivée et d'un schéma X. Pour définir les groupes K_n pour $n > 0$, on considére aussi la catégorie des complexes parfaits de faisceaux sur X. Dans cette catégorie, on a la notion de quasi-isomorphisme et de cofibration, notions que Waldhausen et Thomason-Trobaugh utilisent de manière abstraite pour étendre la définition de Quillen au cadre plus général des schémas. Comme dans le cas des fibrés vectoriels sur les variétés, on associe à une catégorie de complexes parfaits un ensemble simplicial dont les groupes d'homotopie sont (à un décalage près) les groupes $K_n(X)$ recherchés.

A cette occasion, Thomason et Trobaugh répondent à une question laissée ouverte dans SGA 6. Soit X un schéma et $\mathcal{F}$ un complexe parfait au-dessus d'un ouvert affine U de X. Alors on peut l'étendre (à quasi-isomorphisme près) en un complexe parfait sur X si et seulement si sa classe dans $K_0(U)$ appartient à l'image de l'homomorphisme de restriction

$K_0(X) \to K_0(U)$. Ce résultat est une application frappante des méthodes de K-théorie à la géométrie algébrique.

Parallèlement à la K-théorie algébrique qui s'est développée après Grothendieck, sous l'impulsion de Bass, Milnor, Quillen, Waldhausen, Thomason..., la K-théorie topologique ne s'est pas arrêtée à Atiyah, Hirzebruch, Singer et d'autres. Dès la fin des années 60 et surtout dans les années 70, elle a pris un nouveau départ dans le cadre de l'analyse fonctionnelle (les premières amours de Grothendieck !). Ainsi pour toute algèbre de Banach A non nécessairement commutative, on peut définir [**K1**] des groupes de K-théorie topologiques $K_n(A)$ qui sont périodiques de période 8 dans le cas réel et 2 dans le cas complexe (théorèmes analogues à ceux de Bott).

Cette périodicité a beaucoup intrigué les algébristes, car les groupes K_n de Quillen et Thomason sont loin d'être périodiques. Il existe cependant une "périodicité galoisienne" cachée ; cf. par exemple [**FG**], où plusieurs versions algébriques de la périodicité de Bott sont présentées.

Si A et B sont des C^*-algèbres, Kasparov [**Bl**] a même réussi à définir des "bifoncteurs" $KK_n(A, B)$ par des méthodes analytiques et montré que le théorème d'Atiyah-Singer peut se démontrer par un simple formalisme de "KK-théorie". Il n'existe pas d'analogue de ces bifoncteurs pour les schémas ni même les variétés algébriques.

C'est sans aucun doute les travaux de Connes [**C**] qui ont donné à la K-théorie topologique ses lettres de noblesse les plus récentes dans le cadre de la "géométrie non commutative", un concept paradoxalement éloigné de la problématique initiale de Grothendieck.

Faute de place, nous nous limiterons à une seule application de la K-théorie topologique concernant les C^*-algèbres AF (pour "approximativement finies"). Une telle algèbre est l'adhérence dans l'algèbre des opérateurs bornés sur un espace de Hilbert d'une limite inductive d'algèbres complexes semi-simples de dimension finie. Alors deux telles algèbres AF, soient A et B, sont isomorphes si et seulement si les groupes ordonnés $K_0(A)$ et $K_0(B)$ sont isomorphes. Un élément de K_0 est dit "positif" s'il est la classe d'un "vrai" module (pas seulement une différence), notion qui permet de définir une relation d'ordre sur K_0.

On pourra consulter [**Bl**], [**C**] et [**R**] par exemple pour de nombreuses applications de toutes ces idées en analyse et en physique théorique.

Pour conclure ce bref exposé, il convient de mentionner la K-théorie hermitienne, intermédiaire entre la K-théorie algébrique et la K-théorie topologique. On la note $KQ(X)$ ou plus précisément ${}_{\epsilon}KQ(X)$ en tenant du compte du signe de symétrie $\epsilon = \pm 1$. Très sommairement, on la construit comme la K-théorie algébrique des fibrés vectoriels E, en se donnant en plus une forme ϵ-hermitienne non dégénérée sur E. Nous allons donner deux applications frappantes des méthodes de Grothendieck à cette toute nouvelle théorie.

La première est due à Voevodsky [**V**]. Par des méthodes astucieuses combinant les idées de Grothendieck et Verdier sur les catégories dérivées et triangulées et une topologie adéquate sur les schémas (dite de Nisnevich), Voevodsky a réussi à démontrer la conjecture de Milnor [**M2**] que nous allons maintenant décrire.

Soit F un corps de caractéristique différente de 2 et soit $W(F)$ l'anneau de Witt défini à partir des formes quadratiques non dégénérées sur les F-espaces vectoriels E de dimension finie (les espaces hyperboliques étant identifiés à 0). L'idéal fondamental $I(F)$ est le noyau de l'homomorphisme surjectif $W(F) \to \mathbb{Z}/2$, défini par la dimension de E modulo 2. Les puissances $I^n(F)$ de cet idéal définissent alors une filtration décroissante de $W(F)$. La conjecture de Milnor, démontrée par Voevodsky, est la suivante : les quotients successifs $I^n(F)/I^{n+1}(F)$ sont isomorphes par des flèches explicites à certains groupes $k_n^M(F) = K_n^M(F)/2$. Ici les groupes de "K-théorie de Milnor" $K_n^M(F)$ (à ne pas confondre avec ceux de Quillen) sont définis de manière purement algébrique comme dans le théorème de Matsumoto cité plus haut. Le groupe $K_n^M(F)$ est ainsi le quotient du produit tensoriel de F^* n fois par lui-même - en tant que $\mathbb{Z}$-module - par le sous-groupe engendré par les produits tensoriels $x_1 \otimes \cdots \otimes x_n$, lorsque $x_i + x_j = 1$ pour un couple (i, j) avec $i \neq j$.

Dans un autre ordre d'idées, Schlichting [**Sc**] a réussi à étendre la théorie de Grothendieck et Thomason-Trobaugh dans un cadre hermitien. Il a pu ainsi redémontrer de manière purement algébrique les théorèmes de périodicité de Bott sous une forme beaucoup plus générale. Dans ce but, on introduit des théories intermédiaires ${}_\epsilon U$ et ${}_{-\epsilon}V$ comme "fibres" des foncteur "hyperbolique"

$$K \to_\epsilon KQ$$

et "oubli"

$${}_{-\epsilon}KQ \to K$$

respectivement, pour des schémas quasiment arbitraires (cf. [**K2**] et [**Sc**]). Schlichting démontre alors que les théories ${}_{-\epsilon}V$ et ${}_\epsilon U$ sont les mêmes à un décalage de degrés près. Ceci lui permet de redémontrer dans le cadre des groupes classiques un de mes théorèmes [**K2**] : pour une algèbre de Banach involutive réelle ou complexe, on a une équivalence d'homotopie naturelle entre les composantes connexes des espaces suivants (où $GL(A)$ et ${}_\epsilon O(A)$ sont respectivement les groupes linéaire et orthogonal "infinis")

$$GL(A)/{}_\epsilon O(A) \text{ } et \text{ } \Omega({}_{-\epsilon}O(A)/GL(A)) \qquad (H)$$

Pour pouvoir écrire ces quotients de groupes classiques, on remarque qu'on a une suite d'inclusions (la dernière en doublant la taille des matrices)

$${}_\epsilon O(A) \subset GL(A) \subset {}_{-\epsilon}O(A)$$

Il est facile de voir que l'équivalence d'homotopie (H) implique les théorèmes de périodicité de Bott classiques en choisissant pour A les corps $\mathbb{R}$, $\mathbb{C}$ ou $\mathbb{H}$ avec des involutions adéquates. Ce dernier exemple montre à quel point les K-théories algébrique et topologique sont imbriquées et le travail qui reste à faire pour mieux comprendre les relations entre elles.

Quelques souvenirs personnels sur Grothendieck.

Dans un certain sens, mon parcours mathématique a commencé de manière parallèle à celui de Luc Illusie, entré comme moi à l'Ecole Normale Supérieure et qui a été ensuite élève de Grothendieck. Nous participions tous les deux au séminaire Cartan-Schwartz en 1963/64, où nous commencions à comprendre l'utilité de la K-théorie. A cette occasion, je voudrais rendre hommage à Henri Cartan qui a été mon patron de thèse, décédé cette année 2008 à l'âge de 104 ans.

Après ce séminaire de 1963/64, j'avais commencé à avoir des idées sur ma thèse, un peu par hasard, à la suite de nombreux entretiens avec Henri Cartan et Shih Weishu. Quelques éléments étaient déjà écrits en 1965. C'est de cette année que date ma première rencontre avec Grothendieck qui, avec sa générosité habituelle, n'a pas hésité à me prodiguer ses conseils, bien que je n'ai pas été formellement un de ses élèves. Ainsi, par exemple, cette phrase de lui dont je me souviens, après une première rédaction de thèse qu'il avait lue avec soin : "Karoubi, ce n'est pas ainsi qu'on écrit des mathématiques". J'ai eu bien sûr l'occasion de l'écouter à maintes reprises, notamment dans un cours sur le calcul fonctoriel donné à l'Université d'Alger fin 1965, où j'effectuais mon service militaire (sic) dans le cadre de la coopération. Il faut ajouter qu'à cette époque, Grothendieck, avec d'autres mathématiciens français comme Godement, Serre... se déplaçaient plusieurs semaines à Alger pour y donner des cours. Ceux-ci ont contribué au développement d'un Département de Mathématiques, bien nécessaire après la guerre douloureuse (1954-1962) qui s'était conclue par l'indépendance de l'Algérie.

Pendant les années suivantes, Grothendieck est resté attentif à la progression de mes travaux en K-théorie, où je cherchais des applications de ses idées. Par exemple, pour démontrer un théorème, j'avais introduit la notion relativement simple d'enveloppe pseudo-abélienne d'une catégorie additive. Cette définition n'avait pas échappé à la perspicacité de Grothendieck qui, comme chacun sait, l'a utilisée plus tard dans sa théorie des motifs. Je l'ai appris d'ailleurs de manière inopinée lors d'un séjour à l'Institute for Advanced Study à Princeton en 1967 : un visiteur m'a ainsi interpellé, en m'apprenant que Grothendieck utilisait dans ses travaux une notion de catégorie "karoubienne". J'en ai été le premier surpris!

En conclusion, je garde de Grothendieck le souvenir d'une personnalité hors norme sur tous les plans par sa volonté très forte de faire partager ses idées (mathématiques et philosophiques), sa disponibilité sans limite pour les autres et ses conceptions sur ce qu'il est important de faire de notre vie. C'est avec un immense regret que je l'ai vu quitter prématurément la scène scientifique et couper ses relations (pour des raisons qui lui sont propres) avec les nombreux mathématiciens qui l'ont admiré et continuent à le faire, en trouvant tous les jours de nouveaux prolongements et applications de ses idées.

Références (liste non exhaustive)

[A] M..F. Atiyah. K-theory. Benjamin, New York (1967).

[AS] M.F. Atiyah and I.M. Singer. The index of elliptic operators. Ann. of Math. 87 (1968), 484–530, 531–545 (avec G. Segal), 546–604.

[AH] M.F. Atiyah and F. Hirzebruch. Riemann-Roch theorems for differentiable manifolds. Bull. Amer. Math. Soc. 65 (1959) 276–281.

[Ba] H. Bass. Algebraic K-theory, Benjamin, New York (1968).

[Be] J. Berrick. An approach to Algebraic K-theory. Pitman (1982).

[Bl] B.Blackadar. K-theory of operator algebras. Springer (1986).

[BFP] P. Baum, W. Fulton and R. MacPherson. Riemann Roch and topological K-theory for singular varieties. Acta Math 143 (1979), 155–192.

[Bo] A. Borel. Stable real cohomology of arithmetic groups. Ann. Sci. Ec. Norm. Sup. (1974), 235–272.

[BS] A. Borel et J.-P. Serre. Le théorème de Riemann-Roch (d'après Grothendieck). Bull. Soc. Math. France 86 (1958) 97–136.

[C] A. Connes. Noncommutative Geometry. Academic Press, San Diego (1994).

[FG] E. Friedlander and D. Grayson. Handbook of K-theory. Springer (2005).

[FW] E. Friedlander and C. Weibel. School on Algebraic K-theory and applications. An overview of the subject. ICTP Trieste (1997).

[G1] A. Grothendieck. La théorie des classes de Chern. Bull. Soc. Math. France 86 (1958), 137–154.

[G2] A. Grothendieck, P. Berthelot et L. Illusie (SGA6). Théorie des intersections et théorème de Riemann-Roch. Springer Lecture Notes in Maths 225 (1971).

[H] F. Hirzebruch. Topological methods in Algebraic Geometry. Springer 2e édition (1965).

[K1] M. Karoubi. K-theory, an introduction. Grundlehren 226, Springer (1978). Nouvelle édition (2008) dans la série "Classics in Mathematics".

[K2] M. Karoubi. Le théorème fondamental de la K-théorie hermitienne. Annals of Math. 112, (1980) 259–282.

[K3] M. Karoubi. Homologie cyclique et K-théorie. Astérisque 149. Société Mathématique de France (1987).

[M1] J. Milnor. Introduction to Algebraic K-theory. Annals of Math. Studies 197. Princeton University Press (1974).

[M2] J. Milnor. Algebraic K-theory and quadratic forms. Inventiones Math. 9 (1969) 318–344.

[Q] D. Quillen. Higher Algebraic K-theory. Springer Lecture Notes in Math. 341 (1973).

[R] J. Rosenberg. Algebraic K-theory and its applications. Graduate Texts in Maths. 147. Springer (1994).

[Sc] M. Schlichting. The Mayer-Vietoris principle for the Grothendieck-Witt group of schemes. J. K-Theory 5 (2010), 105–165.

[Sr] V. Srinivas. Algebraic K-theory. Birkhauser (1996).

[TT] R. Thomason and T. Trobaugh. Higher Algebraic K-theory of schemes and of derived categories. In the Grothendieck Festshrift, vol. III. Prog. in Maths Vol. 88, Birkhauser (1990) 247–435.

[V] V. Voevodsky. The Milnor conjecture. Preprint 1996, available at http://www.math.uiuc.edu/K-theory/0170/

[W] F. Waldhausen. Algebraic K-theory of spaces. Springer Lecture Notes 1126 (1983) 318–419.

Grothendieck et la théorie des schémas

Michel Raynaud

Avec partialité et néanmoins bonne conscience, je situe le début de la géométrie algébrique moderne en 1955, avec la parution de "Faisceaux algébriques cohérents". Serre y utilise la topologie de Zariski et le langage des faisceaux. Il se limite aux variétés de type fini sur un corps algébriquement clos, mais, peu après, Grothendieck développe la théorie des schémas en toute généralité et en présente les grandes lignes dans sa conférence au congrès ICM de 1958. Avec l'introduction des schémas, la géométrie algébrique a désormais ses objets, mais aussi ses morphismes, ce qui n'allait pas de soi dans les présentations antérieures, où applications rationnelles et correspondances étaient omniprésentes. Grâce aux espaces annelés, à leurs topologies sous-jacentes et leurs faisceaux de fonctions, le géomètre algébriste dispose d'un confort et d'une souplesse d'utilisation en tout point comparables à la situation déjà acquise en topologie générale et en géométrie différentielle.

Le premier coup d'éclat de Grothendieck, en géométrie algébrique, est la démonstration en 57-58, d'un théorème de Riemann-Roch pour un morphisme propre $X \to Y$ entre variétés lisses quasi-projectives. Il fait suite aux travaux de Hirzebruch sur les variétés compactes complexes. Le texte est rédigé par Borel et Serre, dans le langage de FAC. Grothendieck introduit le groupe $K(X)$ des faisceaux cohérents sur X et leurs classes de Chern.

Le point de vue fonctoriel

On considère qu'Eilenberg est le père des catégories. Mais, incontestablement, Grothendieck a beaucoup contribué à les populariser. Elles sont au cœur de son travail. Il montre toute leur richesse et leur souplesse d'utilisation. Grothendieck a une approche très exhaustive des mathématiques. Quand il aborde une théorie, il se place spontanément dans le cadre le plus général et celui-ci s'exprime idéalement dans le langage des catégories. Bien des années plus tard, il écrira fort joliment, dans

Professeur honoraire à l'Université Paris-Sud, Département de Mathématiques, F-91405 Orsay Cedex, France. michel.raynaud@math.u-psud.fr.

"Récoltes et Semailles" : <<. . . , je n'ai pu m'empêcher, au fur et à mesure, de construire des maisons, des très vastes et des moins vastes, et toutes bonnes à être habitées, – des maisons où chaque coin et recoin est destiné à devenir lieu accueillant et familier pour plus d'un. Les portes et fenêtres sont d'aplomb et s'ouvrent et se ferment sans entrebâiller et sans grincer, le toit ne fuit pas et la cheminée tire>>.

Il observe que si C est une catégorie et si $C^\wedge$ désigne la catégorie des foncteurs (contravariants) sur C, à valeurs dans les ensembles, on dispose d'un foncteur $h : C \to C^\wedge$, qui, à tout objet X associe le foncteur h_X qui, sur l'objet Y de C, prend la valeur $\mathrm{Hom}_C(Y, X)$. De plus, ce foncteur h est pleinement fidèle.

Appliqué aux schémas, ce point de vue introduit une véritable révolution: jusque là, la géométrie algébrique mettait l'accent sur les corps, maintenant ce sont les anneaux commutatifs qui sont au cœur de la place. Toutefois le langage géométrique est préservé et $\mathrm{Hom}_C(Y, X)$ va encore s'appeler l'ensemble des "points" de X à valeurs dans Y. Pour connaître un schéma X, il faut connaître ses points à valeurs dans tout schéma Y et beaucoup de propriétés de X se voient sur le foncteur des points.

Certaines notions classiques sont complètement renouvelées. Ainsi on connaissait les variétés lisses (jusque là appelées simples). Maintenant la lissité d'un morphisme se lit sur le foncteur et même sur le foncteur à valeurs dans les anneaux locaux artiniens, une propriété qui n'était pas formulable avec le point de vue classique où l'on ne travaillait qu'avec des variétés réduites. Dans la même veine, les calculs sur les nombres duaux et, plus généralement, les problèmes de déformations infinitésimales s'insèrent parfaitement dans l'approche fonctorielle.

Grothendieck magnifie la trilogie chère à la géométrie différentielle : isomorphismes locaux, immersions et submersions, qui deviennent les morphismes étales, non ramifiés et lisses. De plus, chacune de ces notions se teste agréablement sur le foncteur.

En fait, c'est toute l'approche différentielle qui est revivifiée avec l'utilisation des voisinages infinitésimaux de la diagonale. Quelques années plus tard, Grothendieck franchira une nouvelle étape qui le conduira au site cristallin.

Cette adoption du point de vue fonctoriel ne va pas sans résistance. Les premiers exposés de "Fondements de la géométrie algébrique", au séminaire Bourbaki, se déroulent devant un public quelque peu noyé sous les foncteurs. Bien des géomètres restent indifférents à ce nouveau point de vue : ils ne voient pas l'intérêt de ces développements catégoriels qui, pour eux, sont des "maths molles", par opposition aux "maths dures" qu'ils pratiquent.

Pourtant, Grothendieck va faire des efforts pédagogiques considérables pour expliquer son point de vue. Il convient, certes, qu'un investissement est nécessaire au départ, pour entrer dans "son yoga" ; mais ensuite, la théorie se déroule toute en douceur, pour le plus grand confort de l'utilisateur. Dieudonné insistera sur le caractère naturel de la théorie des schémas.

En 60-61, le séminaire Cartan, porte sur les familles de surfaces de Riemann compactes. Douady donne les premiers exposés en adoptant le point de vue des espaces de Teichmüller. Grothendieck objecte que, selon lui, ce n'est pas la bonne approche. Avec l'accord de Cartan, il prend alors le séminaire en main et expose les bases de la géométrie analytique complexe. C'est l'occasion pour lui, de présenter le point de vue fonctoriel, dans le cadre analytique cette fois. Il décrit quelques-unes des constructions de base, insiste sur la représentabilité et la représentabilité relative. Rodé par la rédaction des EGA, il donne sa pleine mesure et aboutit à un texte magnifique, très didactique, que je ne saurais trop recommander au géomètre débutant.

La géométrie relative

Avec le développement du point de vue fonctoriel, vient tout naturellement la géométrie algébrique relative, sur une base S, et l'opération de changement de base. Le "produit fibré" acquiert une importance primordiale. Il est utilisé de façon dissymétrique : on part de $f : X \to S$, avec, en général, certaines conditions de finitude sur f, puis on fait un changement de base $S' \to S$, plus ou moins arbitraire, pour aboutir à $X \times_S S' \to S'$. La géométrie absolue s'estompe, puisque, même si l'on part d'une variété sur un corps k, il y a lieu de considérer ses points dans tout k-schéma.

Quand Serre arrive en avance à l'IHES, il prend la craie et dessine au tableau

$$\begin{array}{c} X \\ \downarrow \\ S \end{array}$$

qui devient la marque déposée du séminaire de Bures.

Clairement, pour Grothendieck, il y a d'abord les notions qui se comportent bien par changement de base. Elles forment le socle de la théorie et il y a lieu de les étudier en priorité, avec le plus grand soin. Puis, éventuellement, il y a les autres. En fait, il n'y en a pas tellement d'autres, car beaucoup de propriétés autrefois vues comme absolues, admettent une version relative naturelle, sous des hypothèses de finitude et de platitude convenables.

De fait, au fur et à mesure de l'avancement des EGA et des SGA, c'est toute une terminologie cohérente, adaptée à la géométrie algébrique relative, qui se met en place. Grothendieck, tout comme Serre d'ailleurs, accorde beaucoup d'importance au choix du vocabulaire. Il doit être simple, précis et évocateur. L'introduction des mots "lisse" et "étale" est un succès. Par contre, Grothendieck regrettera d'avoir choisi "non ramifié", qu'il juge trop long à écrire et qui présente, de façon négative, une propriété pourtant très naturelle et très utile. Il proposera alors de remplacer "non ramifié" par "net". Dans le même souci de concision, il suggèrera d'appeler le spectre d'un anneau de valuation discrète un "trait". Mais il est alors sur le point de se retirer du devant de la scène et cette terminologie sera peu suivie.

La présentation finie

Faire de la géométrie relative, d'accord, mais dans quel cadre ? Beaucoup des sorites sur les schémas se déroulent sans aucune hypothèse de finitude. Les premiers SGA se situent dans le contexte des schémas localement noethériens.

C'est un peu plus tard, me semble-t-il, peut-être en préparant déjà la rédaction de EGA IV et avec SGA 4, que Grothendieck découvre les charmes de la présentation finie. Il observe, qu'un S-schéma X est localement de présentation finie si et seulement si le S-foncteur de ses points commute aux limites inductives filtrantes d'anneaux. Dès lors, le fait que cette condition de finitude se lise commodément sur le foncteur, est une incitation à l'explorer et, à partir de EGA IV, Grothendieck la considère systématiquement. Certes, travailler avec des S-schémas de (locale) présentation finie est assurément un bon cadre, qui flatte le rédacteur : la base est générale, et les conditions de finitude introduites sur les objets relatifs, sont naturelles. Mais, ne serait-ce que pour des raisons historiques, liées aux références, on doit fréquemment opérer des passages à la limite pour revenir au cas noethérien, ce qui conduit à une rédaction assez lourde et répétitive. Notons toutefois que certains de ces passages à la limite comme ceux concernant la platitude ou la projectivité sont fort instructifs. En quelques années, la technique de passage à la limite inductive filtrante sur les anneaux devient familière, alors qu'auparavant, on ne l'utilisait guère que pour appliquer le "principe de Lefschetz".

Avec la thèse de Monique Hakim, Grothendieck va aller encore plus loin. Pourquoi s'imposer une base qui soit un schéma ? On peut faire de la géométrie algébrique relative sur des espaces topologiques, des variétés différentielles ou analytiques, finalement sur un espace annelé, voire un topos annelé.

La platitude

Dans GAGA, Serre donne un superbe exemple de "couple plat" à propos du passage de l'algébrique à l'analytique. Avec Grothendieck, la platitude va jouer un rôle considérable...

Elle se manifeste de plusieurs façons.

Tout d'abord, elle est au cœur de la géométrie relative. Si l'on pense à un morphisme $f : X \to S$ comme à une famille de schémas sur des corps, paramétrée par les points de S, imposer que f est plat est l'hypothèse la plus commode qui assure qu'une fibre de f au-dessus d'un point s de S, est "exactement" contenue dans l'adhérence schématique d'une fibre au-dessus d'une générisation de s. Ainsi, la platitude apparaît comme un liant naturel entre les fibres, en géométrie relative. Toutefois, la platitude n'est pas une propriété tout à fait "géométrique" ; elle garde, malgré tout, un caractère "algébrique". Par exemple, si S est local artinien, la platitude de f traduit simplement que les anneaux des ouverts affines de X sont des modules libres sur l'anneau de S. Or, dans le contexte noethérien, la platitude se teste précisément, par passage à la limite sur les anneaux locaux artiniens. C'est l'occasion pour Grothendieck, de démontrer un délicat critère de platitude qu'il propose à Bourbaki. Il figure dans Bourbaki Alg. Com. Chap 3.

Précisons les rapports de Grothendieck avec Bourbaki. Dès qu'il entreprend la rédaction des EGA, il n'assiste plus aux congrès, mais reste en bons termes avec le groupe. Comme orateur au séminaire, il jouit d'un statut un peu spécial. Bourbaki est conscient que ses exposés de fondements de la géométrie algébrique seront très utiles et que l'on ne disposera pas d'une autre rédaction avant longtemps. Dès lors Grothendieck est autorisé à proposer des textes écrits plus longs et expose ses propres travaux, ce qui est inhabituel.

Mais revenons à la platitude. Un avantage technique de la notion est que, sous des hypothèses de finitude raisonnables (noethériennes, présentation finie), nombre de propriétés courantes d'un morphisme $f : X \to S$, se lisent sur les fibres, sous l'hypothèse que f est plat. On dispose ainsi d'une méthode générale pour passer de la géométrie sur les corps à la géométrie relative et réciproquement.

Dans un contexte noethérien, les notions de dimension et de profondeur se comportent fort bien lorsque f est plat. Ainsi se développe tout un chapitre d'algèbre commutative relative, utile et facile. Plus tard, dans EGA IV, Grothendieck fera face à des questions autrement délicates d'algèbre commutative, qui lui permettront de compléter les travaux de Nagata : il dégagera la notion de schéma excellent et prouvera de précieux énoncés de permanence.

La platitude, sous des conditions de finitude convenables est idéale pour l'étude des propriétés locales et d'ouverture. Suivant la voie initiée par Serre dans "Algèbre locale et multiplicités", Grothendieck "module" systématiquement ses énoncés. Ainsi un bon cadre est de considérer un morphisme $X \to S$ localement de présentation finie, et un O_X-faisceau de présentation finie M, qui est S-plat.

Mais la platitude apparaît de façon essentielle sous un autre angle : celui de la descente fidèlement plate quasi-compacte (fpqc). C'est là une merveilleuse découverte qui se révèle d'une portée considérable : les propriétés usuelles, utilisées en géométrie algébrique, peuvent se tester après changement de base fpqc. Cette observation, va bien au-delà de la descente galoisienne de Weil et de la descente radicielle de Cartier. Le concept de descente apparaît dès SGA 1. Grothendieck développe un langage et un formalisme pour pouvoir l'énoncer. La notion généralise celle des recouvrements ouverts et préfigure les topologies de Grothendieck. La descente fpqc des propriétés des objets est remarquablement simple à établir ; la descente des objets eux mêmes, est moins automatique. Grothendieck n'est toutefois pas entièrement satisfait par sa propre exposition et renonce finalement à développer les considérations générales sur les catégories fibrées qui lui semblent pourtant nécessaires. Peut être a-t-il déjà en tête la théorie des champs qui sera le sujet de thèse de Jean Giraud.

Ainsi, la géométrie algébrique est profondément renouvelée et s'organise autour de puissantes techniques de localisation. Partant d'un morphisme de type fini $X \to S$ sur une base noethérienne, par passage à la limite sur les voisinages de Zariski d'un point de S, on se ramène au cas d'une base locale. Puis par complétion et descente, au cas où l'anneau local est noethérien complet ; éventuellement, ensuite, par passage à la limite adique, au cas d'un anneau local artinien. Plus tard, avec le développement des techniques d'hensélisation, introduites pour les besoins de la topologie étale, on partira plutôt d'un morphisme de présentation finie, et la complétion pourra être remplacée par l'hensélisation, plus algébrique et qui ne nécessite pas d'hypothèses noethériennes.

Projectivité et propreté

L'étude globale des morphismes est abordée dans EGA II. Grothendieck déroule le sorite des spectres homogènes. Il définit les fibrés projectifs et le point de vue fonctoriel l'amène à privilégier les faisceaux quotients inversibles, donc les hyperplans d'un vectoriel, plutôt que les droites, qui prévalaient dans le point de vue de la géométrie algébrique projective classique.

La notion de morphisme propre, définie à partir de la propriété d'être universellement fermé, s'impose aussi bien en topologie générale (elle est ajoutée dans la seconde édition du chap I de topologie générale de Bourbaki) qu'en géométrie algébrique. L'écart entre propre et projectif se clarifie. Nagata donne le premier exemple d'une surface normale propre, non projective, puis Hironaka le premier exemple de variété lisse (en dimension 3). C'est encore Nagata qui montrera que toute variété de type fini séparée se compactifie en une variété propre, énoncé étendu par Deligne au cas relatif noethérien.

Sous des conditions de finitude convenables, la propreté se teste sur le foncteur grâce aux critères valuatifs de séparation et de propreté. Grothendieck n'aime guère les valuations, hormis les valuations discrètes. Il les utilise au minimum et reproche à Bourbaki de leur avoir consacré tout un chapitre. Sans que cela lui soit spécialement imputable, la génération de Grothendieck voit un effacement de l'usage des valuations, si on la compare à ses devancières.

Avec EGA III, le traité aborde les considérations cohomologiques. Grothendieck généralise les résultats de FAC sur les schémas affines et les schémas propres sur une base noethérienne. Puis vient un réjouissant "GAGA algébrique-formel" qui, d'un point de vue technique, est une brillante utilisation des conditions de Mittag-Leffler et du lemme d'Artin-Rees.

Cette étude globale dans le cas propre, sera complétée dans SGA2 par l'étude de la cohomologie locale et des énoncés du type GAGA, mais cette fois, dans des situations non nécessairement propres. Notons combien la notion de schéma formel est, elle aussi, novatrice. Quant à l'algébrisation des schémas formels, Grothendieck en donne une jolie application, dès SGA 1, avec la comparaison du groupe fondamental modéré d'une courbe lisse en caractéristique $p > 0$ avec son analogue en caractéristique 0, et finalement, avec le groupe fondamental topologique d'une surface de Riemann.

EGA III continue avec un paragraphe, assez indigeste, sur les "Tor". Il faut dire qu'à l'époque où Grothendieck le rédige, il ne dispose pas encore des catégories dérivées, qui seront le sujet de thèse de Jean-Louis Verdier.

EGA III se termine par l'étude du comportement de la cohomologie cohérente par changement de base, dans un contexte propre et plat. La cohomologie ne commute pas nécessairement aux changements de base, mais on peut la représenter universellement par un complexe à objets cohérents et plats sur la base. A partir de SGA 6, on parlera de complexe parfait.

Considérer un S-schéma noethérien et un morphisme propre $X \to S$, muni d'un faisceau cohérent F S-plat est assurément un bon cadre pour faire de la géométrie algébrique globale relative. Grothendieck en donne une illustration saisissante, au séminaire Bourbaki, avec les schémas de Hilbert. Par leur définition naturelle et leur étude différentielle commode, ceux-ci vont vite supplanter les coordonnées de Chow.

Passage au quotient

Passer au quotient, en géométrie algébrique, par exemple sous l'action d'un groupe algébrique G opérant sur une variété X, a été longtemps un obstacle : les géomètres savaient qu'ils manquaient de suffisamment de fonctions algébriques pour séparer les orbites qu'ils pouvaient raisonnablement espérer séparer. Faute de mieux, ils définissaient le quotient X/G par une propriété universelle parmi les variétés équivariantes sous G.

Pour Grothendieck, ce point de vue n'est guère satisfaisant : on prétend définir un schéma par les flèches qui en sont issues et non par celles qui y aboutissent.

Lorsqu'il aborde les schémas en groupes dans son séminaire SGA 3, avec Michel Demazure, il revient au point de vue fonctoriel. Si l'on a un S-schéma en groupes G qui opère sur un S-schéma X, ou plus généralement si l'on a une relation d'équivalence R sur X, voire un groupoïde, il y a lieu de considérer le préfaisceau quotient qui a tout S-schéma T associe l'ensemble quotient naïf de $X(T)$ par $R(T)$. Ensuite, on passe au faisceau associé pour la topologie fpqc, ou une autre topologie de Grothendieck bien choisie. C'est ce faisceau qu'il y a lieu de considérer comme le véritable quotient X/R. Ainsi, tautologiquement, on en connaît ses points et X/R domine tout quotient schématique. Bien sûr, ce quotient est rarement représentable.

Mais avec les exposés de Gabriel dans SGA 3, sont étudiés des cas fort utiles : groupoïdes finis et plats, quotients génériques sous des hypothèses de platitude. Le triomphe survient dans le cas d'une relation d'équivalence projective et plate, annoncé à Bourbaki en 61, qui est le point d'orgue de la construction du schéma de Picard, exposée l'année suivante. Par sa généralité et son élégance, cette construction marque un sommet dans l'utilisation des techniques projectives.

Beaucoup de constructions de schémas de modules conduisent à des problèmes de passage au quotient par le groupe PGL_n. Dans le cas des actions de groupes réductifs, la situation est meilleure, Mumford introduit la notion de stabilité et semi-stabilité dans "Geometric Invariant Theory", qui paraît en 65. Il est aussi le premier à construire, par voie purement algébrique, sur l'anneau des entiers Z, la variété de modules des courbes propres et lisses et celle des variétés abéliennes polarisées.

Représentabilité

Bien sûr, comme on l'a déjà remarqué, une des raisons qui pousse Grothendieck à adopter le point de vue fonctoriel est que nombre de propriétés usuelles des schémas se lisent commodément sur le foncteur. Dès lors, on peut se demander si, partant d'un foncteur contravariant F, sur la catégorie des schémas, à valeurs dans les ensembles, on peut accumuler des informations de diverse nature sur F, jusqu'à pouvoir garantir la représentabilité de F. C'est là une approche entièrement nouvelle que Grothendieck forge dès les premières années où il s'investit en géometrie algébrique et qu'il affine au fur et à mesure de sa progression. Les résultats vont s'avérer assez stupéfiants.

On en a déjà mentionné quelques-uns. Par exemple, le fait que F soit représentable par un S-schéma localement de présentation finie nécessite que F commute aux limites inductives filtrantes d'anneaux. Comme signalé, ce fait apparaît un peu plus tard; au début Grothendieck se limite à des bases localement noethériennes.

Par contre, très tôt, Grothendieck met l'accent sur les propriétés d'exactitude à gauche. Elles résultent, principalement, de la descente fpqc et se traduisent en disant que F doit être un faisceau pour la topologie fpqc. Pour nombre de foncteurs naturels, cette propriété découle automatiquement de la théorie de la descente. Mais quelquefois, on part seulement avec un préfaisceau et il y a lieu de prendre pour F le faisceau associé. C'est le cas, par exemple, pour le foncteur de Picard.

Très vite, Grothendieck dégage la notion de proreprésentabilité. Si F est représentable par un S-schéma localement de type fini, sur une base locale noethérienne de point fermé s, les complétés des anneaux locaux de F, aux points fermés de la fibre spéciale au-dessus de s, sont déterminés par la restriction de F à la catégorie des S-schémas locaux artiniens.

Réciproquement, la connaissance de F sur la catégorie des S-schémas artiniens, jointe à quelques propriétés d'exactitude à droite, détermine les complétés des anneaux locaux de F aux points fermés. La contribution de Schlessinger dégage des critères de proreprésentabilité, très accessibles en pratique.

L'étape suivante, essentielle, est la commutation de F aux limites adiques. Elle nous fait passer des anneaux artiniens aux complétés des futurs anneaux du schéma représentant F. C'est, a priori, une étape délicate. Mais dans un contexte de faisceaux cohérents sur des schémas propres, elle résulte souvent du GAGA algébrique – formel de EGA III.

A-t-on accumulé suffisamment d'informations pour conclure à la représentabilité de F ? Oui, si l'on ajoute que F est séparé sur S et quasi-fini (disons non ramifié), comme en témoigne l'exposé de Murre à Bourbaki en 1966. En effet, dans ce cas, un argument de récurrence sur la dimension de la base et l'utilisation de la descente fpqc, permettent de passer des complétés aux anneaux locaux eux-mêmes.

L'étape finale, dans ces questions de représentabilité, allait être accomplie par Mike Artin avec son merveilleux théorème d'approximation. On est en 67 ; pour les besoins de la topologie et de la cohomologie étale, l'hensélisation bat son plein. L'énoncé d'approximation d'Artin est un peu rébarbatif à formuler. Essayons malgré tout. Soit A un anneau local, essentiellement de type fini sur un anneau de valuation discrète excellent. Notons m son idéal maximal, $A^\wedge$ son complété m-adique et A^h son hensélisé. Considérons un système d'équations dans $A[T_1, \ldots, T_r]$ qui admet une solution s dans $A^\wedge$. Alors il admet une solution dans A^h arbitrairement proche de s, m-adiquement. La démonstration résulte d'une habile application du théorème de préparation de Weierstrass.

Signalons, pour la petite histoire, que Chevalley a, autrefois, démontré la variante analytique complexe, mais que, faute d'en trouver des applications, il n'a pas cru devoir la publier. Artin, au contraire, voit tout l'intérêt de la version hensélienne. Il adapte l'énoncé à l'approximation du foncteur F que l'on souhaite représenter. En le combinant avec les critères patiemment accumulés par Grothendieck, il en conclut que F, à défaut d'être représenté par un schéma, est "représenté" par un "espace algébrique", c'est-à-dire par un faisceau étale, quotient d'un schéma localement de présentation finie par une relation d'équivalence étale.

Je me suis souvent demandé si Grothendieck avait imaginé une conclusion aussi satisfaisante. Oui, peut-être? En tout cas, si Grothendieck en a rêvé, Artin l'a fait.

Autant dire que l'énoncé d'Artin "tue" la représentabilité qui perd tout son mystère : la représentabilité (du moins comme espace algébrique) est maintenant découpée en une somme de propriétés qui, dans la pratique, sont aisées à vérifier séparément. Bien sûr, il faudrait réécrire toute la géométrie algébrique, dans le cadre des espaces algébriques, un programme esquissé par Artin et Knutson.

Aujourd'hui, l'intérêt pour les espaces algébriques a quelque peu faibli et l'accent est plutôt mis sur les champs qui permettent de mieux cerner la réalité géométrique.

The Picard scheme

Steven L. Kleiman

ABSTRACT. This article introduces, informally, the substance and the spirit of Grothendieck's theory of the Picard scheme, highlighting its elegant simplicity, natural generality, and ingenious originality against the larger historical record.

1. Introduction

A scientific biography should be written in which we indicate the "flow" of mathematics... discussing a certain aspect of Grothendieck's work, indicating possible roots, then describing the leap Grothendieck made from those roots to general ideas, and finally setting forth the impact of those ideas.

—*Frans Oort* [**60**, p. 2]

Alexandre Grothendieck sketched his proof of the existence of the Picard scheme in his February 1962 Bourbaki talk. Then, in his May 1962 Bourbaki talk, he sketched his proofs of various general properties of the scheme. Shortly afterwards, these two talks were reprinted in [**31**], commonly known as **FGA**, along with his commentaries, which included statements of nine finiteness theorems that refine the single finiteness theorem in his May talk and answer several related questions.

However, Grothendieck had already defined the Picard scheme, via the functor it represents, on pp. 195-15, 16 of his February 1960 Bourbaki talk. Furthermore, on p. 212-01 of his February 1961 Bourbaki talk, he had announced that the scheme can be constructed by combining results on quotients sketched in that talk along with results on the Hilbert scheme to be sketched in his forthcoming May 1961 Bourbaki talk. Those three talks plus three earlier talks, which prepare the way, were also reprinted in [**31**].

Department of Mathematics, Massachusetts Institute of Technology, Cambridge, MA 02139-4307, USA. kleiman@math.mit.edu.

Moreover, Grothendieck noted in [**31**, p. C-01] that, during the fall of 1961, he had discussed his theory of the Picard scheme in some detail at Harvard in his term-long seminar, which David Mumford and John Tate continued in the spring. In November 2003, Mumford kindly lent me his own folder of notes from talks given by each of the three, and notes written by each of them. Virtually all the content was published long ago.

Those notes contain a rudimentary form of the tool now known as *Castelnuovo–Mumford regularity.* Grothendieck mentions this tool in his commentaries [**31**, p. C-10], praising it as the basis for an "extremely simple" proof of a bit weaker version of his third finiteness theorem. Mumford sharpened the tool in his book [**46**, Lect. 14], so that it yields the finiteness of the open subscheme of the Hilbert scheme that parameterizes all closed subschemes with given Hilbert polynomial.

Grothendieck [**31**, p. 221-1] correctly foresaw that the Hilbert scheme is "destined to replace" Chow coordinates. As he [**31**, p. 195-14] put it, they are "irremediably insufficient," because they "destroy the nilpotent elements in parameter varieties." Nevertheless, he [**31**, p. 221-7] had to appeal to the theory of Chow coordinates to prove the finiteness of the Hilbert scheme. So after he received a prepublication edition of [**46**], he wrote a letter on 31 August 1964 to Mumford in which he [**48**, p. 692] praised the theory in Lecture 14 as "a significant amelioration" of his own.

Mumford made use of the finiteness of the Hilbert scheme in his construction of the Picard scheme over an algebraically closed field in [**46**, Lect. 19], whereas Grothendieck took care to separate existence from finiteness, giving an example in [**31**, Rem. 3.3, p. 232-07] over a base curve of a Picard scheme with connected components that are not of finite type.

Mumford's book [**46**] was based closely on the lovely course he gave at Harvard in the spring of 1964. It was by far the most important course I ever took, due to the knowledge it gave me and the doors it opened for me. During the academic year of 1966–67, I was a Postdoc under Grothendieck at the IHES (Institut des Hautes Études Scientifiques). When he learned from me that I had taken that course and had advanced some of the finiteness theory in my thesis [**35**, Ch. II], he asked me to write up proofs of his nine finiteness theorems for SGA6 [**8**, Exps. XII, XIII].

Grothendieck, perhaps, figured that I had learned how to prove his nine theorems at Harvard, but in fact I had not even heard of them. At any rate, he told me very little about his original proofs, and left me to devise my own, which I was happy to do. There is one exception: the first theorem, which concerns generic relative representability of the Picard scheme. Its proof has a very different flavor, as it involves nonflat descent, Oort dévissage, and representability of unramified functors. Grothendieck asked Michel Raynaud to lecture on this theorem and to send me his lecture notes, which I wrote up in [**8**, Exp. XII].

My experience led me to study Grothendieck's construction of the Picard scheme, and to teach the whole theory a number of times. Further, in collaboration with Allen Altman, Mathieu Gagné, Eduardo Esteves, Tony Iarrabino, and Hans Kleppe, I extended some of Grothendieck's theory to the compactified Picard scheme. The underlying variety had been introduced via Geometric Invariant Theory in 1964 by Alan Mayer and Mumford in [**44**, § 4]. The scheme has been studied and used by many others ever since then.

Thus my experience is like the experiences of Nicholas Katz and Barry Mazur, which were described by Allyn Jackson in [**33**, p. 1054]. Katz said that Grothendieck assigned him the topic of Lefschetz pencils, which was new to him, but "he learned a tremendous amount from it, and it had a big effect on my future." Mazur said that Grothendieck asked him this question posed earlier by Gerard Washnitzer: "Can the topology of an algebraic variety vary with the complex embedding of its field of definition?" Mazur, then a differential topologist, added, "But for me, it was precisely the right kind of motivation to get me to begin to think about algebra."

Both Katz and Mazur then confirmed my impression that our experiences were typical. Jackson quotes Katz as saying that Grothendieck got visitors interested in something, but with "a kind of amazing insight into what was a good problem to give to that particular person to think about. And he was somehow mathematically incredibly charismatic, so that it seemed like people felt it was almost a privilege to be asked to do something that was part of Grothendieck's long range vision of the future." Similarly, Mazur said that Grothendieck had an instinct for "matching people with open problems. He would size you up and pose a problem that would be just the thing to illuminate the world for you. It's a mode of perceptiveness that's quite wonderful and rare."

I spent the summer of 1968 at the IHES. Grothendieck invited me to his home in Massy-Verrières to discuss my drafts for my contributions to SGA 6. His comments ranged from providing insight into the theory of bounded families of sheaves to criticizing my starting sentences with symbols.[1] Again, my experience was typical:[2] Jackson [**33**, p. 1054] quotes Luc Illusie as saying,

[1]Many years later, Jean-Pierre Serre told me that he had taught Grothendieck not to start sentences with symbols.

[2]Grothendieck gave me another project during my Postdoc. On April 18 and 25 that year, he talked in his seminar at the IHES on his Standard Conjectures and Theory of Motives. He asked me to write up his talks, gave me copies of his notes on related matters, and invited me to his home a year later, in the summer of 1968, to discuss my draft. That work too led me to learn some good mathematics and to write several articles, although they are more expository. Also, it led to my co-chairing an organizing committee for an AMS summer research conference in 1991.

However, my experience was the exception that proved the rule: Grothendieck had already asked others to write up his talks; they tried, and gave up! Also, curiously he never told me about his talk on the Standard Conjectures at a conference in Bombay, India, in January 1968, let alone offer me his notes. Moreover, in the conference proceedings, his writeup cites a talk of mine at the IHES, which I never gave, crediting me for an observation; but it is due to Saul Lubkin, and credited to him in my writeup [**36**, p. 361], which Grothendieck critiqued in his home that summer.

that Grothendieck often worked at home with colleagues and students, making a wide range of apposite comments on their manuscripts.

One time, Grothendieck found that I didn't know some result treated in EGA ([**29**] and sequels). So he gently advised me, for my own good, to read a little EGA every day, in order to familiarize myself with its content. After all, he pointed out, he had been writing EGA as a service to people like me; now it was up to us to take advantage of this resource. That experience supports a statement Leila Schneps made in [**64**, p. 16]: "The foundational work that Grothendieck and [Jean] Dieudonné were undertaking [was] in the service of all mathematicians, of mathematics itself. The strong sense of duty and public service was felt by everyone around Grothendieck."

As Grothendieck stated on p. 6 of [**29**], he planned to develop in EGA the ideas he sketched in [**31**]. He did not succeed. Nevertheless, those ideas have become a basic part of Algebraic Geometry. So they were chosen as the subject of a summer school held 7–18 July 2003 at the ICTP (International Center for Theoretical Physics) in Trieste, Italy. The first Bourbaki talk reprinted in [**25**] was not covered; it treats Grothendieck's generalization of Serre duality for coherent sheaves, so is somewhat apart and was already amply developed in the literature.

The lectures were written up, and published in [**25**]. As stated on p. viii, "this book fills in Grothendieck's outline. Furthermore, it introduces newer ideas whenever they promote understanding, and it draws connections to subsequent developments." In particular, I wrote about the Picard scheme, beginning with a 14-page historical introduction, which served as a first draft for the present article.

Mumford stated the goal of his book [**46**] on pp. vii–viii: "a complete clarification of... the so-called[3] Completeness of the Characteristic Linear System of a good complete algebraic system of curves on a surface.... Until about 1960, no algebraic proof of this purely algebraic theorem was known.... [Then] a truly amazing development occurred:" by combining his results on the Hilbert scheme and the Picard scheme with Cartier's result, "that group schemes in characteristic 0 are reduced," Grothendieck [**31**, pp. 221-23,24] obtained in February 1961 an enlightening, purely algebraic proof. "The key... is the systematic use of nilpotent elements."

Grothendieck had, moreover, reversed history: he proved Completeness via the Picard scheme. By contrast, in December 1904 Federigo Enriques and sometime in 1905 Francesco Severi gave algebraic proofs of Completeness from scratch. In the first half of 1905, on the basis of Enriques's work, Guido Castelnuovo **introduced** the Picard variety in order to prove the

[3]On p. 8, the theorem is formulated as "problem (B)," and two analytic solutions are outlined. On p. 157, a more precise version is formulated as the "Fundamental Theorem," and given its first algebraic proof. On p. 169, an important special case is proved, following Grothendieck's somewhat different algebraic treatment. However, none of those is called the "Theorem of Completeness."

Fundamental Theorem of Irregular Surfaces. It asserts the surprising equality of the four basic invariants: the dimension of the Picard variety, the irregularity, the number of independent Picard integrals of the first kind, and half the first Betti number. Grothendieck's theory, without reference to Completeness, also yields the first part of the Fundamental Theorem, that the dimension of the Picard variety is equal to the irregularity in characteristic 0.

Both Enriques's and Severi's proofs have serious gaps, as Severi himself noted in 1921.[4] Severi then proved a more restricted version of Completeness, but one sufficient for Castelnuovo's work. Severi's proof was based on Henri Poincaré's construction of a key system of curves. That construction appeared in 1910, 1911; it is rigorous, but analytic. After 1921, finding a fully rigorous, purely algebraic proof of a suitable version of Completeness became a major endeavor — undertaken by Enriques, Severi, and others — until Grothendieck finally settled the matter. Section 2 explains more fully the history and meaning of Completeness and of the Fundamental Theorem; Section 5 elaborates on Grothendieck's proof.

When Grothendieck worked on his theory of the Picard scheme, the general algebro-geometric theory of the Picard variety had been under active development for nearly fifteen years. More than twenty mathematicians had worked on various aspects. Grothendieck clarified and settled a number of issues. Section 3 explains those issues in chronological order. Sections 4 and 5 give more detail about Grothendieck's advances, which involve many great innovations.

One issue was a topic of conversation between Grothendieck and Jacob Murre sometime in the academic year 1960/61. Murre told Schneps about it, and she [**64**, pp. 1–2] quoted him as saying, "A very important unsolved question ... [was] the behavior of the Picard variety if the original variety ... moved in a system and moreover — and worse — in[to] characteristic $p > 0$. ... I asked Grothendieck whether he could explain this behavior. ... He said he would certainly [do so]. ... Then, in 1962, Grothendieck completely solved the question. ... I attended his Bourbaki lectures, and needless to say, I was very impressed!"

As it happens, much earlier, in his 1958 talk [**28**, p. 118] at the ICM (International Congress of Mathematicians), Grothendieck said, "We shall not give here the precise definition of a 'relative Picard schema', but ... if this schema exists then it behaves in the simplest conceivable way with respect to change of base-space." In his February 1960 Bourbaki talk [**31**, pp. 195-15, 16], he added that "in particular, the Picard schemes of the fibers" of a system are the fibers of the relative Picard scheme, once existence is proved. In his February 1961 talk [**31**, pp. 212-01], as noted above, he announced his proof of existence.

[4]In 1949, Severi [**72**, p. 40] lamented the fact that "this annoying episode was taken as an article of indictment for the [crime of] lack of rigor in Italian algebraic geometry!"

Thus when Grothendieck had succeeded in settling a major issue, such as the Behavior of the Picard Variety in a Family or the Completeness of the Characteristic System, he noted the advance, but did not tout it. Cartier [**12**, p. 17] describes Grothendieck's philosophy as follows: "Grothendieck was convinced that if one has a sufficiently unifying vision of mathematics, if one can sufficiently penetrate the essence of mathematics and the strategies of its concepts, then particular problems are nothing but a test; they do not need to be solved for their own sake."

One beautiful illustration of Grothendieck's "unifying vision" is provided by his theory of the Picard functor. It is the *functor of points* of the Picard scheme — that is, the functor whose values are the sets of maps from a variable source into the scheme. Often, a functor of points is said to provide nothing more than another way of expressing the universal property of a fine moduli scheme. That statement is true for the Hilbert scheme, but a half-truth for the Picard scheme.

What is the universal property of the Picard scheme? The naive answer falls short! However, Grothendieck saw the hidden common thread in descent of the base field, Galois cohomology, and sheaf theory; he concluded that any functor of points must be a sheaf for the fpqc Grothendieck topology. Thus the right Picard functor has to be the sheaf associated to the naive Picard functor, regarded as a presheaf. More work with the functor leads to the construction of the Picard scheme. It is automatically compatible with base change, because the Picard functor is so. Sections 4 and 5 explain all that theory.

In short, Section 2 gives a historical introduction to two venerable theorems: the Theorem of Completeness of the Characteristic System, and the Fundamental Theorem of Irregular Surfaces. Section 3 gives a historical introduction to the inadequate algebro-geometric theory of the Picard variety. Please note: these two introductions are **not** meant to be either serious historical studies or rigorous mathematical surveys, but simply fascinating informal accounts, providing background material for comprehending the nature and extent of Grothendieck's advances.

Section 4 explains Grothendieck's innovative theory of the Picard functor, culminating in his main construction of the Picard scheme. Finally, Section 5 explains how the theory of the Picard scheme enabled Grothendieck and others to provide enlightening treatments of the issues discussed in Sections 2 and 3. The discussions in Section 4 and 5 are mathematically rigorous, but just introductory. Sources for more information are given at the beginning of each of Sections 2–5.

There are three minor mathematical novelties below: (1) the proof on p. 46 of the equivalence of the 19th century definition of the arithmetic genus of a surface and the modern definition, (2) the algebro-geometric treatment on p. 64 of Severi's 1921 version of Completeness, and (3) the "nearly formal" treatment on p. 67 of the Albanese variety, including duality, intriguingly announced by Grothendieck on p. 232-14 of his February 1962 Bourbaki talk [**31**].

Acknowledgments. It is a great pleasure to thank my good friend Luc Illusie for carefully reading an earlier version of this paper, and making numerous comments, which led to many significant improvements. It is a pleasure as well to thank Michael Artin, Matthias Kuenzer, Arthur Mattuck, Barry Mazur, David Mumford, Brian Osserman, Ragni Piene, Michel Raynaud, and the referee for their comments, which led to many other important improvements.

2. Irregular Surfaces

But to demonstrate the power of modern abstract ideas
to solve older very concrete problems,
I think that this example is unmatched.
David Mumford [**50**, p. 7]

In the quotation above, *this example* refers to Grothendieck's treatment of the Theorem of Completeness of the Characteristic System. In fact, the example is the centerpiece of Mumford's article [**50**] in this volume. Moreover, Mumford notes that Completeness yields the Fundamental Theorem of Irregular Surfaces. Thus if we are to appreciate the full significance of Grothendieck's contribution, then we must review the history of those two main theorems. We do so in this section. First we pursue, intuitively, the spirit of the original work. Then we treat that work rigorously, beginning at the end of this section and continuing in Section 5.

A number of historical reviews are already available, and served as a basis for the account here. Notably, in 1906, Castelnuovo and Enriques wrote[5] one [**14**] at the request of Emile Picard to be an appendix to Tome II of his book [**62**] with Georges Simart. In 1934, Oscar Zariski reviewed various aspects of the development in different places in his celebrated book [**78**]. Those reviews are fairly technical. In 1994, Fabio Bardelli [**7**] wrote a more informal review of the developments through 1934. In 1974, Dieudonné published a masterful history of algebraic geometry, which touches on these theorems in particular. The book was translated as [**21**] by Judith Sally, and supplemented with an extensive annotated bibliography.

In 2011, Mumford [**49**] carefully analyzed the mathematics in a 1936 paper by Enriques on Completeness. Mumford, in his introduction, stated his conclusion: "Enriques must be credited with a nearly complete [algebro-]geometric proof using, as did Grothendieck, higher order infinitesimal deformations.... Let's be careful: he certainly had the correct ideas about infinitesimal geometry, though he had no idea at all about how to make precise

[5]Please also see their encyclopedia article [**15**] and Castelnuovo's historical note [**23**, pp. 339–353].

definitions."[6] Mumford's article is preceded by a lovely article by Donald Babbitt and Judith Goodstein [**6**], which focuses on the times, lives, and personalities of Enriques and his colleagues; please also see their related articles [**5**] and [**26**]. All the articles mentioned above give many precise references, which are not repeated here.

Around 1865, Alfred Clebsch caused a sea change in algebraic geometry, turning it away from the concrete study of particular curves and surfaces, and toward the abstract study of their *birational invariants* — the numbers that depend only on their field of rational (or global meromorphic) functions.

In 1868, Clebsch considered a connected *smooth*[7] complex projective algebraic surface $\widetilde{X}$ of large degree n. He studied it via its general projection in 3-space, which is a surface

$$X : f(x, y, z) = 0 \quad \text{and} \quad n := \deg f$$

with "ordinary" singularities, none at infinity, and no point at infinity on the z-axis.

Clebsch found the algebraic double integrals on $\widetilde{X}$ of the *first kind* — that is, those finite on any bounded analytic domain of integration — to be of the form

$$\iint \frac{h(x, y, z)}{\partial f/\partial z}\, dx\, dy$$

where h is a polynomial of degree at most $n - 4$ vanishing on the singular locus,

$$\Gamma : f,\ \partial f/\partial x,\ \partial f/\partial z,\ \partial f/\partial z = 0,$$

a curve of double points. The number of linearly independent such integrals became known as the *geometric genus* and denoted by p_g.

Clebsch asserted without proof that p_g is a birational invariant. In 1870, his student, Max Noether, gave an algebraic proof. In 1869, Arthur Cayley worked out a formula for the number of independent h; essentially, he found an explicit expression for $F(n-4)$ where F is the Hilbert polynomial of the homogeneous ideal of Γ. The value $F(n - 4)$ was later called the *arithmetic genus* and denoted by p_a.[8]

In 1871, Hieronymous Zeuthen used Cayley's formula to prove algebraically that p_a too is a birational invariant. Also in 1871, Cayley observed that, if X is a ruled surface with plane section of genus g, then $p_a = -g \leq 0$, although $p_g = 0$.

[6] Mumford elaborated in his resumeé at the end: "Although [Enriques] gave [infinitesimal deformations] names, they remained in limbo, without substance, because he did not think of what it meant to have a function on them. Grothendieck realized that functions on such objects should be rings with nilpotent elements, and this gave life to these infinitesimal deformations."

[7] Also called *nonsingular,* $\widetilde{X}$ is defined by polynomials with Jacobian matrix of maximal rank.

[8] Cayley [**16**] denoted it by D, and called it the *deficiency.* Picard and Simart [**62**, p. 88] denoted it by p_n, and called it the *numerical genus.* Those definitions soon fell into disuse.

The disagreement between p_g and p_a came as a surprise. In 1875, Noether explained it: $F(n-4)$ is the number of independent h only if n is suitably large. In any case, $p_g \geq p_a$. Moreover, if X is smooth or rational, then $p_g = p_a$. It was thought that, as a rule, $p_g = p_a$, and when so, X was dubbed *regular*. The failure of X to be regular is quantified by the difference $p_g - p_a$; so it became known as the *irregularity*. Zariski [**78**, p. 75] denoted it by q; following suit, set

$$q := p_g - p_a.$$

In 1884, Picard initiated the study of algebraic simple integrals

$$\int P(x,y,z)\,dx + Q(x,y,z)\,dy$$

that are *closed*, or $\partial P/\partial y = \partial Q/\partial x$; they became known as *Picard integrals*. He proved that there are only finitely many independent such integrals of the *first kind*, those finite on any bounded analytic path of integration; use[9] s to denote their number. Picard noted that, if X is smooth, then $s = 0$.

In 1894, Georges Humbert considered an *algebraic system*, or *algebraic family*, of curves. Its members are the zeros on X of a polynomial

$$\varphi(x,y,z;\lambda_0,\ldots,\lambda_t)$$

where the λ_i satisfy polynomial equations, which define the *parameter variety* Λ. The curves can all contain common subcurves; some of them are included as *fixed components* of the system, and the others, omitted. The system is said to be *linear* if there are homogeneous polynomials $\varphi_i(x,y,z)$ of the same degree with

$$\varphi = \lambda_0\varphi_0 + \cdots + \lambda_t\varphi_t.$$

Humbert proved a remarkable result: if $s = 0$, then every algebraic system is a subsystem of a linear system. That result inspired Castelnuovo to prove in 1896 that, if $q = 0$, then again every algebraic system is a subsystem of a linear system under a certain restriction, which Enriques removed in 1899.

In 1897, Castelnuovo fixed a linear system of curves on X. Let r be its *dimension*, the number of linearly independent restrictions $\varphi_i|X$ diminished by 1. Castelnuovo studied its *characteristic* linear system, the system cut out by the φ_i on a general member curve D_η, assuming D_η is *irreducible*, that is, not the union of two smaller curves. The characteristic system has dimension has $r-1$.

Castelnuovo formed the *complete*, or largest, linear system on D_η containing the characteristic system. Let δ be the amount, termed the *deficiency*, by which the dimension of the characteristic system falls short of the dimension of its complete linear system. Castelnuovo proved that

$$\delta \leq q, \tag{1}$$

[9]Castelnuovo and Enriques [**14**, p. 495] used q, whereas Zariski [**78**, p. 162] used r_0.

with equality if the linear system consists of all *hypersurface sections* of high degree, namely if the $\varphi_i(x, y, z)$ generate all homogeneous polynomials of that degree.

In February 1904, Severi extended Castelnuovo's work. Severi fixed an algebraic system of curves on X, and a general member D_η. He assumed that D_η is irreducible and that $D_\lambda \neq D_\mu$ for all distinct λ, $\mu \in \Lambda$. As λ approaches η along a path in Λ, the intersections $D_\lambda \cap D_\eta$ approach a limit, which depends only the tangent vector at η to the path. The various limits form a linear system on D_η, parameterized by the projectivized tangent space to Λ at η. Thus Severi constructed the *characteristic linear system* of the algebraic system. Set $R := \dim \Lambda$. Then this characteristic system is of dimension $R - 1$.

In the algebraic system, form the largest linear subsystem containing D_η. Denote its dimension by r. Form its characteristic linear system. Let δ be its deficiency. Then its complete linear system has dimension $r - 1 + \delta$, and it also contains the characteristic system of the algebraic system. Thus Severi proved that

$$R \leq r + \delta, \tag{2}$$

with equality if and only if the latter characteristic system is complete.

In December 1904 Enriques and sometime in 1905 Severi each constructed an algebraic system with $R = r + q$. Both constructions are short and delicate. Both rely on the completeness of the characteristic system of certain[10] algebraic systems. Both are flawed, as Severi himself pointed out in 1921. In 1934, Zariski [**78**, pp. 99–102] reviewed those constructions, "in order to analyze the assumption on which they are based and for which as yet an algebro-geometric proof is not available."

In 1910 and 1911 using a new method of "normal functions," Poincaré gave a rigorous analytic construction of an algebraic system with $r = 0$ and $R = q$. His construction was simplified and developed by Severi in 1921 and Solomon Lefschetz in 1921 and 1924. In 1934, Zariski [**78**, pp. 169–173] reviewed that work too.[11] He [**78**, p. 102] noted that "the value of the construction of such a system is greater than that of mere example; indeed it is an essential step in the theory."

Zariski then derived Severi's May 1905 theorem[12] that, if there is one system with $R = r + q$, then $R = r + q$ holds for every complete system whose general member D_η is *arithmetically effective*; namely, a certain lower semi-continuous combination of its numerical characters is nonnegative,

---

[10]Severi [**72**, p. 41] noted that both he and Enriques believed at the time that they had proved every complete algebraic system with irreducible general member has a complete characteristic system!

[11]Please also see the reviews of Mumford [**46**, pp. 9–10] and Dieudonné [**21**, p. 53].

[12]In December 1904, Enriques proved the theorem under the more stringent, but still sufficient, hypothesis that D_η is "regular," later renamed "regular and nonspecial." Please see Fn.43 on p. 63.

a common condition (please see p. 64). Hence, by (1) and (2), if D_η is irreducible too, then its characteristic system is complete and $\delta = q$. By Bertini's Theorem, usually D_η is irreducible.

The **Theorem of Completeness** came[13] to mean the following assertion:

(3) Every complete algebraic system whose general member is arithmetically effective and irreducible has a complete characteristic system.

Moreover, (3) is equivalent to the existence of at least one system with $R = r + q$, and in turn to the existence of suitably many such systems.

On 16 January 1905 in the C. R. Paris, Enriques [**22**, pp. 134–135] announced that Severi had just proved $q \geq s$ and $q = b - s$, where b is the number of independent Picard integrals of the *second kind*, those with polar singularities. It was known before 1897 that b is equal to the first Betti number; please see [**78**, p. 157]. In the same issue of the C. R., Picard [**61**] proved $q = b - s$ independently.[14]

In the next issue on 23 January, Castelnuovo [**13**] outlined the last step in this direction. He gave the details in three notes in the Rend. Accad. Lincei of 21 May and 4 and 8 June 1905. Specifically, he took a complete algebraic system with arithmetically effective (in fact, regular) general member, fibered it into linear systems, and formed the quotient, P say. Then P is projective, and P is of dimension q as $R = r + q$, an equation he considered proved. Morover, P is, up to isomorphism, independent of the choice of algebraic system, and sum (union) of curves induces an addition of points of P, turning P into a commutative group variety.

Hence, by a general 1895 theorem of Picard, completed in 1901 by Painlevé, P is an *Abelian variety*: P is parameterized by q *Abelian functions*, or $2q$-ply periodic functions of q variables, with a common lattice of periods. Castelnuovo proved that these functions induce independent Picard integrals on X. Therefore, $q \leq s$. Thus Castelnuovo obtained the **Fundamental Theorem of Irregular Surfaces**:

$$\dim P = q = s = b/2.$$

In 1905, the term "Abelian variety" was not yet in use. So naturally enough, Castelnuovo termed P the *Picard variety* of X.[15]

[13] According to Severi [**72**, p. 42], in 1921 he derived the theorem essentially in this form from Poincaré's construction.

[14] Picard presented Enriques's note to the Academy, but explained in Fn. ([1]) on p. 122 of [**61**] that he had completed his own note before receiving Enriques's.

[15] Castelnuovo [**13**, p. 221] explained that "out of respect for Picard's profound research on surfaces [sic] admitting a group of birational automorphisms, [he] proposes calling the variety P (and [a certain] group G_d) *the Picard variety (and Picard group) associated to the surface* X."

Andre Weil [**76**, I, p. 572] discussed his own use of the term "Picard variety" in his commentary on his 1950 paper on Abelian varieties. Weil said, "Historically speaking, it

In 1903, Severi [**69**, § 26] discovered a remarkable expression for p_a in terms of a different Hilbert polynomial. Say the smooth surface $\widetilde{X}$ is a subvariety of some higher dimensional projective space $\mathbb{P}^N$. Form the Hilbert polynomial $\widetilde{F}(\nu)$ of the homogeneous ideal of $\widetilde{X}$. Then $\widetilde{F}(0) - 1 = p_a$.

Serre, in his 1954 ICM talk [**66**, pp. 286–291], announced a theory of coherent algebraic sheaves, inspired by the analytic work of Friedrich Hirzebruch, Kunihiko Kodaira, and Donald Spencer. In particular, Serre proved the Euler characteristic of the twisted structure sheaf $\chi(\mathcal{O}_{\widetilde{X}}(\nu))$ is equal to $\widetilde{F}(\nu)$. Thus $p_a = \chi(\mathcal{O}_{\widetilde{X}}) - 1$;[16] so p_a is independent of the embedding of $\widetilde{X} \subset \mathbb{P}^N$ and of the projection $\widetilde{X} \to \mathbb{P}^3$.

Further, Serre Duality yields this equality of dimensions of cohomology groups: $h^i(\mathcal{O}_{\widetilde{X}}) = h^{2-i}(\Omega^2_{\widetilde{X}})$ for all i where $\Omega^2_{\widetilde{X}}$ is the sheaf of algebraic 2-forms. However, $p_g = h^0(\Omega^2_{\widetilde{X}})$, essentially by definition, and $h^0(\mathcal{O}_{\widetilde{X}}) = 1$ as $\widetilde{X}$ is connected. Thus

$$p_a = \chi(\Omega^2_{\widetilde{X}}) - 1, \quad p_g = h^2(\mathcal{O}_{\widetilde{X}}), \quad q = h^1(\mathcal{O}_{\widetilde{X}}).$$

Above, the first equation is a form of Severi's discovery. Here is a proof of it using Grothendieck's generalization of Serre Duality. Let ω_X be the dualizing sheaf. Since $X \subset \mathbb{P}^3$ and $\Omega^3_{\mathbb{P}^3} = \mathcal{O}_{\mathbb{P}^3}(-4)$, duality theory and simple computation yield

$$\omega_X = \mathit{Ext}^1(\mathcal{O}_X\,,\ \Omega^3_{\mathbb{P}^3}) = \mathcal{O}_X(n-4).$$

Hence, duality for the finite map $\pi\colon \widetilde{X} \to X$ and elementary manipulation yield

$$\pi_*\Omega^2_{\widetilde{X}} = \mathit{Hom}(\pi_*\mathcal{O}_{\widetilde{X}}\,,\ \omega_X) = \mathfrak{C}(n-4) \quad \text{where} \quad \mathfrak{C} := \mathit{Hom}(\pi_*\mathcal{O}_{\widetilde{X}}\,,\ \mathcal{O}_X).$$

Thus $\chi(\Omega^2_{\widetilde{X}}) = \chi(\mathfrak{C}(n-4))$.

Here, $\mathfrak{C}$ is the conductor; it is the ideal sheaf on X of the curve Γ of double points. Let $\mathfrak{C}_0$ be the ideal sheaf on $\mathbb{P}^3$ of Γ. Then $p_a = \chi(\mathfrak{C}_0(n-4))$, essentially by definition. Form the standard exact sequence

$$0 \to \mathcal{O}_{\mathbb{P}^3}(-4) \xrightarrow{\times f} \mathfrak{C}_0(n-4) \to \mathfrak{C}(n-4) \to 0.$$

By Serre's Computation, $\chi(\mathcal{O}_{\mathbb{P}^3}(-4)) = -1$. Thus $p_a = \chi(\mathfrak{C}(n-4)) - 1$, as desired.

Recall $q = h^1(\mathcal{O}_{\widetilde{X}})$. Also, $s = h^0(\Omega^1_{\widetilde{X}})$ essentially by definition. So Hodge Theory yields $q = s$ and $q = b/2$, but Hodge Theory is not algebraic. However, a p-adic algebraic proof that $q = s$ was given by Kirti Joshi [**34**]. Further, if by b is meant the dimension of the first Grothendieck étale cohomology group, then a standard algebraic argument yields $\dim P = b/2$; see [**36**, Lem. 2A7, p. 375] for example. The latter argument also works in positive characteristic, but the equations $\dim P = q$ and $q = s$ may fail. In

would have been justified to give it Castelnuovo's name, but it was a matter of tampering as little as possible with common usage rather than rendering due homage unto this master."

[16] Afterwards, it became common to define p_a by this formula.

1955, Jun-ichi Igusa gave an example with $\dim P = 1$ but $q = s = 2$; in 1958, Serre [**66**, p. 529] gave one with $\dim P = s = 0$ but $q = 1$.

Finally, $H^1(\mathcal{O}_{\tilde{X}})$ is always the Zariski tangent space at 0 to the Picard scheme by Grothendieck's theory, and over $\mathbb{C}$ the Picard scheme is smooth by Cartier's theorem; so $\dim P = q$. Thus the Fundamental Theorem of Irregular Surfaces can be proved algebraically over $\mathbb{C}$, and the proof does not involve the Theorem of Completeness of the Characteristic System. Yet, the latter theorem has taken on a life of its own, and Grothendieck's work is heavily involved in proving both theorems algebraically. All that work is discussed further in Section 5.

3. The Picard Variety

Ever since 1949, I considered the construction
of an algebraic theory of the Picard variety
as the task of greatest urgency in abstract algebraic geometry.
André Weil [**76**, II, p. 537]

Up to 1949, Weil worked primarily in Number Theory and Algebraic Geometry.[17] That work culminated in proofs of the Riemann hypothesis for curves in 1948 and in the formulation of his celebrated conjectures for arbitrary dimension in 1949. Next, he led "the construction of an algebraic theory of the Picard variety." In turn, that theory led Grothendieck to develop his theory of the Picard scheme. However, the Weil Conjectures themselves motivated much of Grothendieck's work. In particular, they led to the notion of a Grothendieck topology, which, as noted in the introduction, is fundamental for the very definition of the Picard functor; that functor is the subject of Section 4.

In the present section, so that we may better appreciate Grothendieck's advances, let us consider in chronological order up to 1962, what was sought and what was proved about the Weil conjectures and the Picard variety. Good secondary sources include Dieudonné's history [**21**] for all of it, Mazur's 1974 expository article [**43**] for the Weil conjectures, and the explanatory comments and historical notes in Serge Lang's 1959 book [**40**] for the Picard variety. Again, as those sources contain many primary references, those references are not always repeated here.

In his 1921 thesis, which was published in 1924, Emil Artin developed an analogue of the classical Riemann hypothesis, in effect, for a hyperelliptic curve over a prime field of odd characteristic. In 1929, Friedrich Karl Schmidt generalized Artin's work to all curves over all finite fields, recasting it in the algebro-geometric style of Richard Dedekind and Heinrich Weber. In 1882, they had viewed a curve as the set of discrete valuation rings in a finitely

[17] "Weil was far from confining himself to" those subjects, as Serre [**67**, pp. 523–526] noted, citing Weil's work in real and complex analysis, representation theory, and differential geometry.

generated field of transcendence degree 1 over $\mathbb{C}$, but their approach works in any characteristic. In particular, Schmidt ported their proof of the Riemann–Roch theorem, and used it to prove that Artin's Zeta Function satisfies a natural functional equation.

In 1936, Helmut Hasse proved Artin's Riemann hypothesis in genus 1 via an analogue over finite fields for the theory of elliptic functions. Then he and Max Deuring noted that to extend the proof to higher genus would require developing a similar analogue for the nineteenth century theory of correspondences between complex curves.

Their work inspired Weil.[18] In each of two notes, [**76**, I, pp. 257–259] of 1940 and [**76**, I, pp. 277–279] of 1941, he sketched a different proof of the Riemann hypothesis in any genus. In both, the key is a certain positivity theorem for correspondences. It was found over $\mathbb{C}$ by Castelnuovo[19] in 1906, and proved over a field of any characteristic by Weil in two ways: in 1940 by algebraizing Adolf Hurwitz's transcendental theory of 1886, and in 1941 by porting to positive characteristic the algebro-geometric theory in Severi's textbook [**70**] of 1926.

To provide the details, Weil had to redo the foundations of Algebraic Geometry over a field of arbitrary characteristic. The first instalment [**73**] appeared in 1946. Building on ideas of Emmy Noether, Bartel van der Waerden, and Schmidt from the 1920s, Weil fixed a *universal domain* Ω, a field of infinite transcendence degree over the prime field. Then a *projective variety* X is the locus of zeros with coordinates in Ω of homogeneous polynomials with coefficients in a variable *coefficient field* or *field of definition* k, a subfield of Ω over which Ω has infinite transcendence degree. Also X is *absolutely irreducible*, not the union of two smaller such loci. Then Weil formed *abstract* varieties by patching pieces of projective varieties.

Finally, Weil treated *cycles*. They are the formal $\mathbb{Z}$-linear combinations of subvarieties. Those of codimension 1 are called (*Weil*) *divisors*, and play a major role in the theory of the Picard variety. Weil developed a calculus of cycles, including intersection products, inverse images, and direct images.

In 1948, Weil published two books. In the first [**74**], he completed his note of 1941. He reproved the Riemann–Roch theorem, and developed an elementary theory of correspondences for curves. To prove Castelnuovo's theorem, he used his full calculus of cycles on products of numerous copies of the curve. The proof is "the most complicated part of the" book, as Otto

---

[18]The interaction among the three and others has attracted a lot of study. One delightful and well-documented report was published by Michèle Audin in 2012 as [**4**]. It describes the political, social, and personal circumstances at the time, while focusing on three reviews of Weil's first note.

[19]In both notes, Weil cites only Severi. In his commentary on the second note [**76**, I, p. 553], Weil wrote, "it's one of Castelnuovo's most beautiful discoveries (see his *Memorie Scelete,* no. XXVIII, pp. 509–517). But I didn't read Castelnuovo until 1945 in Brazil; then I realized that Severi in the *Trattato* ([**70**, pp. 286–287]) had not given his elder due credit."

Schilling observed in his Math Review [MR0027151]. Then Weil proved the Riemann hypothesis.[20]

Weil's proof inspired three others. First, in his 1953 thesis under Hasse, Peter Roquette translated it into the more arithmetic language of Schmidt, and simplified it to involve the product of just two different curves. Second, in 1958, Arthur Mattuck and John Tate applied the Riemann–Roch theorem for surfaces, which had been proved in any characteristic by Zariski in 1952 and by Serre in 1956. Mattuck and Tate proved the version of Castelnuovo's theorem for the product of two curves that Severi [**70**] gave on p. 265. They dubbed it the *inequality of Castelnuovo–Severi.* Then they rederived the Riemann hypothesis, thus showing that it is a fairly simple consequence of the general theory of algebraic surfaces.

Third, right as Mattuck and Tate finished their paper, Grothendieck [**27**, p. 208], "attempting to understand the full import of their method," found that it produces an index theorem on any surface, which yields the Castelnuovo–Severi inequality. According to Grothendieck however, Serre pointed out to him that he had proved an algebraic version of William Hodge's 1937 analytic index theorem, and moreover that the same version had already been proved the same way by Beniamino Segre in 1937 and independently by Jacob Bronowski in 1938.

In Weil's second book [**75**] of 1948, he completed his note of 1940. He developed the abstract theory of *Abelian varieties*, which he defined as the group varieties that are *complete*, the abstract equivalent of "compact." He proved that they are commutative, and that a map between two is a homomorphism plus a translation.

Weil constructed the Jacobian J of a smooth curve C of genus g by patching together copies of an open subset of the symmetric product $C^{(g)}$. Given a prime l different from the characteristic, he constructed, out of the points on J of order l^n for all $n \geq 1$, an l-adic representation of the ring of correspondences, which, over $\mathbb{C}$, is equivalent to the representation on the first cohomology group. He proved that the trace of this representation is positive definite, and recovered Castelnuovo's theorem. Finally, he reproved the Riemann hypothesis for curves.

Weil left open, as Lang [**40**, p. 17] noted, two important questions: (i) Is J defined over the given coefficient field of C? (ii) Is every Abelian variety projective? Both questions were soon answered affirmatively: (i) by Wei-Liang Chow, who announced his answer in 1949 but published it in 1954, and (ii) by Teruhisa Matsusaka in 1953. In 1954, Weil gave a much simpler and more direct answer to (ii).

---

[20] It is extraordinarily important. Dieudonné [**21**, p. 83] gave one reason why: it "allows proofs, in analytic [sic] number theory, of 'the best possible' upper bounds, inaccessible" by other means, such as this bound on a Kloosterman sum: $\left|\sum_{x=1}^{p-1} \exp\left(\frac{2\pi i}{p}(x+x^*)\right)\right| \leq cp^{1/2}$ for any prime p, where x^* is the inverse of x modulo p and c is a constant independent of p.

In 1956, in order to handle (i), Weil addressed the general question of finding a smaller coefficient field, but only in the case where the resulting field extension is finitely generated and separable. In turn, Weil's work inspired Grothendieck to develop his general Descent Theory, which he then sketched in his December 1959 Bourbaki talk [**31**, 190]. Grothendieck said on p. 190-1 that he was also inspired by Cartier's subsequent treatment [**10**, §4] of purely inseparable extensions, but that "due to the lack of the language of schemes, and especially the lack of nilpotents, Cartier could not express the basic commonality of the two cases."

In 1949, Weil published his celebrated conjectures about the zeta function of a variety of arbitrary dimension. He did not involve a hypothetical cohomology theory outright, but one is implicit. Moreover, one was credited to him explicitly in Serre's 1956 "Mexico paper" [**66**, p. 502] and in Grothendieck's 1958 ICM talk [**28**, p. 103], where the term "Weil cohomology" appears, likely for the first time.

Grothendieck then announced "an approach [to Weil cohomology]... suggested to [him] by the connections between sheaf-theoretic cohomology and cohomology of Galois groups on the one hand, and the classification of unramified coverings of a variety on the other..., and by Serre's idea that a 'reasonable' algebraic principal fiber space... should become locally trivial on some covering unramified over a given point." Thus, on p. 104, he announced the birth of *Grothendieck topology.*

In 1950, Weil published a remarkably prescient note [**76**, I, pp. 437–440] on Abelian varieties. For each normal[21] projective variety X of any dimension in any characteristic, he said that there ought to be two associated Abelian varieties, the *Picard variety* P and the *Albanese*[22] *variety* A, with the following properties:

Universality: The Picard variety P parameterizes the linear equivalence classes of all divisors on X algebraically equivalent[23] to 0. There is a rational map[24] from X into A, defined wherever X is smooth, such that every rational map from X into an Abelian variety factors uniquely, up to translation, through it.

[21] *Normal* means the singular locus has codimension at least 2, and (Zariski's Main Theorem) a rational map is defined everywhere if its graph projects finite-to-one onto X (so isomorphically).

[22] In 1913, Severi introduced and studied A over the complex numbers. Nevertheless, much to Severi's dismay, Weil [**76**, I, p. 571] named A after Severi's student, Giacomo Albanese, ostensibly because, in 1934, Albanese viewed A as a quotient of a symmetric power of X. However, Weil [**76**, I, p. 562] left the impression that rather it is because he owed a debt of gratitude to Albanese for enriching the library of the University of Saõ Paulo, Brazil, with works of Castelnuovo, Torelli and others, which were new to Weil and from which he "profited amply."

[23] *Algebraic equivalence* and *linear equivalence* are just the equivalence relations generated by the algebraic systems and the linear systems.

[24] A *rational map* is given by the ratio of two polynomials, and is *defined* at a point if, for some choice of the two, the denominator does not vanish there.

Duality: If X is an Abelian variety, so that $X = A$, then A is the Picard variety of P; such a pair, A and P, are called *dual* Abelian varieties. Also, A and P are *isogenous*, or finite coverings of each other, and of dimension equal to the irregularity [sic]. If X is arbitrary, then A and P are dual; in fact, the universal map $X \to A$ induces the canonical isomorphism from the Picard variety of A to P. If X is a curve, then both A and P coincide with the Jacobian.

In the note, Weil said that he had complete treatments of P and A for a smooth complex X, and sketches in general. The sketches rest on two criteria for linear equivalence of divisors in terms of linear-space sections. The criteria were found in 1906 by Severi, and reformulated in the note by Weil, who referred to pp. 104–105, 164–165 in Zariski's book [**78**]; please see Mumford's comments [**78**, p. 120] as well. Weil announced proofs of the criteria in 1952, and gave the details in 1954.

In 1951, Matsusaka gave the first algebraic construction of P. He extended the coefficient field k in order to apply two of Weil's results: one of the equivalence criteria and the construction of the Jacobian. Both applications involve the *generic curve*, the section of X by a linear space of appropriate dimension defined over a transcendental extension of k. In 1952, Matsusaka gave a second construction; it does not require extending k, but does require X to be smooth.

Both of Matsusaka's constructions are like Castelnuovo's: first Matsusaka constructed a complete algebraic system of sufficiently positive divisors, and then he formed the quotient modulo linear equivalence. To parameterize the divisors, he used the theory of "Chow coordinates," which was developed in 1938 by Chow and van der Waerden and was under refinement by Chow. In 1952, Matsusaka also used Chow coordinates to form the quotient. Further, he made the first construction of A, again using the Jacobian of the generic curve, but he did not relate A and P.

Also in 1952, in § II of his paper *On Picard Varieties* [**76**, II, pp. 73–102], Weil refined the sense in which P parameterizes classes of divisors. Working complex analytically, he constructed "an algebraic family of divisors on X, parameterized by P, containing one and only one representative of each class."

Weil did not name that family of divisors. However, the same year, André Néron and Pierre Samuel [**56**] constructed,[25] in any characteristic, a similar family, which they named a *Poincaré family* citing [**76**, II, pp. 73–102] in a way suggesting the name[26] is due to Weil. The family is defined by a divisor D on $X \times P$, which is called a *Poincaré divisor* by Lang [**40**, p. 114].

[25]Unfortunately, Lang [**40**, p. 175] felt that he had to write: "It can not be said that the Picard variety is constructed in [**56**] because this paper begins by a false statement concerning the birational invariance. This is a delicate point, when the varieties involved have singularities."

[26]Of course, here and below, Poincaré's name is used to honor his work mentioned above.

Moreover, Lang showed that the pair (P, D) is unique, P up to isomorphism and D up to addition of a "trivial" divisor.

In 1955, Chow constructed A and P in a new way, as what he called respectively the "image" and the "trace" of the Jacobian of a generic curve on X. Also, he proved that, indeed, the universal map $X \to A$ induces an isomorphism of the Picard variety of A onto P.

In a course at the University of Chicago, 1954–55, Weil gave a more complete and elegant treatment, based on the "see-saw principle," which he adapted from Severi, and on his own Theorem of the Square and Theorem of the Cube. This treatment became the core of Lang's 1959 book [**40**]. The idea is to construct A first using the generic curve, and then to construct P as a quotient of A modulo a finite subgroup. Thus there is no need for Chow coordinates.

In 1958, Serre [**66**, p. 555] worked over a fixed algebraically closed coefficient field k of any characteristic. He reproved Igusa's 1955 bound $\dim A \leq h^0(\Omega^1_X)$, and obtained a simple direct construction of A over k, not using the generic curve.

In 1958 Cartier [**10**] and in 1959 Nishi [**57**] independently proved Weil's conjectured duality of A and P: in any characteristic, each is the others Picard variety.

Between 1952 and 1957, Maxwell Rosenlicht published a remarkable series of papers, inspired by Severi's 1947 monograph [**71**], which treated curves with double points. Treating a curve with arbitrary singularities, Rosenlicht generalized the notions of linear equivalence and differentials of the first kind. Then he constructed a *generalized Jacobian* J over $\mathbb{C}$ by integrating and in arbitrary characteristic by patching. It is not an Abelian variety, but an extension of the Jacobian J_0 of the desingularized curve by an affine algebraic group. In 1962, Frans Oort [**59**] gave another construction, which gives J as a successive extension of J_0 by additive and multiplicative groups.

For arithmetic applications, Tate suggested, according to Lang [**40**, p. 176], doing this: Given finitely many simple points on X, consider the divisors avoiding them. Form linear equivalence classes via functions congruent to 1 to given multiplicities at the points. Finally, seek a generalized Picard variety to parameterize these classes.

In 1959, Serre published a textbook [**65**] on the case $\dim X = 1$ and its application to Lang's class field theory over function fields. In particular, Serre recovered Rosenlicht's generalized Jacobian of an X with one singular point[27], constructed by identifying given points with given multiplicities on a given smooth curve.

In 1962, Murre [**52**] constructed Tate's generalized Picard variety P by adapting Matsusaka's two constructions. Thus Murre obtained P for any

[27]The singularity cannot be arbitrary; for example, it cannot be a planar triple point.

(normal) X via patching and for any smooth X directly over the same ground field.

In 1956, Igusa established the compatibility of specializing a curve with specializing its generalized Jacobian, possibly under reduction mod p, provided the general curve is smooth and the special curve has at most one node. Igusa explained that, in 1952, Néron had studied the total space of such a family of Jacobians, but had not explicitly analyzed the special fiber.[28] Igusa's approach is, in spirit, like Castelnuovo's, Chow's, and Matsusaka's before him.

In 1960, Claude Chevalley [**19**] constructed a Picard variety using Weil divisors locally defined by one equation; they are called *Cartier divisors* in honor of Cartier's 1958 Paris thesis [**11**]. First, Chevalley constructed a *strict* Albanese variety; it is universal for *regular* maps, ones defined everywhere, into Abelian varieties. Then he took its Picard variety to be that of X. He noted his Picard and Albanese varieties need not be equal to those of a desingularization of X. By contrast, Weil's P and A are birational invariants, and his universal map $X \to A$ is a rational map. In 1962, Conjeerveram Seshadri [**68**] generalized Chevalley's construction to an X with arbitrary singularities, recovering Rosenlicht's generalized Jacobian.

In 1961, Mattuck [**42**] took, on a smooth X over an algebraically closed field, a complete algebraic system Σ of suitably positive divisors C. He parameterized Σ by the *Chow variety* H, the locus of points given by the Chow coordinates of the $C \in \Sigma$. He fixed a $D \in \Sigma$, and took the class of the difference $C - D$ for $C \in \Sigma$, to get a rational map $\alpha \colon H \to P$. In order that α be defined everywhere, he reembedded X in another projective space, because Murre [**51**] had just proved that then H is smooth, so normal.

Mattuck proved that α is a projective bundle[29] and has a section. The section corresponds to a refined Poincaré divisor: not only does it define a Poincaré family, but it contains no fiber of α, so cuts each fiber in a divisor. Finally, Mattuck studied the case that X is a curve of genus $g > 1$, so that H is the n-th symmetric power of X where n is the degree of the divisors. He proved that α is a bundle[30] if $n > 2g - 2$, and that α has a section[31] if $n > 4g$, but no section if $n = 2g - 1$ and the divisor classes modulo algebraic equivalence on P form a group of $\mathbb{Q}$-rank 1. So it seems unlikely that α has a section if $n = 2g - 1$ and X has general moduli.

Thus, in 1962, the algebraic theory of the Picard variety was indirect, involved, and incomplete. There were competing definitions and constructions, each with advantages and disadvantages. There was a lot of fussing

---

[28]For a comprehensive discussion of the "Néron model" and its connection to the Picard scheme (and Picard algebraic space) along with historical notes and references to the original sources, please see the textbook [**9**] of Siegfried Bosch, Werner Lütkebohmert, and Raynaud.

[29]Earlier, in 1956, Kodaira [**38**] obtained a similar result analytically.

[30]Atiyah [**3**, p. 451] wrote in 1957 that this case is "well known."

[31]In fact, $n > 2g - 1$ suffices; please see p. 66 below.

with fields of definition. There were loose ends. Notably, there was no fully satisfactory way to parameterize divisors or to construct quotients. So there was not enough machinery to prove the Completeness of the Characteristic System or to treat, in general, the behavior of the (generalized) Picard variety in a family. Grothendieck brilliantly handled those issues in the way explained in the next two sections.

4. The Picard Functor

Grothendieck certainly did not feel
that he was attempting to use powerful techniques
in order to obtain stronger results by generalizing.
What he perceived himself as doing was simplifying situations and objects,
by extracting the fundamental essence of their structure.
Leila Schneps [**64**, p. 3]

Many times, Grothendieck proceeded "by extracting the fundamental essence" of existing theories, and then developing his own versions, for example his own theories of schemes, representable functors, the Hilbert scheme, and the Picard scheme. Parts of those theories belong to the theory of the Picard functor. Those parts are treated in depth in [**32**] and [**25**], and they are introduced in this section.

To begin, here is a bit more informal history. Starting in 1937, Zariski made deep use of the local ring of all rational functions that are finite at a given point of a variety with a fixed field of definition. Inspired by Zariski's work, Chevalley developed an intrinsic theory of abstract varieties V in his paper [**17**], submitted on 2 July 1954. On p. 2, he called the set of all the local rings of the points of V its *model,* and then developed a theory of models. Earlier, in January 1954, he lectured on his theory at Kyoto University, according to Masayoshi Nagata [**55**, p. 78], who then generalized it by replacing the field of definition by a Dedekind domain.

In 1944, Zariski topologized the set of all valuation rings of the field of rational functions of a variety in order to use the finite-covering property to pass from local uniformization to global desingularization. In 1949, Weil [**76**, I, pp. 411–413] observed that his abstract varieties support what he called the *Zariski Topology,* whose closed sets are the subvarieties and their finite unions.

Weil used the Zariski topology to define locally trivial fiber spaces. He discussed the natural bijective correspondence between the line bundles on a smooth variety and its linear equivalence classes of divisors. Then, in his 1950 paper on Abelian varieties [**76**, I, pp. 438–439], he suggested that those line bundles might be used to develop, for any abstract variety, a version of Severi's generalized Jacobian.

As already noted in Section 2, Serre, in his 1954 ICM talk, announced a theory of coherent algebraic sheaves. In fact, he worked over an arbitrary algebraically closed field k, of any transcendence degree over the prime field, and used k both as the field of definition and as the field of coordinates.[32] Moreover, he worked only with projective space and its subvarieties, which he allowed to be reducible, and he viewed as the closed sets of a topology, which he too called the "Zariski topology."

In 1955, Serre presented the details in his celebrated paper *Faisceaux algébriques cohérents* [**66**, pp. 310–391]. He also generalized the notion of abstract variety via Henri Cartan's notion[33] of *ringed space*. It is a topological space X endowed with a *sheaf of rings* $\mathcal{O}_X$, called the *structure sheaf*: over each open set, its sections form a ring; for each smaller open set, the restriction map is a ring homomorphism. To be a variety, X must be covered by finitely many open subsets, each of which, when endowed with the restriction of $\mathcal{O}_X$, is isomorphic to an *affine variety*; the latter's space is a closed subspace of an affine space, and its structure sheaf has, as its sections over an open set U, the rational functions defined everywhere on U.

Both Serre and Chevalley downgraded rational maps, preferring maps[34] that are defined everywhere. For Serre, a map of varieties $\varphi\colon X \to Y$ is a map of ringed spaces: φ is a continuous map equipped with a map φ^* relating the two structure sheaves, so that a section f of $\mathcal{O}_Y$ over an open set V yields a section $\varphi^* f$ of $\mathcal{O}_X$ over $f^{-1}V$ in a way respecting addition, multiplication, and restriction.

Grothendieck "extracted the fundamental essence of" those ideas, and developed a theory of *schemes*.[35] By February 1956 (see [**20**, p. 32]), he was working with ringed spaces that have an open covering by *affine schemes*, or *ring spectra*.[36] The spectrum of a ring R is the set of all its prime ideals $\mathfrak{p}$. Its topology is generated by its *principal open subsets* $D(f)$ for all $f \in R$, where $D(f) := \{\mathfrak{p} \not\ni f\}$. Over $D(f)$, the sections of the structure sheaf are the fractions a/f^n for all $a \in R$ and $n \geq 0$. In 1956, Grothendieck took R to be Noetherian, but in EGA I [**29**], which appeared in 1961, R is an arbitrary commutative ring. In any case, R may have nilpotents.

Moreover, Grothendieck worked with Cartier's generalization in [**11**, p. 206] of coherent sheaves, the *quasi-coherent* sheaves. He [**20**, p. 32] told

---

[32]Cartier [**11**] generalized Serre's theory to an arbitrary field of definition, and studied the effect of extending it. He took the field of coordinates to be a universal domain.

[33]Cartan had used the notion to define C^∞-manifolds. He and Serre had used it to define complex analytic manifolds.

[34]Serre called them *regular maps,* a traditional term; Chevalley [**18**, p. 219] used *morphisms.*

[35]Cartier [**12**, Fn. 8] noted, "This word results from a typical epistemological shift from one thing to another: for Chevalley, who invented the name in 1955, it indicated the 'scheme' or 'skeleton' of an algebraic variety, which itself remained the central object. For Grothendieck, the 'scheme' is the focal point, the source of all the projections and all the incarnations."

[36]Cartier [**12**, Fn. 29] explained, "It was [Israel] Gelfand's fundamental idea [of 1938] to associate a normed commutative algebra to a space. . . . The term 'spectrum' comes directly from Gelfand."

Serre why: they are "technically very convenient because they have the relevant properties of coherent sheaves without requiring the finiteness (on an affine, they correspond to all the modules over the coordinate ring, and not just the finitely generated modules)."

The generality is vast, but not idle. Murre put it as follows, according to Schneps [**64**, p. 2]: "Undoubtedly, people did see in the mid 50's that one could generalize a lot of things to schemes, but Grothendieck saw that such a generalization was not only possible and natural, but necessary to explain what was going on, even if one started with varieties."

The spectrum S of a field k is a single point, but S has k as structure sheaf. Call k the *field of definition* of a scheme X if there is a distinguished map $X \to S$. On the other hand, given a universal domain Ω extending k, let T be its spectrum. Then a *point* of X with "coordinates" in Ω, or an Ω-*point* of X, is just a map $T \to X$ that respects the distinguished maps $X \to S$ and $T \to S$. Cartier [**12**, p. 20] described this new situation as being of "admirable simplicity — and a very fruitful point of view — but a complete change of paradigm!"

For example, say X is the spectrum of the residue ring R of the polynomial ring $k[u_1, \ldots, u_n]$ modulo the ideal generated by polynomials f_λ. Then the inclusion $k \to R$ defines the map $X \to S$. Further, a zero $(c_1, \ldots, c_n)$ of the f_λ in Ω^n amounts exactly to a k-algebra map $\gamma\colon R \to \Omega$ that carries the residue of u_i to c_i. In turn, γ amounts exactly to an S-map $T \to X$.

However, there's no need to restrict S and T to be the spectra of fields, and good reason not to. Moreover, the right setting for this theory, as for any theory whose principals are objects and maps, is Category Theory. It is not simply a convenient language for expressing abstract ideas, but more importantly, an effective tool, which eases the work at hand and affords new possibilities. Grothendieck recognized as much, and promoted Category Theory.

Thus, given a *base* scheme S, an S-*scheme* is a scheme equipped with a map to S, its *structure map*. An S-*map* is a map between S-schemes that commutes with the two structure maps. The category of S-schemes has products:[37] the product of X and Y is an S-scheme $X \times_S Y$ with a distinguished pair of S-maps to X and Y, the *projections*, such that composition with them sets up a bijection from the S-maps $T \to X \times_S Y$ to the pairs of S-maps $T \to X$ and $T \to Y$.

The product $X \times_S Y$ is determined, formally, up to unique isomorphism. It is constructed by patching together the spectra of the tensor products of the rings of affines that cover S, X, and Y. It can also be viewed as the result of *base change* of X when Y is viewed as a new base. By contrast, Cartier [**12**, p. 20] noted that, "in both [Serre's and Chevalley's] cases,

[37]Mac Lane [**41**, p. 76] said that he himself, in 1948 and 1950, formulated "the idea that... products could be described by universal properties of their projections."

the two fundamental problems of products and base change could only be approached indirectly."

Another way to view a map $X \to S$ is as a family X/S with S as parameter space and X as total space. Its members are the *fibers*, the preimages X_s of the points $s \in S$. More precisely, $X_s := X \times_S Y$ where Y is the spectrum of the residue field k_s of the stalk $\mathcal{O}_{S,s}$, which is the local ring of all functions that are each defined on some neighborhood of s. Grothendieck discovered that, many properties of the X_s vary continuously when X is *S-flat*; that is, $\mathcal{O}_{X,x}$ is $\mathcal{O}_{S,s}$-flat for all $s \in S$ and $x \in X_s$. Often, he considered what he called a *geometric fiber*, which is the product $X \times_S Y'$ where Y' is the spectrum of an algebraically closed field containing k_s.

An S-map $T \to X$ is called a *T-point* of X, and the set of all of them is denoted by $X(T)$ or $h_X(T)$. An S-map $T' \to T$ induces a set map $h_X(T) \to h_X(T')$. Thus h_X is a contravariant functor from the category of S-schemes to the category of sets; it is called the *functor of points* of X.

The contravariant functors H from S-schemes to sets themselves form a category. The assignment $X \mapsto h_X$ is a functor from S-schemes into the latter category. This functor is an embedding by Yoneda's Lemma. Given an H, if an X is found with $h_X = H$, then H is said to be *representable* by X. If so, then $H(X)$ contains a *universal* element W, which corresponds to the identity map of X; namely, each element $Y \in H(T)$ defines a unique S-map $\varphi\colon T \to X$ with $H(\varphi)W = Y$. In other words, the T-points of X *represent* the elements of $H(T)$. Conversely, if an X is found with a universal $W \in H(X)$, then there's a canonical isomorphism $h_X = H$.

The first important example is the functor $P(\mathcal{E})$, where $\mathcal{E}$ is an arbitrary quasi-coherent sheaf on S. For each S-scheme T, the set $P(\mathcal{E})(T)$ is the set of invertible quotients $\mathcal{L}$ of the pullback $\mathcal{E}_T$; *invertible* means that $\mathcal{L}$ is the sheaf of sections of a line bundle. The functor $P(\mathcal{E})$ is representable by an S-scheme $\mathbf{P}(\mathcal{E})$. Automatically, $\mathbf{P}(\mathcal{E})$ carries a *universal* invertible quotient of $\mathcal{E}_{\mathbf{P}(\mathcal{E})}$, denoted $\mathcal{O}_{\mathbf{P}(\mathcal{E})}(1)$; namely, each invertible quotient $\mathcal{L}$ of $\mathcal{E}_T$ defines a unique S-map $\varphi\colon T \to \mathbf{P}(\mathcal{E})$ with $\varphi^*\mathcal{O}_{\mathbf{P}(\mathcal{E})}(1) = \mathcal{L}$. Moreover, $\mathbf{P}(\mathcal{E}) \times_S Y = \mathbf{P}(\mathcal{E}_Y)$ for any S-scheme Y. In particular, the fiber $\mathbf{P}(\mathcal{E})_s$ over $s \in S$ is the projective space of 1-dimensional quotients of the vector space $\mathcal{E}_s \otimes k_s$ where $\mathcal{E}_s$ is the stalk of $\mathcal{E}$ at s.

For convenience, **assume from now on** that all schemes are *locally Noetherian*, or covered by affine schemes on Noetherian rings, and that S is *Noetherian*, or covered by finitely many such. Assume **also** that X is projective over S; that is, X can be embedded as a closed subscheme of $\mathbf{P}(\mathcal{E})$ for some coherent sheaf $\mathcal{E}$ on S.

Consider the *Hilbert functor* $\mathrm{Hilb}_{X/S}$, treated by Grothendieck in his May 1961 Bourbaki talk [**31**, Exp. 221]. For each S-scheme T, the set $\mathrm{Hilb}_{X/S}(T)$ is the set of T-flat closed subschemes Y of $X \times_S T$. Moreover, for each polynomial $F(\nu)$ with rational coefficients, $\mathrm{Hilb}_{X/S}$ has a subfunctor $\mathrm{Hilb}^F_{X/S}$; namely, for all T, the set $\mathrm{Hilb}^F_{X/S}(T)$ is the set of Y such that Y_t has

Hilbert polynomial F for all $t \in T$, that is, $F(\nu) = \chi(\mathcal{O}_{Y_t}(\nu))$ where $\mathcal{O}_{Y_t}(\nu)$ is the pullback to Y_t of the νth tensor power of $\mathcal{O}_{\mathbf{P}(\mathcal{E})}(1)$. Grothendieck proved that $\mathrm{Hilb}_{X/S}$ is representable by a locally Noetherian S-scheme $\mathbf{Hilb}_{X/S}$, the *Hilbert scheme.* In fact, $\mathbf{Hilb}_{X/S}$ is the disjoint union of projective S-schemes $\mathbf{Hilb}^F_{X/S}$, which represent the functors $\mathrm{Hilb}^F_{X/S}$.

Automatically, $X \times_S \mathbf{Hilb}_{X/S}$ has a *universal* subscheme W; namely, each T-flat closed subscheme Y of $X \times_S T$ defines a unique S-map $T \to \mathbf{Hilb}_{X/S}$ with $W \times_{\mathbf{Hilb}_{X/S}} T = Y$. Note that the $\mathbf{Hilb}^F_{X/S}$ depend on the choice of embedding of X in some $P(\mathcal{E})$, but $\mathbf{Hilb}_{X/S}$ does not. Thus the Hilbert scheme is a noble replacement for Chow coordinates; the latter only parameterize the cycles on a variety V, and depend on the choice of embedding of V in projective space.

A subscheme R of $X \times_S X$ defines a *flat and projective equivalence relation* if each projection $R \to X$ is flat and projective and if, for each S-scheme T, the subset $h_R(T)$ of $h_X(T) \times h_X(T)$ defines a set-theoretic equivalence relation. Grothendieck found two constructions of a *quotient* X/R in the strongest sense of the term. Namely, first, an S-map $X \to Z$ factors through X/R if and only if the two compositions $R \rightrightarrows X \to Z$ are equal; if so, then $X/R \to Z$ is unique. So by "abstract nonsense," X/R is determined up to unique isomorphism. Second, the quotient map $X \to X/R$ is flat and projective, and the canonical map $R \to X \times_{X/R} X$ is an isomorphism.[38]

Grothendieck's first construction [**31**, p. 212-15] uses quasi-sections to reduce to the case where X is affine and each $R \to X$ has finite fibers. However, his second construction [**31**, p. 232-13] is easier and more elegant. It proceeds as follows: R lies in $\mathrm{Hilb}_{X/S}(X)$, so defines a map $\varphi\colon X \to \mathbf{Hilb}_{X/S}$; the graph Γ_φ is a closed subscheme of the universal subscheme W; finally, by Grothendieck's Descent Theory, Γ_φ descends to a closed subscheme of $\mathbf{Hilb}_{X/S}$, which is the desired X/R.

Assume from now on that X is **also** S-flat. Then $\mathrm{Hilb}_{X/S}$ has an important subfunctor $\mathrm{Div}_{X/S}$; namely, for each S-scheme T, let $\mathrm{Div}_{X/S}(T)$ consist of the *effective Cartier divisors* D in $\mathrm{Hilb}_{X/S}(T)$, that is, the flat subschemes D whose ideal $\mathcal{I}_D$ is locally generated by one nonzerodivisor; equivalently, $\mathcal{I}_D$ is invertible as an abstract sheaf. Grothendieck [**31**, p. 232-10] proved that $\mathrm{Div}_{X/S}$ is representable by an open subscheme $\mathbf{Div}_{X/S}$ of $\mathbf{Hilb}_{X/S}$.

Given an invertible sheaf $\mathcal{L}$ on X, define[39] a subfunctor $\mathrm{LinSys}_{\mathcal{L}/X/S}$ of $\mathrm{Div}_{X/S}$: for each S-scheme T, let $\mathrm{LinSys}_{\mathcal{L}/X/S}(T)$ consist of the D in $\mathrm{Div}_{X/S}(T)$ for which there is an invertible sheaf $\mathcal{M}$ on T such that the

[38]It follows that $h_{X/R}$ is the fpqc sheaf associated to h_X/h_R in the sense discussed below.

[39]This subfunctor appears in [**31**, p. 232-10], but the notation for it comes from [**46**, p. 93].

inverse $\mathcal{I}_D^{-1}$ is isomorphic to the tensor product on $X \times_S T$ of the pullbacks of $\mathcal{L}$ and $\mathcal{M}$.

Assume in addition from now on the geometric fibers of X/S are *integral*; that is, each affine ring of each geometric fiber is an integral domain. Grothendieck [**31**, p. 232-11] proved that then there is a coherent sheaf $\mathcal{Q}$ on S, determined up to unique isomorphism, such that $\mathrm{LinSys}_{\mathcal{L}/X/S}$ is representable by $\mathbf{P}(\mathcal{Q})$. Hence, $\mathbf{P}(\mathcal{Q})$ is equal to a closed subscheme of $\mathbf{Div}_{X/S}$. Also, if $H^1(\mathcal{L}|X_s) = 0$ at $s \in S$, then s has a neighborhood on which $\mathcal{Q}$ is *free*, or isomorphic to $\mathcal{O}_S^r$ for some r.

In general, what makes a functor H representable? Say $H = h_X$. Then given any S-scheme T and any open covering $\{T_\lambda\}$ of T, two maps $T \to X$ are equal if their restrictions to each T_λ are equal. Furthermore, maps $\varphi_\lambda \colon T_\lambda \to X$ are the restrictions of a single map $T \to X$ if, for all λ and μ, the restrictions of φ_λ and φ_μ to $T_\lambda \cap T_\mu$ are equal. In other words, as U ranges over the open sets of T, the $H(U)$ form a sheaf. The latter condition does not explicitly involve X. So it makes sense for any H, representable or not. If it is satisfied, H is called a *Zariski sheaf.*

Here's another formulation. Let T' be the disjoint union of the T_λ, and consider the induced map $T' \to T$. Then $T' \times_T T'$ is the disjoint union of the $T_\lambda \cap T_\mu$. So the condition to be a Zariski sheaf just means that the induced sequence of sets

$$H(T) \to H(T') \rightrightarrows H(T' \times_T T') \tag{4}$$

is *exact*; that is, the first map is injective, and its image consists of the elements of $H(T')$ whose two images are equal in $H(T' \times_T T')$.

Grothendieck's Descent Theory yields more. Let $T' \to T$ be an *fpqc* map; namely, it is flat and surjective, and the preimage of any affine open subscheme is a finite union of affine open subschemes. If H is representable, remarkably (4) is still exact; in other words, H is an *fpqc sheaf.* Indeed, the *fpqc Grothendieck topology* may be defined as the refinement of the Zariski topology with the fpqc maps as additional generalized open coverings. In particular, H is an *étale sheaf,* the notion obtained by requiring the maps to be *étale*, that is, flat, unramified and locally of finite type.

The *Picard group* $\mathrm{Pic}(X)$ is the group, under tensor product, of isomorphism classes of invertible sheaves on X. The *absolute Picard functor* Pic_X is defined by $\mathrm{Pic}_X(T) := \mathrm{Pic}(X \times_S T)$. It is never a Zariski sheaf, so never representable.

There is a sequence of ever more promising "Picard functors." First comes the *relative Picard functor* $\mathrm{Pic}_{X/S}$ defined by

$$\mathrm{Pic}_{X/S}(T) := \mathrm{Pic}(X \times_S T)/\operatorname{Pic}(T)$$

where $\mathrm{Pic}(T)$ acts via pullback. Following it are its *associated sheaves* in the Zariski, étale and fpqc topologies: $\mathrm{Pic}_{(X/S)(\mathrm{Zar})}$, $\mathrm{Pic}_{(X/S)(\text{ét})}$, and $\mathrm{Pic}_{(X/S)(\mathrm{fpqc})}$. Grothendieck [**31**, p. 232-03, (1.6)] formed them as direct limits. For example,

$$\mathrm{Pic}_{(X/S)(\mathrm{Zar})}(T) := \varinjlim_{T'} \mathrm{Pic}_{X/S}(T')$$

where T' ranges over the small category of all open coverings of T.

Recall that S is Noetherian and that X is a flat and projective S-scheme with integral geometric fibers. Grothendieck [**31**, pp. 232-4–6] proved[40] that *then the three canonical comparison maps are, respectively, injective, injective, and bijective:*

$$\mathrm{Pic}_{X/S} \hookrightarrow \mathrm{Pic}_{(X/S)(\mathrm{Zar})} \hookrightarrow \mathrm{Pic}_{(X/S)(\text{ét})} \xrightarrow{\sim} \mathrm{Pic}_{(X/S)(\mathrm{fpqc})} .$$

Moreover, the first two maps are bijective if $X \to S$ has a section; the middle map is bijective if it just has a section on a Zariski neighborhood of each point of S.

A simple example shows that, in general, we must pass to the étale sheaf. Namely, in the real projective plane, consider the conic $X : u^2+v^2+w^2 = 0$. Let S and T be the spectra of $\mathbb{R}$ and $\mathbb{C}$. Then $T \to S$ is an étale covering. Moreover, $X \times_S T$ is the complex conic with the same equation; so X is isomorphic to the complex projective line. The latter's universal sheaf $\mathcal{O}(1)$ defines an element $\tau \in \mathrm{Pic}_{(X/S)(\text{ét})}(S)$, as the two pullbacks of $\mathcal{O}(1)$ to $X \times_S T \times_S T$ are isomorphic. And τ is not in the image of $\mathrm{Pic}_{(X/S)(\mathrm{Zar})}(S)$, as X has no S-points.

Grothendieck's main existence theorem [**31**, p. 232-06] says that $\mathrm{Pic}_{(X/S)(\text{ét})}$ *is representable by a scheme* $\mathbf{Pic}_{X/S}$. It is called the *Picard scheme.* Of course, if $\mathrm{Pic}_{(X/S)(\mathrm{Zar})}$ is representable, then it is an étale sheaf, so equal to $\mathrm{Pic}_{(X/S)(\text{ét})}$, and representable by $\mathbf{Pic}_{X/S}$. Similarly, if $\mathrm{Pic}_{X/S}$ is representable, then all four functors are equal, and representable by $\mathbf{Pic}_{X/S}$.

Grothendieck's proof is fairly simple at this point. Here is the idea. Fix an embedding of X in a $\mathbf{P}(\mathcal{E})$. Given any S-scheme T and any quasi-coherent sheaf $\mathcal{F}$ on $\mathbf{P}(\mathcal{E}) \times_S T$, let $\mathcal{F}(n)$ denote the tensor product of $\mathcal{F}$ and the pullback of the nth tensor power of the universal sheaf $\mathcal{O}_{\mathbf{P}(\mathcal{E})}(1)$. Let $\mathcal{I}$ be the ideal of the universal divisor on $X \times_S \mathbf{Div}_{X/S}$, and $\mathcal{I}^{-1}$ its inverse. Form the open subscheme $\mathbf{D}^+$ of $\mathbf{Div}_{X/S}$ on which all the higher direct images of $\mathcal{I}^{-1}(n)$ vanish for all $n \geq 0$.

Set $\mathcal{L} := \mathcal{I}^{-1}|(X \times_S \mathbf{D}^+)$. Then $\mathrm{LinSys}_{\mathcal{L}/X\times_S\mathbf{D}^+/\mathbf{D}^+}$ is representable by $\mathbf{P}(\mathcal{Q})$ where $\mathcal{Q}$ is a coherent sheaf on $\mathbf{D}^+$. Moreover, $\mathcal{Q}$ is *locally free*; that is, each point of $\mathbf{D}^+$ has neighborhood on which $\mathcal{Q}$ is free. Thus $\mathbf{P}(\mathcal{Q})$ is flat over $\mathbf{D}^+$.

Set $R := \mathbf{P}(\mathcal{Q})$. Then R is a closed subscheme of $\mathbf{D}^+ \times_S \mathbf{D}^+$. Moreover, for each S-scheme T, the subset $h_R(T)$ of $h_{\mathbf{D}^+}(T) \times h_{\mathbf{D}^+}(T)$ consists of the pairs of D, $D' \in \mathbf{D}^+(T)$ for which there is an invertible sheaf $\mathcal{M}$ on T such that the ideal $\mathcal{I}_D$ is isomorphic to the tensor product of $\mathcal{I}_{D'}$ and the pullback of $\mathcal{M}$. Thus $h_R(T)$ is a set-theoretic equivalence relation.

---

[40] In [**31**], Grothendieck did not consider $\mathrm{Pic}_{(X/S)(\text{ét})}$. However, his methods apply to it, and show that it is equal to $\mathrm{Pic}_{(X/S)(\mathrm{fpqc})}$, because in the case at hand, there exists an *étale quasi-section*, an S-map $S' \to X$ for which the structure map $S' \to S$ is étale.

Although $\mathbf{D}^+$ isn't Noetherian, nevertheless it is the disjoint union of Noetherian subschemes, as $\mathbf{Hilb}_{X/S}$ is the disjoint union of the projective S-schemes $\mathbf{Hilb}^F_{X/S}$, and R decomposes compatibly. Consequently, the quotient $\mathbf{D}^+/R$ exists.

For each $m \geq 0$, let P_m be the fpqc subsheaf of $\mathrm{Pic}_{(X/S)(\mathrm{fpqc})}$ associated to the subfunctor of $\mathrm{Pic}_{X/S}$ whose value at T consists of the classes of invertible sheaves $\mathcal{L}$ on $X \times_S T$ for which all the higher direct images of $\mathcal{L}(n)$ vanish on T for all $n \geq m$, but the direct image doesn't vanish. Then P_0 is representable by $\mathbf{D}^+/R$.

Tensor product with the pullback of $\mathcal{O}_X(1)$ defines an isomorphism $P_{m+1} \xrightarrow{\sim} P_m$ for all $m \geq 0$. So the P_m are representable by (isomorphic) schemes U_m. Each inclusion $P_m \hookrightarrow P_{m+1}$ is representable by an open embedding $U_m \hookrightarrow U_{m+1}$. Finally, $\mathrm{Pic}_{(X/S)(\mathrm{fpqc})}$ is the "union" of the P_m; so is representable by the union, or rather direct limit, of the U_m. Thus Grothendieck proved his main existence theorem.

Commenting on his proof, Grothendieck [**31**, p. 232-13] noted that "the approach is basically the one followed by Matsusaka" (so by Igusa, Chow and Castelnuovo). Further, he [**31**, p. 232-14] noted that "the proof appeals neither to the preliminary construction of the Jacobians of curves... nor to the theory of Abelian varieties, and thus differs in an essential way from the 'traditional' treatments in Lang's book [**40**] and Chevalley's paper [**19**]....That the construction of the Picard scheme ought to precede and not follow the theory of Abelian varieties is clear a priori from the fact that...Rosenlicht's 'generalized Jacobians' are not Abelian varieties." More of Grothendieck's advances are highlighted in the next section.

5. The Picard Scheme

His [Grothendieck's] feeling was that "those people" made too strict assumptions and tried to prove too little.
Jacob Murre, quoted in [**64**, p. 2]

Above, Murre describes Grothendieck's feeling about the theory of the Picard variety: it was hampered by its developers' narrow vision. This section explains how Grothendieck's broader vision led to clarifying and settling a number of issues. Primarily, we focus on the two major issues: Behavior in a Family and Completeness of the Characteristic System. In addition, we consider some other issues mentioned earlier, especially Poincaré divisors and the Albanese variety. And we consider some other ways that other mathematicians enhanced Grothendieck's theory between 1962 and 1973, especially ways of generalizing his main existence theorem. For more discussion of those issues and some discussion of a lot of other issues of the same sort, please see [**31**], [**25**], [**9**], and [**8**].

As noted in the Introduction, Grothendieck explained the behavior of the Picard schemes of the members of a family as compatibility with base change. More precisely, if the functor $\mathrm{Pic}_{(X/S)(\mathrm{fpqc})}$ is representable by an S-scheme $\mathbf{Pic}_{X/S}$, then for any S-scheme S', the functor $\mathrm{Pic}_{(X\times S'/S')(\mathrm{fpqc})}$ is representable by the S'-scheme $\mathbf{Pic}_{X/S}\times_S S'$. In particular, if S' is the spectrum of the residue field k_s of $s \in S$, then the Picard scheme of the fiber X_s of X/S is just the fiber of $\mathbf{Pic}_{X/S}$.

Compatibility holds for this reason. For any S'-scheme T, the equation

$$\mathrm{Pic}_{X\times S'/S'}(T) = \mathrm{Pic}_{X/S}(T)$$

results from the definitions, because $(X \times_S S') \times_{S'} T = X \times_S T$. So the equation

$$\mathrm{Pic}_{(X\times S'/S')(\mathrm{fpqc})}(T) = \mathrm{Pic}_{(X/S)(\mathrm{fpqc})}(T)$$

follows, because a map of S'-schemes $T' \to T$ is a covering if and only if it is a covering when viewed as a map of S-schemes. However, the equation

$$\big(\mathbf{Pic}_{X/S}\times_S S'\big)(T) = \mathbf{Pic}_{X/S}(T)$$

holds, because the structure map $T \to S'$ is fixed. Since the right sides of the last two equations are equal, so are their left sides, as desired.

Until otherwise said near the end of the section, assume that S is the spectrum of an algebraically closed field k and that X is an integral and projective S-scheme. As is common, write "k-scheme," $\mathbf{Div}_{X/k}$, etc. for "S-scheme," $\mathbf{Div}_{X/S}$, etc.

In order to complete the discussion in Section 2 of the algebraic proofs of the Theorem of Completeness of the Characteristic System and of the Fundamental Theorem of Irregular Surfaces, we must discuss what's called[41] the *Zariski tangent space* $\mathbb{T}_z(Z)$ to a k-scheme Z at a *rational point* z, a point whose residue field k_z is k. Let $\mathfrak{m}$ be the maximal ideal, and set $\mathbb{T}_z(Z) := \mathrm{Hom}_k(\mathfrak{m}/\mathfrak{m}^2, k)$.

Then $\mathbb{T}_z(Z)$ can be viewed as the vector space of k-derivations $\delta\colon \mathcal{O}_{Z,z} \to k$; indeed, $\delta(\mathfrak{m}^2) = 0$, and so δ corresponds to a linear map $\mathfrak{m}/\mathfrak{m}^2 \to k$. Let k_ε be the ring of *dual numbers,* the ring obtained by adjoining an element ε with $\varepsilon^2 = 0$. Then δ corresponds to the map of k-algebras $u\colon \mathcal{O}_{Z,z} \to k_\varepsilon$ given by $u(a) := \overline{a} + \delta(a)\varepsilon$ where $\overline{a} \in k$ is the residue of a. Finally, let S_ε be the spectrum of k_ε; it is the *free tangent vector.* Then u corresponds to a k-map $S_\varepsilon \to Z$, whose image is supported at z. Denote the set of such k-maps by $h_Z(S_\varepsilon)_z$. Then in summary $\mathbb{T}_z(Z) = h_Z(S_\varepsilon)_z$.

Often, if Z represents a given functor H, that is $h_Z = H$, then we can work out a useful description of $h_Z(S_\varepsilon)_z$ by viewing it as the subset of $H(S_\varepsilon)$ of elements whose image in $H(S)$ is z. For example, say $Z = \mathbf{Hilb}_{X/k}$ and $z \in Z$ represents $Y \subset X$. Then $h_Z(S_\varepsilon)_z$ is the set of S_ε-flat closed subschemes of $X \times_k S_\varepsilon$ whose fiber over S is Y. Say the ideal of Y is $\mathcal{I}_Y$. Working it out, Grothendieck [**31**, pp. 221-21–23] found $h_Z(S_\varepsilon)_z = H^0(\mathcal{N}_Y)$

[41] In 1947, Zariski [**77**] introduced and studied $\mathfrak{m}/\mathfrak{m}^2$ for a variety in any characteristic, and called it the "local vector space." In Weil's math review of Zarisk's paper, Weil wrote: "the dual vector-space ... seems to deserve to be called the 'tangent vector-space'."

where $\mathcal{N}_Y := \mathcal{H}om(\mathcal{I}_Y/\mathcal{I}_Y^2,\ \mathcal{O}_Y)$. If Y is a Cartier divisor D, then $\mathcal{N}_D$ is invertible on D, and

$$\mathbb{T}_z(\mathbf{Div}_{X/k}) = H^0(\mathcal{N}_D). \tag{5}$$

Let Λ parameterize a system of divisors on X including D, and say $\lambda \in \Lambda$ represents D; in other words, there is a map $\Lambda \to \mathbf{Div}_{X/k}$ carrying λ to z. It induces a map of vector spaces $\theta\colon \mathbb{T}_\lambda(\Lambda) \to \mathbb{T}_z(\mathbf{Div}_{X/k})$. If D is integral, then, owing to (5), the image of θ defines a linear system on D, the storied *characteristic linear system.* When is it complete? More generally, for any D, when is θ surjective?

Each version of the Theorem of Completeness provides conditions guaranteeing the existence of a Λ that is smooth at λ and for which θ is surjective. But then $\mathbf{Div}_{X/k}$ is smooth at z, owing to a simple general observation [**37**, p. 305]. Thus the conditions in question just guarantee that $\Lambda := \mathbf{Div}_{X/k}$ is smooth at z.

Some conditions are necessary. Indeed, in 1943 Severi's student, Guido Zappa, found a smooth complex surface X such that $\mathbf{Div}_{X/k}$ has an isolated point z with $\dim \mathbb{T}_z(\mathbf{Div}_{X/k}) = 1$; so $\mathbf{Div}_{X/k}$ has nilpotents; for details, please see [**46**, pp. 155–156] or [**25**, p. 285]. Commenting, Grothendieck [**31**, pp. 221-24] wrote that this example "shows in a particularly striking way how varieties with nilpotents are needed to understand the phenomena of the most classical theory of surfaces."

Grothendieck then gave an enlightening proof of Kodaira's 1956 version [**38**] of Completeness. As to Kodaira's own proof, Kodaira and Spencer [**39**, p. 477] said that it's "based essentially on the theory of harmonic differential forms" [so not algebraic]; it's "indirect and does not reveal the real nature of the theorem."

Kodaira proved that, *if X and D are smooth and complex, and*[42] *if $h^1(\mathcal{I}_D^{-1}) = 0$, then $\mathbf{Div}_{X/k}$ is smooth at z, where $\mathcal{I}_D$ is the ideal of D.* For example, $h^1(\mathcal{I}_D^{-1}) = 0$ by Serre's computation if D is a hypersurface section of large degree. In 1904, Enriques studied the case that X is a surface and D is *regular* and *nonspecial,* meaning[43] $h^1(\mathcal{I}_D^{-1}) = h^2(\mathcal{I}_D^{-1})$ and $h^2(\mathcal{I}_D^{-1}) = 0$. Thus then Completeness holds.

Grothendieck proved Kodaira's theorem for any X and D as follows. Let $\mathcal{I}$ denote the ideal of the universal divisor on $X \times_k \mathbf{Div}_{X/k}$. Then $\mathcal{I}^{-1}$ is invertible. So it defines a map $\boldsymbol{\alpha}_{X/k}\colon \mathbf{Div}_{X/k} \to \mathbf{Pic}_{X/k}$, called the *Abel map.* Assume $H^1(\mathcal{I}_D^{-1}) = 0$. Then $\boldsymbol{\alpha}_{X/k}$ is smooth at z; see below. Hence $\mathbf{Div}_{X/k}$ is smooth at z if and only if $\mathbf{Pic}_{X/k}$ is smooth at $\boldsymbol{\alpha}_{X/k}(z)$, or equivalently by translation, everywhere. In characteristic zero, $\mathbf{Pic}_{X/k}$ is smooth by Cartier's Theorem [**46**, p. 167]. Thus $\mathbf{Div}_{X/k}$ is smooth at z in characteristic zero.

---

[42] It is now common to let $\mathcal{O}_X(D)$ stand for $\mathcal{I}_D^{-1}$, but that practice is not followed here.

[43] At first, "regular" alone was used to mean $h^1(\mathcal{I}_D^{-1}) = 0$ and $h^2(\mathcal{I}_D^{-1}) = 0$.

In positive characteristic, $\mathbf{Pic}_{X/k}$ can be nonreduced everywhere even if X is a smooth surface; see below. If so and $h^1(\mathcal{I}_D^{-1}) = 0$, then $\mathbf{Div}_{X/k}$ is nonreduced at z. Thus Completeness fails, even if D is a general hypersurface section of large degree.

Since k is algebraically closed, X has a rational point, so that $X \to S$ has a section. Set $P := \mathbf{Pic}_{X/k}$. Then $\mathrm{Pic}_{X/k}$ is representable by P. So $X \times_k P$ carries an invertible sheaf $\mathcal{P}$, called a *Poincaré sheaf*, whose class modulo $\mathrm{Pic}(P)$ is universal. In particular, there is an invertible sheaf $\mathcal{M}$ on P such that $\mathcal{I}^{-1}$ is isomorphic to the tensor product on $X \times_k \mathbf{Div}_{X/k}$ of the pullbacks of $\mathcal{P}$ and $\mathcal{M}$. Then $\mathrm{LinSys}_{\mathcal{P}/X\times P/P}$ is representable, on the one hand, by $\mathbf{Div}_{X/k}$ regarded as a P-scheme via $\boldsymbol{\alpha}_{X/k}$, and on the other, by $\mathbf{P}(\mathcal{Q})$ for some coherent sheaf $\mathcal{Q}$ on P. So $\mathbf{Div}_{X/k}$ and $\mathbf{P}(\mathcal{Q})$ are canonically isomorphic P-schemes. If $H^1(\mathcal{I}_D^{-1}) = 0$, then $\mathcal{Q}$ is free at $\boldsymbol{\alpha}_{X/k}(z)$, and so $\boldsymbol{\alpha}_{X/k}$ is smooth at z, as desired.

Grothendieck [**31**, pp. 236-16] asserted $\mathbb{T}_0(\mathbf{Pic}_{X/k}) = H^1(\mathcal{O}_X)$; for proofs, please see [**46**, pp. 163–164] and [**25**, pp. 281–282]. So $\dim \mathbf{Pic}_{X/k} \le h^1(\mathcal{O}_X)$, with equality if and only if $\mathbf{Pic}_{X/k}$ is smooth. That result is part of Grothendieck's contribution to the proof of the Fundamental Theorem of Irregular Surfaces. In the examples of Igusa and Serre recalled in Section 2, we have $\dim \mathbf{Pic}_{X/k} < h^1(\mathcal{O}_X)$; hence, $\mathbf{Pic}_{X/k}$ is not smooth at 0, so nonreduced everywhere.

Grothendieck noted smoothness holds if $H^2(\mathcal{O}_X) = 0$, owing to the Infinitesimal Criterion for Smoothness and a well-known computation; for details, please see [**25**, pp. 285–286]. For example, if X is a curve, then $\mathbf{Pic}_{X/k}$ is smooth, so of dimension g where $g := h^1(O_X)$. In positive characteristic, Mumford [**46**, pp. 193–198] proved $\mathbf{Pic}_{X/k}$ is smooth if and only if Serre's Bockstein operations [**66**, p. 505] all vanish.

Grothendieck did not consider the other versions of Completeness, but his work does provide a basis for proving them algebraically. First consider Severi's 1921 version (3) on p. 45. Given an invertible sheaf $\mathcal{L}$ on X, set $e(\mathcal{L}) := \chi(\mathcal{L}) - 1 - h^2(\mathcal{L})$. Call $\mathcal{L}$ *arithmetically effective* if $e(\mathcal{L}) \ge 0$. Vary $\mathcal{L}$. Then $\chi(\mathcal{L})$ is locally constant, and $h^2(\mathcal{L})$ is upper semi-continuous. So $e(\mathcal{L})$ is lower semi-continuous.

Hence there is an open subscheme of $\mathbf{Pic}_{X/k}$, say $\mathbf{P}_{\mathrm{ae}}$, that parameterizes the arithmetically effective $\mathcal{L}$ on X. Set $\mathbf{D}_{\mathrm{ae}} := \boldsymbol{\alpha}_{X/k}^{-1}\mathbf{P}_{\mathrm{ae}}$. Assume $\dim X = 2$. Then $\mathbf{D}_{\mathrm{ae}}$ surjects onto $\mathbf{P}_{\mathrm{ae}}$, since over a point representing an $\mathcal{L}$, the fiber has dimension $e(\mathcal{L}) + h^1(\mathcal{L})$, which is nonnegative. In these terms, a refined Version (3) says that, *if* $\mathbf{Pic}_{X/k}$ *is smooth too, then* $\mathbf{D}_{\mathrm{ae}}$ *is smooth on a dense open subset* U.

Since $h^0(\mathcal{L})$ is upper semi-continuous in $\mathcal{L}$, there is an open subscheme $V \subset \mathbf{P}_{\mathrm{ae}}$ that parameterizes the $\mathcal{L}$ where $h^0(\mathcal{L})$ has a local minimum. Set $U := \boldsymbol{\alpha}_{X/k}^{-1}V$. Recall that $\mathbf{Pic}_{X/k}$ carries a coherent sheaf $\mathcal{Q}$ such that $\mathbf{P}(\mathcal{Q}) = \mathbf{Div}_{X/k}$. Then V is precisely the set of points of $\mathbf{P}_{\mathrm{ae}}$ at which the rank of $\mathcal{Q}$ has a local minimum. Suppose $\mathbf{Pic}_{X/k}$ is smooth. Then the restriction $\mathcal{Q}|V$ is locally free. Hence $U \to V$ is smooth. So U is smooth. Thus we have refined and proved Severi's version (3).

In 1944, Severi discovered another condition on D for Completeness to hold if $\mathbf{Pic}_{X/k}$ is smooth. The condition requires D to be *semi-regular*; namely, in the standard long exact sequence of cohomology

$$\begin{aligned} &0 \longrightarrow H^0(\mathcal{O}_X) \longrightarrow H^0(\mathcal{I}_D^{-1}) \longrightarrow H^0(\mathcal{N}_D) \\ &\xrightarrow{\partial^0} H^1(\mathcal{O}_X) \longrightarrow H^1(\mathcal{I}_D^{-1}) \xrightarrow{u} H^1(\mathcal{N}_D) \xrightarrow{\partial^1} H^2(\mathcal{O}_X), \end{aligned} \tag{6}$$

the map u is 0, or equivalently ∂^1 is injective. In particular, D is semi-regular if either $H^1(\mathcal{I}_D^{-1}) = 0$ or $H^1(\mathcal{N}_D) = 0$.

Severi worked with an integral D on a smooth surface X, and he formulated the condition in its dual form: the restriction $H^0(\Omega_X^2) \to H^0(\Omega_X^2|D)$ is surjective; in other words, the canonical system on X cuts out a complete system on D. In 1959, Kodaira and Spencer [**39**, p. 481] reformulated Severi's condition as $u = 0$ in any dimension. Then they proved that, *in the complex analytic case, if X and D are smooth and if $u = 0$, then $\mathbf{Div}_{X/k}$ is smooth at z.*

Grothendieck did not consider semi-regularity per se, but he [**31**, pp. 221-23] did observe that $H^1(\mathcal{N}_D)$ houses the obstruction to deforming D in X. Thus[44] *if $H^1(\mathcal{N}_D) = 0$, then $\mathbf{Div}_{X/k}$ is smooth at z in any characteristic* whether X and D are smooth or not. For example, if X is a curve, then $\mathbf{Div}_{X/k}$ is smooth everywhere; however, $\boldsymbol{\alpha}_{X/k}$ is not smooth at z if $\deg D < g$ where $g := h^1(\mathcal{O}_X)$, since $\dim \mathbf{Div}_{X/k} = \deg D$ by (5) and $\dim \mathbf{Pic}_{X/k} = g$ as noted above.

Mumford [**46**, pp. 157–159] explicitly computed the obstruction to deforming D, as well as its image under ∂^1. He proved that this image vanishes in characteristic 0 using an exponential. Therefore, if ∂^1 is injective, then $\mathbf{Div}_{X/k}$ is smooth at z. Cartier's Theorem is not involved, but recovered. Thus in 1966 Mumford gave the first algebraic proof that semi-regularity yields Completeness in characteristic 0.

In 1973, I [**37**] gave another algebraic proof, yielding a more refined statement: *assume $\mathbf{Pic}_{X/k}$ is smooth; then $\mathbf{Div}_{X/k}$ is smooth at z of dimension ρ where*

$$\rho := h^1(\mathcal{O}_X) - 1 + h^0(\mathcal{I}_D^{-1}) - h^1(\mathcal{I}_D^{-1})$$

[44]Perhaps not surprisingly, the condition $H^1(\mathcal{N}_D) = 0$ is related to the flaw in the constructions of a good algebraic system made in 1904 by Enriques and in 1905 by Severi. In 1934, Zariski [**78**, p. 100] noted that both constructions rely on a certain assumption and that Severi's 1921 "criticism is to the effect that the available algebro-geometric proof of this assumption fails if" $H^1(\mathcal{N}_D) \neq 0$.

if and only if D is semi-regular. My proof[45] does not use obstruction theory, but a short formal analysis, essentially due to George Kempf, of the scheme $\mathbf{P}(\mathcal{Q})$ above.

In passing, set $R := \dim_z \mathbf{Div}_{X/k}$ and note that (5) yields $R \leq h^0(\mathcal{N}_D)$, with equality if and only if $\mathbf{Div}_{X/k}$ is smooth at z. Also, (6) yields $\rho \leq h^0(\mathcal{N}_D)$, with equality if and only if D is semi-regular. Thus if $\mathbf{Div}_{X/k}$ is smooth at z, then D is semi-regular if and only if $R = \rho$.

Generalizing more of Section 2, set $\delta := \dim \operatorname{Coker}(\partial^0)$ and $q := h^1(\mathcal{O}_X)$. Then (6) yields $\delta \leq q$, with equality if $H^1(\mathcal{I}_D^{-1}) = 0$. As $H^1(\mathcal{I}_D^{-1}) = 0$ if D is a hypersurface section of large degree, we have generalized Castelnuovo's result (1). Next, set $r := h^0(\mathcal{I}_D^{-1}) - 1$. Then (6) yields $h^0(\mathcal{N}_D) = r + \delta$. Hence $R \leq r + \delta$, with equality if and only if $\mathbf{Div}_{X/k}$ is smooth at z. Thus we have generalized Severi's result (2). Finally, *if X is a surface, $\mathbf{Pic}_{X/k}$ is smooth and z lies in the open subset $U \subset \mathbf{D}_{\mathrm{ae}}$, then $\mathbf{Div}_{X/k}$ is smooth at z, and so* $R = r + \delta$, just as Severi discovered.

Grothendieck [**31**, pp. 2-12] proved the following basic properties of the connected component of 0 in $\mathbf{Pic}_{X/k}$, denoted $\mathbf{Pic}^0_{X/k}$. It is open and closed. It is irreducible. Forming it commutes with base change. It is *quasi-projective*; that is, it is an open subscheme of a projective k-scheme. Moreover, it is projective if X is normal. Of course, the *Picard variety* of X is the set of points of $\mathbf{Pic}^0_{X/k}$ with coordinates in a given universal domain. If X is a curve, then $\mathbf{Pic}^0_{X/k}$ is its *generalized Jacobian.*

Define a *Poincaré divisor* to be a Cartier divisor Δ on $X \times P$, where P is some connected component of $\mathbf{Pic}_{X/k}$, such that Δ yields a section $P \to \mathbf{Div}_{X/k}$. Such Δ abound, as is shown next, generalizing Mattuck's work mentioned on p. 53.

Given any connected component P' of $\mathbf{Pic}_{X/k}$, notice it's a translate of $\mathbf{Pic}^0_{X/k}$, so quasi-projective. Hence there's an n such that, for any invertible sheaf $\mathcal{L}$ on X represented by a point of P', we have $h^0(\mathcal{L}(n)) > \dim \mathbf{Pic}_{X/k}$ and $h^1(\mathcal{L}(n)) = 0$ where $\mathcal{L}(n)$ is the nth twist by $\mathcal{O}_X(1)$. Let P be the translate of P' defined by $\mathcal{O}_X(n)$. Since P is quasi-projective, it has a universal sheaf $\mathcal{O}_P(1)$.

Recall $\mathbf{Pic}_{X/k}$ carries a coherent sheaf $\mathcal{Q}$ such that $\mathbf{P}(\mathcal{Q}) = \mathbf{Div}_{X/k}$. Notice the restriction $\mathcal{Q}|P$ is locally free of rank $h^0(\mathcal{L}(n))$, so of rank more than $\dim P$. Take m so that $\mathcal{H}om(\mathcal{Q}|P, \mathcal{O}_P)(m)$ is generated by its global sections, so by finitely many. Then a general linear combination of the latter vanishes nowhere on P by a well-known lemma, [**3**, p. 426] or [**46**, p. 148], due to Serre. So there's a surjection $\mathcal{Q}|P \twoheadrightarrow \mathcal{O}_P(m)$. It defines a section $P \to \mathbf{P}(\mathcal{Q})$, and so a Poincaré divisor Δ.

[45]The proof works over any Noetherian S, and yields this more general result: *let D be a divisor on an (integral) geometric fiber of X/S, and assume $\mathbf{Pic}_{X/S}$ is smooth; then $\mathbf{Div}_{X/S}$ is smooth of relative dimension ρ at the point representing D if and only if D is semi-regular.*

Suppose also that X is a curve. Set $g := h^1(\mathcal{O}_X)$ and recall $g = \dim \mathbf{Pic}_{X/k}$. Take P to be any connected component of $\mathbf{Pic}_{X/k}$ that parameterizes invertible sheaves $\mathcal{L}$ on X of $\deg \mathcal{L} > 2g-1$. Then $h^0(\mathcal{L}) > g$ and $h^1(\mathcal{L}) = 0$. So similarly there is a section $P \to \mathbf{P}(\mathcal{Q})$, and so a Poincaré divisor Δ.

Here's an introduction[46] to the scheme-theoretic theory of the Albanese variety. Assume X is normal. Then $\mathbf{Pic}^0_{X/k}$ is projective. Let P denote its *reduction*, namely, the subscheme defined by the nilradical of the structure sheaf of $\mathbf{Pic}^0_{X/k}$. It too is a *group scheme*; that is, its T-points form a group for all T. So P is smooth. Call any such connected smooth projective group scheme an *Abelian variety.*

If X is an Abelian variety, then $\mathbf{Pic}^0_{X/k}$ is already reduced. Mumford [**45**] gave a proof on pp. 117–118, which he attributed to Grothendieck on p. 115. Then $\mathbf{Pic}^0_{X/k}$ is denoted by $\widehat{X}$ or X^*, and called the *dual Abelian variety.*

In general, let Y be another integral and projective k-scheme, and fix rational points $x \in X$ and $y \in Y$. Then a k-map $f\colon Y \to P$ with $f(y) = 0$ is defined by an invertible sheaf $\mathcal{L}$ on $X \times_k Y$ whose restriction to $X \times_k y$ is $\mathcal{O}_X$. At first, $\mathcal{L}$ is only determined modulo $\mathrm{Pic}(Y)$, but normalize $\mathcal{L}$ as follows: restrict it to $x \times_k Y$, pull the restriction back to $X \times_k Y$ via the projection, and replace $\mathcal{L}$ by its tensor product with the inverse of the pullback.

Let Q be the reduction of $\mathbf{Pic}_{Y/k}$. By symmetry, $\mathcal{L}$ corresponds to a k-map $g\colon X \to Q$ with $g(x) = 0$. Plainly, this correspondence $f \longleftrightarrow g$ is *functorial*: given a similar triple (Z, z, R) and a k-map $h\colon Z \to Y$ with $g(z) = y$, the composition $f \circ h\colon Z \to P$ corresponds to the composition $h^* \circ g\colon X \to R$ where $h^*\colon P \to R$ is the k-map induced by pullback of invertible sheaves, which is a group homomorphism.

For a moment, take $Y := P$ and $y := 0$ and $f := 1_p$. Since P is an Abelian variety, Q is its dual P^*. Set $A := P^* = Q$ and $a := g$. Then A is called the *Albanese variety* of X, and there is a canonical k-map $a\colon X \to A$.

The map $a\colon X \to A$ is the *universal example* of a map $g\colon X \to Q$ where Q is the reduction of $\mathbf{Pic}_{Y/k}$ for some integral and projective k-scheme Y; that is, any such g factors uniquely through a. Here's why. Say g corresponds to $h\colon Y \to P$. Then by functoriality, $1_P \circ h$ corresponds to $h^* \circ a$.

By definition, A^* is the Albanese of P. Moreover, the canonical map $p\colon P \to A^*$ is an isomorphism, since by "abstract nonsense," a universal example is determined up to unique isomorphism, and $1_P\colon P \to P$ is trivially

[46] Doubtless, Grothendieck had something similar in mind when he [**31**, p. 232-14] wrote, "the theory of Abelian varieties, and more generally of Abelian schemes, becomes much simpler once we have a general theory of the Picard scheme at our disposal. In particular, the theory of duality for Abelian schemes, and notably results like Cartier's, thus become nearly formal (cf. for example the forthcoming notes to the Mumford–Tate seminar at Harvard in the spring term of 1962)." However, it seems that nothing like the present introduction appears in Mumford's personal notes to the seminar, or has already appeared in print.

another universal example. In fact, p^{-1} is just the map a^* induced by the canonical map $a\colon X \to A$, because by functoriality, $1_A \circ a$ corresponds to $a^* \circ p$. Thus A and P are dual to each other.

Suppose X is an Abelian variety; take $x := 0$. Mumford [**47**, p. 125] constructed X^* as a quotient of X by a finite subgroup; so X and X^* are isogenous. Mumford [**47**, Cor., p. 43, 132] proved that $a : X \to X^{**}$ is an isomorphism of groups and of schemes. Hence, for any X, the map $a\colon X \to A$ is the *universal example* of a map $X \to B$ where B is any Abelian variety, because $B = Q$ if $Y = B^*$.

Suppose finally that X is a smooth curve of genus $g > 0$. Then X is a component of $\mathbf{Div}_{X/k}$. So the Abel map restricts to a k-map $X \to \mathbf{Pic}_{X/k}$. Its image lies in the connected component parameterizing the sheaves $\mathcal{L}$ of degree 1. Fix an $\mathcal{L}$. Translating by $\mathcal{L}^{-1}$ yields a map $\alpha\colon X \to P$. It is an embedding by general principles, since its fibers are finite and $X = \mathbf{P}(\mathcal{Q})$ for some coherent sheaf $\mathcal{Q}$ on P. It is proved (in a more general form) in [**24**, Thm. 2.1, p. 595] that $\alpha^*\colon P^* \to P$ is an isomorphism, which is independent of the choice of $\mathcal{L}$.

To end this article, let's consider some important ways in which Grothendieck and others generalized the existence theorem culminating Section 4. First, let k be an arbitrary field. On pp. 232-15–17 in [**31**], Grothendieck outlined a construction of $\mathbf{Pic}_{X/k}$ for any projective k-scheme X. He used that earlier theorem plus a method of *relative representability*, by which $\mathbf{Pic}_{X/k}$ is constructed from $\mathbf{Pic}_{X'/k}$ for a suitable surjective k-map $X' \twoheadrightarrow X$. The method employs two main tools: nonflat descent and Oort dévissage. The former refers to descent along maps not required to be flat; however, key objects are flat. The second tool was introduced by Oort in [**58**] to construct $\mathbf{Pic}_{X/k}$ from $\mathbf{Pic}_{X'/k}$ where X' is the reduction of X.

On p. 232-17 in [**31**], Grothendieck, in effect, made two conjectures: first, $\mathbf{Pic}_{X/k}$ exists for any proper k-scheme X; second, given any surjective k-map between proper k-schemes, the induced map on Picard schemes is affine.

The first conjecture was proved in 1964 by Murre [**53**], who thanked Grothendieck for help. However, instead of using relative representability, Murre identified seven conditions that are necessary and sufficient for the representability of a functor from schemes over a field to Abelian groups. Then he checked the seven for $\mathrm{Pic}_{(X/k)(\mathrm{fpqc})}$.

From now on, assume S is Noetherian and X is a flat and proper S-scheme.

Murre [**53**, p. 5] said that Grothendieck too proved the first conjecture. In 1965, Murre [**54**] sketched Grothendieck's proof of the following key intermediate result: *let $\mathcal{F}$ be a coherent sheaf on X, and $S_{\mathcal{F}}$ the functor of all S-schemes T such that the pullback $\mathcal{F}_T$ is T-flat; then $S_{\mathcal{F}}$ is representable by an unramified S-scheme of finite type.* The proof involves identifying and checking eight conditions that are necessary and sufficient for representability by a scheme of the desired sort.

In 1966, Raynaud [**8**, Exp. XII] gave Grothendieck's proof of another key intermediate result: *assume S is integral and let $X' \twoheadrightarrow X$ be a surjective map of proper S-schemes; then there's a nonempty open subscheme $V \subset S$ such that $\mathbf{Pic}_{X' \times V/V}$ and $\mathbf{Pic}_{X \times V/V}$ exist, and the induced map between them is quasi-affine.* The proof does indeed involve suitably general versions of nonflat descent and Oort dévissage. As corollaries, that result yields Grothendieck's two conjectures.

If the geometric fibers of X/S are not all integral, then $\mathbf{Pic}_{X/S}$ need not exist. On p. 236-01 in [**31**], Grothendieck described an example of Mumford's; one geometric fiber is integral, but another is a pair of conjugate lines. On the other hand, on p. viii in [**46**], Mumford asserted this theorem: *Assume X/S is projective, and its geometric fibers are reduced and connected; assume the irreducible components of its ordinary fibers are geometrically irreducible; then $\mathbf{Pic}_{X/S}$ exists.* Mumford said the proof is like the one on pp. 133–149, involving his theory of independent 0-cycles.

On p. 236-01, Grothendieck attributed a slightly different theorem to Mumford, and referred to the Mumford–Tate seminar. Mumford's seminar notes contain a precise statement of the theorem and a rough sketch of the proof. However, he crossed out the hypothesis that the geometric fibers are connected, and made the weaker assumption that the ordinary fibers are connected.

On p. 236-13, Grothendieck wrote that "it is not ruled out that $\mathbf{Pic}_{X/S}$ exists" whenever[47] the direct image of $\mathcal{O}_{X \times T}$ is $\mathcal{O}_T$ for any T. "At least, this statement is proved for analytic spaces when X/S is also projective." Mumford's example shows the statement is false for schemes. Michael Artin's work shows it holds for *algebraic spaces*, which he introduced in 1968 in [**1**]. They are formed by gluing together schemes along open subsets that are isomorphic locally in the étale topology. Over $\mathbb{C}$, these open sets are locally analytically isomorphic; so a separated algebraic space is a kind of complex analytic space.

In 1969, Artin [**2**], inspired by Grothendieck and Murre, found five conditions on a functor that are necessary and sufficient for it to be representable by a well-behaved algebraic space. A key new ingredient is Artin's Approximation Theorem; it facilitates the passage from formal power series to polynomials. By checking that the conditions hold if the direct image of $\mathcal{O}_{X \times T}$ is always $\mathcal{O}_T$, Artin [**2**, Thm. 7.3, p. 67] proved[48] $\mathbf{Pic}_{X/S}$ exists as

[47]This condition holds if the geometric fibers of X/S are integral by [**30**, Prp. (7.8.6), p. 74]. For more about its significance when S is the spectrum of a discrete valuation ring, please see [**63**].

[48]In fact, he proved a more general theorem, in which S and X are algebraic spaces. That theorem and Grothendieck's theorem in Section 4 are the two main representability theorems for the Picard functor. Grothendieck used projective methods. Artin's work has a very different flavor. Moreover, it yields a major improvement of Murre's representability theorem stated above, and it yields the representability of the Hilbert functor and related functors in algebraic spaces.

an algebraic space, a magnificent achievement. Also, he [**2**, Lem. 4.2, p. 43] proved that, if S is the spectrum of a field, then $\mathbf{Pic}_{X/S}$ is a scheme. Thus he obtained a third proof of Grothendieck's first conjecture.

As S and X are schemes, so are the fibers of X/S. Hence their Picard schemes exist. Furthermore, if the direct image of $\mathcal{O}_{X\times T}$ is always $\mathcal{O}_T$, then these Picard schemes form a family; its total space $\mathbf{Pic}_{X/S}$ is an algebraic space, but need not be a scheme. Thus Artin proved the definitive statement explaining the behavior of the Picard schemes of the members of a family.[49]

References

[1] Artin, Michael. *The implicit function theorem in algebraic geometry.* 1969 Algebraic Geometry (Internat. Colloq., Tata Inst. Fund. Res., Bombay, 1968) pp. 13–34. Oxford Univ. Press, London. MR0262237 (41 #6847)

[2] Artin, Michael. *Algebraization of formal moduli. I.* 1969 Global Analysis (Papers in Honor of K. Kodaira) pp. 21–71. Univ. Tokyo Press, Tokyo. MR0260746 (41 #5369)

[3] Atiyah, Michael. *Vector bundles over an elliptic curve.* Proc. London Math. Soc. (3) **7** (1957), 414–452. MR0131423 (24 #A1274)

[4] Audin, Michèle. *La guerre des recensions. Autour d'une note d'André Weil en 1940.* J Math. Semesterber. **59** (2012), no. 2, 243–260. MR2970483

[5] Babbitt, Donald; Goodstein, Judith. *Guido Castelnuovo and Francesco Severi: two personalities, two letters.* Notices Amer. Math. Soc. **56** (2009), no. 7, 800–808. MR2546822

[6] Babbitt, Donald; Goodstein, Judith. *Federigo Enriques's quest to prove the "completeness theorem."* Notices Amer. Math. Soc. **58** (2011), no. 2, 240–249. MR2768117 (2012a:14001)

[7] Bardelli, Fabio. *On the origins of the concept of irregularity of an algebraic surface.* in "Algebra and geometry (1860–1940): the Italian contribution (Cortona, 1992)." Rend. Circ. Mat. Palermo (2) Suppl. No. *36* (1994), 11–26. MR1308338

[8] Berthelot, Pierre; Grothendieck, Alexander; Illusie, Luc. Avec la collaboration de D. Ferrand, J.-P. Jouanolou, O. Jussila, S. Kleiman, M. Raynaud, J.-P. Serre. "Théorie des intersections et théorème de Riemann–Roch." Séminaire de géométrie algébrique du Bois Marie (Series); SGA 6. Lecture notes in mathematics (Springer-Verlag) **225** (1971).

[9] Bosch, Siegfried; Lütkebohmert, Werner; Raynaud, Michel. "Néron models." Ergebnisse der Mathematik und ihrer Grenzgebiete (3) [Results in Mathematics and Related Areas (3)] (Springer-Verlag) **21** 1990. MR1045822 (91i:14034)

[10] Cartier, Pierre. *Dualité des variétés abéliennes.* Séminaire Bourbaki; 10e année: 1957/1958. Exposé 164, May 1958.

[11] Cartier, Pierre. *Questions de rationalité des diviseurs en géométrie algébrique* Bull. Soc. Math. France. **86** 1958, 177–251. MR0106223 (21 #4957)

[12] Cartier, Pierre. *A country of which nothing is known but the name, Grothendieck and "motives."* This volume.

[13] Castelnuovo, Guido. *Sur les intègrales de différentielles totales appartenant à une surface irrégulère.* Comptes rendus hebdomadaires des séances de l'Académie des sciences, Paris. **140**, 23 Jan 1905. pp. 220–222.

[49]As the Picard varieties in the family are the points of the component $\mathbf{Pic}^0_{X_s/k_s}$ for $s \in S$, their behavior is explained by an open subspace $\mathbf{Pic}^0_{X/S}$ of $\mathbf{Pic}_{X/S}$ whose fibers are the $\mathbf{Pic}^0_{X_s/k_s}$. Such a $\mathbf{Pic}^0_{X/S}$ is observed in [**9**, p. 233] to exist when $\mathbf{Pic}_{X/S}$ is S-smooth along the 0-section.

[14] Castelnuovo, Guido; Enriques Federigo. *Sur quelques résultats nouveaux dans la théorie des surfaces algébriques.* Note V in [**62**], pp. 485–522.

[15] Castelnuovo, Guido; Enriques Federigo. "Die algebraischen Flächen vom Gesichtspunkte der birationalen Transformationen," Enzyklop. d. math. Wiss. III_2 6b, 1914.

[16] Cayley, Arthur. "On the deficiency of certain surfaces." Math. Ann. **3** no. 4 (1871), 526–529.

[17] Chevalley, Claude. *Sur la théorie des variétés algébriques.* Nagoya Math. J. **8** (1955), 1–43. MR0069544 (16,1048d)

[18] Chevalley, Claude. *La notion de correspondance propre en géométrie algébrique.* Séminaire Bourbaki; 10e année: 1957/1958. Textes des conférences; Exposés 152 à 168; 2e éd. corrigée, Exposé 152, pp. 219–229, Dec. 1957 Vol. **11** Secrétariat mathématique, Paris. 1958. MR0108495 (21 #7211)

[19] Chevalley, Claude. *Sur la théorie de la variété de Picard.* (French) Amer. J. Math. **82** (1960), 435–490. MR0118723 (22 #9494)

[20] Colmez, Pierre; Serre, Jean-Pierre (eds.) "Correspondance Grothendieck-Serre." (French) [Grothendieck-Serre correspondence] Documents Mathématiques (Paris) [Mathematical Documents (Paris)], 2. Société Mathématique de France, Paris, 2001. MR1942134 (2003k:14002)

[21] Dieudonné, Jean. "History of algebraic geometry. An outline of the history and development of algebraic geometry." Translated from the 1974 French version by Judith D. Sally. Wadsworth Mathematics Series. Wadsworth International Group, Belmont, CA, 1985. MR0780183 (86h:01004)

[22] Enriques Federigo. *Sur les surfaces algébriques irregulières.* Comptes rendus hebdomadaires des séances de l'Académie des sciences, Paris. **140**, 16 Jan 1905. pp. 133–135.

[23] Enriques Federigo. "Le superficie algebriche." Zanichelli, Bologna, 1949.

[24] Esteves, Eduardo; Gagné, Mathieu; Kleiman, Steven. *Autoduality of the compactified Jacobian.* J. London Math. Soc. (2) **65** (2002), no. 3, 591–610. MR1895735 (2003d:14038)

[25] Fantechi, Barbara; Göttsche, Lothar; Illusie, Luc; Kleiman, Steven; Nitsure, Nitin; Vistoli, Angelo. "Fundamental algebraic geometry. Grothendieck's FGA explained." Mathematical Surveys and Monographs, 123. American Mathematical Society, Providence, RI, 2005. MR2222646 (2007f:14001)

[26] Goodstein, Judith; Babbitt, Donald. *A fresh look at Francesco Severi.* Notices Amer. Math. Soc. **59** (2012), no. 8, 1064–1075. MR2985808

[27] Grothendieck, Alexander. *Sur une note de Mattuck-Tate.* (French) J. Reine Angew. Math. **200** (1958), 208–215. MR0136607 (25 #75)

[28] Grothendieck, Alexander. *The cohomology theory of abstract algebraic varieties.* Proc. Internat. Congress Math. (Edinburgh, 1958) Cambridge Univ. Press, New York 1960. pp. 103–118. MR0130879 (24 #A733)

[29] Grothendieck, Alexander. "Éléments de géométrie algébrique. I. Le langage des schémas." (French) Inst. Hautes Études Sci. Publ. Math. **4** (1960). MR0217083 (36 #177a); MR0163908 (29 #1207)

[30] Grothendieck, Alexander. "Éléments de géométrie algébrique. III. Étude cohomologique des faisceaux cohérents. II. (French) Inst. Hautes Études Sci. Publ. Math. **17** (1963). MR0163911 (29 #1210)

[31] Grothendieck, Alexander. "Fondements de la géométrie algébrique. [Extraits du Séminaire Bourbaki, 1957–1962.]" (French) Secrétariat mathématique, Paris 1962. MR0146040 (26 #3566)

[32] Grothendieck, Alexander; Dieudonné, Jean. "Éléments de géométrie algébrique. I." Grundlehren Math. Wiss., vol. 166, Springer-Verlag, 1971.

[33] Jackson, Allyn. *Comme appelé du néant—as if summoned from the void: the life of Alexandre Grothendieck.* Notices Amer. Math. Soc. **51** (2004), no. 9, 1038–1056. MR2089093

[34] Joshi, Kirti. *A p-adic proof of Hodge symmetry for threefolds.* C. R. Math. Acad. Sci. Paris **338** (2004), no. 10, 781–786. MR2059487 (2005e:14028)

[35] Kleiman, Steven. *Toward a numerical theory of ampleness.* Ann. of Math. (2) **84** (1966), 293–344. MR0206009 (34 #5834)

[36] Kleiman, Steven. *Algebraic cycles and the Weil conjectures.* in "Dix exposés sur la cohomologie des schémas." North-Holland, Amsterdam; Masson, Paris, 1968. Pages 359–386. MR0292838 (45 #1920)

[37] Kleiman, Steven. *Completeness of the characteristic system.* Advances in Math. **11** (1973), 304–310. MR0357397 (50 #9865)

[38] Kodaira, Kunihiko. *Characteristic linear systems of complete continuous systems.* Amer. J. Math. **78** (1956), 716–744. MR0090096 (19,767c)

[39] Kodaira, Kunihiko; Spencer, Donald. *A theorem of completeness of characteristic systems of complete continuous systems.* Amer. J. Math. **81** (1959), 477–500. MR0112156 (22 #3011)

[40] Lang, Serge. "Abelian varieties." Interscience Tracts in Pure and Applied Mathematics. No. 7 Interscience Publishers, Inc., New York; Interscience Publishers Ltd., London 1959. MR0106225 (21 #4959)

[41] Mac Lane, Saunders. "Categories for the working mathematician." Graduate Texts in Mathematics, Vol. 5. Springer-Verlag, New York-Berlin, 1971. MR0354798 (50 #7275)

[42] Mattuck, Arthur. *Picard bundles.* Illinois J. Math. **5** (1961), 550–564. MR0150141 (27 #144)

[43] Mazur, Barry. *Eigenvalues of Frobenius acting on algebraic varieties over finite fields.* in "Symposium in Pure Mathematics (29th Humboldt State University). Algebraic geometry, Arcata, 1974" American Mathematical Society. Providence, 1975. Pages 231–261.

[44] Mumford, David *Further comments on boundary points.* in "Lecture notes prepare in connection the summer institute on algebraic geometry held at the Whitney estate, Woods Hole, MA." Amer. Math. Soc. 1964.

[45] Mumford, David. "Geometric invariant theory." Ergebnisse der Mathematik und ihrer Grenzgebiete, Neue Folge, Band **34**, Springer-Verlag, Berlin-New York 1965. MR0214602 (35 #5451)

[46] Mumford, David. "Lectures on curves on an algebraic surface. With a section by G. M. Bergman." Annals of Mathematics Studies, No. **59** Princeton University Press, Princeton, N.J. 1966. MR0209285 (35 #187)

[47] Mumford, David. "Abelian varieties." Tata Institute of Fundamental Research Studies in Mathematics, No. **5** Published for the Tata Institute of Fundamental Research, Bombay; Oxford University Press, London 1970. MR0282985 (44 #219)

[48] Mumford, David. "Selected Papers, V. II: On Algebraic Geometry, including Correspondence with Grothendieck." Neeman, Amnon; Chai, Ching-Li; Shiota, Takahiro (Eds.) 2010.

[49] Mumford, David. *Intuition and rigor and Enriques's quest.* Notices Amer. Math. Soc. **58** (2011), no. 2, 250–260. MR2768118 (2012c:14001)

[50] Mumford, David. *My introduction to Schemes and Functors.* This volume.

[51] Murre, Jacob. *On Chow varieties of maximal, total, regular families of positive divisors.* Amer. J. Math. **83.1** (1961) 99–110. MR0122818 (23 #A151)

[52] Murre, Jacob. *On generalized Picard varieties.* Math. Ann. **145** (1962), 334–353. MR0152529 (27 #2507)

[53] Murre, Jacob. *On contravariant functors from the category of pre-schemes over a field into the category of abelian groups (with an application to the Picard functor).* Inst. Hautes Études Sci. Publ. Math. **23** (1964), 5–43. MR0206011 (34 #5836)

[54] Murre, Jacob. *Representation of unramified functors. Applications (according to unpublished results of A. Grothendieck).* Séminaire Bourbaki, May 1965, Exp. No. 294; Vol. 9, 243–261, Soc. Math. France, Paris, 1995. MR1608802

[55] Nagata, Masayoshi. *A general theory of algebraic geometry over Dedekind domains. I. The notion of models.* Amer. J. Math. **78** (1956), 78–116. MR0082725 (18,600e)

[56] Néron, André; Samuel, Pierre. La variété de Picard d'une variété normale. (French) Ann. Inst. Fourier Grenoble **4** (1952), 1—30. MR0061858 (15,896e)

[57] Nishi, Mieo. *The Frobenius theorem and the duality theorem on an abelian variety.* Mem. Coll. Sci. Univ. Kyoto. Ser. A. Math. **32** (1959), 333–350. MR0116020 (22 #6815)

[58] Oort, Frans. *Sur le schéma de Picard.* (French) Bull. Soc. Math. France **90** (1962), 1–14. MR0138627 (25 #2070)

[59] Oort, Frans. *A construction of generalized Jacobian varieties by group extensions.* Math. Ann. **147** (1962), 277–286. MR0141667 (25 #5064)

[60] Oort, Frans. *Did earlier thoughts inspire Grothendieck?* This volume.

[61] Picard, Emile. *Sur quelques théorèmes relatifs aux surfaces algébriques de connexion linéaire supérieure à l'unité.* Comptes rendus hebdomadaires des séances de l'Académie des sciences, Paris. **140**, 16 Jan 1905. pp. 117–122.

[62] Picard, Emile; Simart, Georges. "Théorie des fonctions algébriques de deux variables indépendantes." Tome I, II (French) Réimpression corrigée (en un volume) de l'édition en deux volumes de 1897 et 1906. Chelsea Publishing Co., Bronx, N.Y., 1971. MR0392468 (52 #13285)

[63] Raynaud, Michel. *Spécialisation du foncteur de Picard.* (French) Inst. Hautes Études Sci. Publ. Math. **38** (1970), 27–76. MR0282993 (44 #227)

[64] Schneps, Leila. "Chapter 1. Mathematics and creativity." `http://www.math.jussieu.fr/~leila/grothendieckcircle/chap1.pdf`. accessed 22 Oct 2008.

[65] Serre, Jean-Pierre. "Groupes algébriques et corps de classes." (French) Publ. de l'Institut de Math. de l'Université de Nancago, VII. Hermann, Paris 1959. MR0103191 (21 #1973)

[66] Serre, Jean-Pierre. "Œuvres. Vol. I. (French) [Collected papers. Vol. I] 1949–1959." Springer-Verlag, Berlin, 1986. MR0926689 (89h:01109a)

[67] Serre, Jean-Pierre. *André Weil. 6 May 1906–6 August 1998.* Biographical Memoirs of Fellows of the Royal Society. Vol. **45** (nov., 199), pp. 520–529.

[68] Seshadri, C. S. *Variété de Picard d'une variété complète.* (French) Ann. Mat. Pura Appl. (4) **57** (1962), 117–142. MR0138623 (25 #2066)

[69] Severi, Francesco. *Su alcune questioni di postulazione.* Rend. del Circ. Mat. di Palermo. **17** (1903), 73–103.

[70] Severi, Francesco. "Trattato di geometria algebrica." Vol. 1. N. Zanichelli, 1926.

[71] Severi, Francesco. "Funzioni quasi abeliane." (Italian) Pontificiae Academiae Scientiarum Scripta Varia, **4** Vatican City, 1947. MR0024985 (9,578a)

[72] Severi, Francesco. *La géométrie algébrique italienne. Sa rigeur, ses méthodes, ses problèmes.* (French) Colloque de géométrie algébrique, Liège, 1949, pp. 9–55. Georges Thone, Liège; Masson et Cie., Paris, 1950. MR0038094 (12,353f)

[73] Weil, André. "Foundations of Algebraic Geometry." American Mathematical Society Colloquium Publications, vol. **29**, 1946. MR0023093 (9,303c)

[74] Weil, André. "Sur les courbes algébriques et les variétés qui s'en déduisent." Actualités Sci. Ind., no. 1041 = Publ. Inst. Math. Univ. Strasbourg. **7** (1945). Hermann et Cie., Paris. 1948. MR0027151 (10,262c)

[75] Weil, André. "Variétés abéliennes et courbes algébriques." Actualités Sci. Ind., no. 1064 = Publ. Inst. Math. Univ. Strasbourg **8** (1946) Hermann et Cie., Paris. 1948. MR0029522 (10,621d)

[76] Weil, André. "Scientific works. Collected papers. Vol. I (1926–1951). Vol. II (1951–1964). Vol. III (1964–1978)." (French) Springer-Verlag, New York-Heidelberg, 1979. MR0537937 (80k:01067a). MR0537935 (80k:01067b). MR0537936 (80k:01067c)

[77] Zariski, Oscar. *The concept of a simple point of an abstract algebraic variety.* Trans. Amer. Math. Soc. **62** (1947), 1–52. MR0021694 (9,99j)

[78] Zariski, Oscar. "Algebraic surfaces. Second supplemented edition." With appendices by S. S. Abhyankar, J. Lipman, and D. Mumford. Ergebnisse der Mathematik und ihrer Grenzgebiete, Band **61**. Springer-Verlag, New York-Heidelberg, 1971. MR0469915 (57 #9695)

My introduction to schemes and functors

David Mumford

1. Some recollections

I want to talk about how Grothendieck's revolution profoundly affected my own understanding of algebraic geometry. But to do that, I need to reconstruct for the reader the mathematical environment in which I grew up. When I started studying algebraic geometry around 1956, the Italian school was no longer active. André Weil and Oscar Zariski were carrying the ball and had together pioneered the extension of algebraic geometry to characteristic p. Weil was motivated by the idea of creating a merger of algebraic geometry with number theory to solve, e.g. Mordell's conjecture. Zariski was motivated by the need to make the work of the Italian school rigorous by using the new methods of commutative algebra. Everyone realized that the field needed better foundations to handle these new ideas, in which the explicit traditional geometry of complex varieties was replaced by an abstract geometry based on algebra and number theory. Both Weil and Zariski cobbled together some tentative definitions to make discussions and papers possible, both written as books for the AMS colloquium series, though only Weil's was published. But their "Foundations" did not have the feeling of inevitability that one associated, for example, with Bourbaki and his treatise and were never widely used.

Zariski was highly conscious of the fact that the Italian literature was filled with deep ideas and that it was essential to mine them, to update them using the new perspectives. He was much more open to new techniques than Weil, who radiated cynicism about anyone else's abstractions. A key breakthrough occurred when Serre and Zariski's interests met in the attempt to understand the Italian work on calculating the dimensions of complete linear systems. In modern terms, this means calculating the dimension of $\Gamma(X, L)$, the global sections of a line bundle L on a variety X. The Italians had worked from curves to surfaces to three-folds and so on, at each stage

Division of Applied Mathematics, Brown University, 182 George St., Providence, RI 02912, U.S.A. david_mumford@brown.edu.

comparing the linear system on a variety X with its trace on a hypersurface section H. Again, putting this in modern terms means considering the map:

$$\Gamma(X, L) \to \Gamma(H, L \otimes o_H).$$

They wanted criteria for this map to be surjective and, if not, to know the dimension of the cokernel. Zariski published a paper *Complete Linear Systems on Normal Varieties and a Generalization of a Lemma of Enriques-Severi* in the 1952 Annals in which he began to reexamine their analysis. He proved that, when X is a normal variety, the above map was surjective if the degree of H is sufficiently big.

In many ways, the cohomology of sheaves was implicit in all the calculations of this sort made by the Italian school. Cokernels, like the one above, were the classical way of dealing with the first cohomology group H^1. H^2's came up also, for example in the Cayley-Bacharach theorem that a plane curve C of degree $n+m-3$ which passes through all but one of the nm points $A = D \cap E$ of intersection of curves D, E of degrees n and m must pass through the last point. This is a geometric way of saying that $H^2(o_{P^2}(-3))$ is one-dimensional: tensor the exact sequence:

$$0 \to o_{\mathbb{P}^2}(-D-E) \to o_{\mathbb{P}^2}(-D) \oplus o_{\mathbb{P}^2}(-E) \to o_{\mathbb{P}^2} \to o_A \to 0$$

with $o_{\mathbb{P}^2}(n+m-3)$ and work out the usual cohomology. A careful examination of the Italian work noting, line by line, the equivalent cohomological calculations would be an interesting study, but this has never been done.

It was J.-P. Serre who made the cohomological approach to algebro-geometric questions explicit. Sheaves and cohomology were the latest hot technique in the Cartan Seminar and Serre saw that this was exactly the formalism which made sense of the Italian work. His work appeared in his famous paper *Faisceaux Algébrique Cohérents* (FAC) in the 1955 Annals and had previously been the subject of his 1954 talk at the Amsterdam International Congress. Zariski embraced these ideas instantly and talked on *Algebraic Sheaf Theory* in the 1954 AMS Summer Institute. The main theorem of Zariski's 1952 paper now became $H^1(X, L(-H)) = (0)$ if the degree of H is large enough, which Serre reproved near the end of his paper FAC.

2. Continuous systems of curves on surfaces

Much of the Italian work on linear systems fell into place and could be extended amazingly once it was recast in terms of sheaves and cohomology. As students of Zariski, many of us exploited this golden opportunity over the next decade. Zariski asked us, in particular, if we could reprove and extend to characteristic p the wonderful synthesis of the Italian work on algebraic surfaces and their classification contained in Enriques's posthumous 1949 book *Le Superficie Algebriche.* But one fundamental issue remained mysterious.

I have to digress to introduce the Italian way of describing the key invariants of surfaces. In the early days of surface theory, everyone sought the natural generalization of idea of *genus* from curves C to surfaces F. There were two natural definitions. Since the genus of a curve was the dimension of $\Gamma(\Omega^1_C)$ the space of 1-forms with no poles, one could take on a surface the dimension of $\Gamma(\Omega^2_F)$, the space of 2-forms with no poles. This they called p_g, the *geometric genus.* But one can also take the polynomial giving $\dim \Gamma(F, o_F(nH))$ for large n and evaluate it at $n = 0$. In modern terms this is the Euler characteristic $\chi(o_F)$ and, subtracting 1, the dimension of $\Gamma(o_F)$, they got the *arithmetic genus* p_a. In cohomology terms, it equals $h^2(o_F) - h^1(o_F)$. For the simplest examples, e.g. non-singular complete intersections and rational surfaces, they found that $p_g = p_a$.

But not always. Going back again to curves of genus g, the set of 0-cycles of degree n on C breaks up into a g-dimensional family of linear systems, the family being the Jacobian. In a similar way, Picard and the Italian school considered 'complete continuous systems of curves on the surface F', e.g. the biggest family of curves on F containing the hypersurface sections H of some high degree. Then they broke this family up into its component linear systems and defined what we call the *Picard variety* as the set of these. Although they avoided using divisors with negative coefficients, in effect, they defined the Picard variety $\mathrm{Pic}(X)$ as the group of divisors mod linear equivalence and called its dimension q, the *irregularity.* And then, experimentally, they found that $q = p_g - p_a$ always seemed to hold. They conjectured this must always be true.

Speaking loosely, they 'knew' a version of the fact that $\Gamma(\Omega^2_F)$ and $H^2(o_F)$ had the same dimension. Their argument can be caricatured by saying that for a high degree hypersurface H:

$$\begin{aligned} H^2(o_F) &\cong H^1(o_H(H.H)) \\ &\cong H^0(\Omega^1_H(-H.H))^* && \text{via Riemann-Roch on curves} \\ &\cong H^0(\Omega^2_F)^* && \text{via residues} \end{aligned}$$

Thus $p_g = h^2(o_F)$ and $p_g - p_a = h^1(o_F)$. Therefore their conjecture was that $h^1(o_F) = \dim \mathrm{Pic}(F)$.

Not only did these numbers always seem equal, there was a direct way of showing why they *ought* to be equal. Take a complete continuous system of curves $\{H_t\}$ of high degree on F and intersect them with one member H_0 of the family, getting divisors $H_t.H_0$ on H_0. Let t tend to 0. Then these intersections tend to a linear system of divisors on H_0, which they called the *characteristic linear system* of the family $\{H_t\}$. If this was complete, i.e. equal to the full space of sections $\Gamma(o_{H_0}(H.H))$, then the desired equality could be proven. In modern terms, this is a consequence of the exact sequence:

$$0 \to \Gamma(o_F) \to \Gamma(o_F(H_0)) \to \Gamma(o_{H_0}(H.H)) \to H^1(o_F) \to 0$$

because, if the characteristic linear system is complete, the dimension of the continuous system divided by linear equivalence will be

$$\dim\left(\Gamma(o_{H_0}(H.H))/\Gamma(o_F(H_0))\right)$$

or $h^1(o_F)$.

Enriques and Severi argued back and forth about whether they had a proof for this, but somehow each paper purporting to have a complete proof was answered with a *dubbio critico.* In fact, the equality was established first by Poincaré using his analytic tool of normal functions. But the intervention of analysis was seen as a blemish on the Italian theory and, moreover, once Weil and Zariski constructed the theory in characteristic p, the question in the characteristic p case remained open. A good deal of the history can be found in some half dozen papers all in the *Comptes Rendus de l'Academie des Sciences*, volume 140, 1905. The key paper of Enriques is in the *Rendiconti dell'Accademia delle Scienze di Bologna*, volume 9, 1904. However Enriques eventually hit on the right track and he published his final argument for the completeness theorem in the *Rendiconti del seminario matematico della universita di Roma*, series 4, volume 1, 1936, pp. 1-9 (see also Enriques's 1938 paper *Sulla proprietà caratteristica delle superficie algebriche irregolari* in the *Rendiconti della Accademia dei Lincei*, volume 27, pp.493-498). I have analyzed this paper in detail in the *Notices of the Amer. Math. Soc.*, volume 58, 2011,pp. 249-260. He speaks here of the exponential map in the Picard variety and asserts that analogously higher order infinitely near curves can be generated from first order infinitely near ones. Unfortunately, he possesses no tools whatsoever for going beyond an intuitive description of why the method of higher order infinitesimals should work: the theory of schemes was clearly what he lacked. My conclusion is that he had the germ of the right idea but he could only express it in a hand-waving sort of way. My paper follows a fascinating paper by Donald Babbitt and Judith Goodstein in same issue of the Notices based on newly discovered correspondence concerning this controversy.

3. Enter Grothendieck

Just as Zariski had welcomed Serre's introduction of sheaf cohomology, he welcomed Grothendieck's new schemes. He invited Grothendieck to Harvard in 1958 and tried to set up a regular visiting appointment for Grothendieck. He didn't exactly tear up his foundational colloquium manuscript but he was deeply impressed by Grothendieck's new way of setting up algebraic geometry via schemes. One of Zariski's deepest theorems was that the inverse image of every normal point under a proper birational morphism from one variety onto another is connected. Then Grothendieck came along and he reproved this result now by a *descending* induction on an assertion on the higher cohomology groups with Zariski's theorem resulting from the H^0 case: this seemed like black magic.

Grothendieck was a hypnotizing presence at Harvard. He seemed to have infinite energy and was always willing to schedule another lecture, to explain yet another facet of his theory. It seemed to be advancing like a tidal wave. In staid God-fearing Christian Yankee country, when no other time could be found, he created consternation by proposing to hold a seminar at 11 o'clock on Sunday. He wrote so fast and fluidly on the blackboard, I thought it resembled the 'grass writing' that I had heard about in a lecture on Chinese calligraphy: writing like the waves in the grass as a gust of wind sweeps over it. The web with which Zariski had ensnared his students was now itself ensnared in a larger, stranger one.

My fellow student Michael Artin took up Grothendieck's still vague ideas on what was to become étale cohomology, whereas the then post-doc Heisuke Hironaka hewed closer to Zariski's traditional interests pursuing birational geometry. My involvement came about because I had been studying the construction of varieties classifying families of algebraic structures, especially moduli spaces of vector bundles and of curves. Whereas I had thought loosely of such a classifying space as having a 'natural' one-one correspondence with the set of objects in question (just as Riemann and Picard had), Grothendieck expressed the relationship much more precisely with *functors*. This was clearly the right perspective. There were 'fine' moduli spaces which carried a universal family of objects, e.g. a universal family of curves from which all other families (over any base scheme) were unique pull-backs. Therefore they represented the functor of all such families. And there were also 'coarse' moduli spaces, the best possible representable approximation to the desired functor (the approximation being caused e.g. by the fact that some curves had automorphisms).

The most beautiful part of his formulation, however, seemed to me to be his 'reification' of infinitesimal deformations. In Kodaira and Spencer's work on analytic moduli spaces, they had introduced $H^1(\Theta_X)$, Θ_X the tangent bundle to X, to describe first order deformations of a compact complex analytic manifold. But now Grothendieck was saying these first order deformations *were actual families*, families whose base space was the *embodied* tangent vector $\mathrm{Spec}(k[\epsilon]/(\epsilon^2))$. And spanning the gap between families whose parameter space was a true variety and these first order families over the dual numbers were a whole stable of families over one point bases, spectra of all possible Artin rings. These were the families which Enriques had not been able to define clearly but which he seemed to intuit clearly. Not being a number theorist and worried about $\mathrm{Spec}(\mathbb{Z})$, for me these sorts of objects were the real argument for schemes, the really new thing.

Grothendieck came back to Harvard in 1961 and he, John Tate and I ran a seminar on existence theorems and the representability of various functors, especially Picard schemes. At about this time, I believe, the more comprehensive category of stacks emerged. Because, for instance, Hironaka had discovered that the quotient of a non-projective variety by a finite group need not exist as a scheme, it became clear that general existence theorems could only be true in a bigger category and stacks were the natural candidate. The definitive existence theorem which was the culmination of this part of Grothendieck's program turned out to be a theorem for stacks. It was discovered by Michael Artin a bit later as a Corollary of his remarkable Approximation theorem. It can be found in his 1971 book *Algebraic Spaces*.

Having an existence theorem for a Picard scheme of a variety X, which represented the functor of all families of line bundles on X, instantly solved the above completeness problem, the main conjecture of Italian algebraic geometry. It is immediate that the tangent space to the Picard scheme at the identity is $H^1(o_X)$ because, by the functorial definition of Pic, its tangent space gives the space of all line bundles over the base scheme $\text{Spec}(k[\epsilon]/(\epsilon^2))$ (or of infinitesimal deformations of any sufficiently ample divisor mod linearly equivalent deformations). So the key equality $q = h^1(o_X)$ is simply the statement that the Picard scheme is *reduced*: its dimension equals the dimension of its tangent space at the identity. In characteristic zero, the exponential map shows immediately that *all* group schemes are reduced – consider the restriction maps:

$$\text{Pic}(X \times \text{Spec}(k[t]/(t^{n+1}))) \to \text{Pic}(X \times \text{Spec}(k[t]/(t^2))) \to \text{Pic}(X).$$

Then, in characteristic zero, the exponential map defines the lifting

$$\{1 + ta_{ij}\} \mapsto \{1 + \cdots + t^n a_{ij}^n/n!\}$$

of the kernel of the right hand arrow to the big Pic on the left. Thus Grothendieck's idea of *representing* the functor of families of line bundles over *all schemes* immediately gives a purely algebraic proof that $h^1(o_X) = q$.

It is tantalizing to see how close Enriques was to this proof. In his paper that I referred to above, he said:

Re-examining the same question ... I observed, however, that the conclusions of my treatment would keep their validity if one admitted that curves infinitely close to a given curve on a surface (meaning n^{th} order deformations) *had an effective existence and that one could operate on them as on finite curves, by adding and subtracting. Thus, letting C_1 be a curve infinitely close to C and inequivalent to it, the operation $+C_1 - C$, successively repeated, serves to define, in the neighborhood of any curve K whatsoever, a series of infinitely close curves $K_1, K_2, K_3, \cdots$ belonging to a suitably high order neighborhood and this leads to the conclusion that this K should belong to a continuous non-linear series in which that K_1 would be close to K.*

I show how this can be made into a rigorous scheme-theoretic argument in my Notices paper referred to above. He goes on to give a different and reasonable argument for this based ultimately on the existence of 1-parameter subgroups of a complex torus through any tangent vector at the origin. What he never had was a definition of infinitely close curves of higher order (i.e. curves over $\text{Spec}(k[t]/(t^{n+1}))$) without which any argument is subject to what the Italians called a *dubbio critico*.

This proof seems to me that this is an example par excellence of Grothendieck's basic philosophy – that if you analyze a question down to its simplest and most abstract components, answers to the most puzzling questions should fall out. Even nicer, the theorem turned out to be false in characteristic p and necessary and sufficient conditions for its truth can be given by asking that certain 'Bockstein operators' from $H^1(o_X)$ to $H^2(o_X)$ must be zero (see my 1966 book *Lectures on Curves on an Algebraic Surface* where I discuss many of Grothendieck's existence theorems in most constructive possible fashion). It shows the power of nilpotent schemes and the functorial point of view in the clearest possible light. Grothendieck, of course, went on to construct and prove many much sexier things for which he is better known. But to demonstrate the power of modern abstract ideas to solve older very concrete problems, I think that this example is unmatched.

Can we summarize how Grothendieck transformed algebraic geometry? First of all, there was simply the *definition* of a scheme. The ingredients, in some sense, had all been there: Zariski's concept of a generic point; Zariski's use of complete local rings, especially in the context of completing at a point of the image variety of a morphism; Weil's concept of 'specialization' with respect to a valuation including from characteristic zero to characteristic $p > 0$; Enriques's idea of higher order deformations of a curve on a surface; the idea of higher order jets and jet-spaces in differential geometry; all the analogies between Dedekind rings of algebraic integers and rings of functions on affine curves over finite fields. When he made the definition, its logic and its unifying power were immediately obvious. But no-one else had taken this ultimate yet simple, Bourbaki-style step. In the same vein, there was his relativization and functorization of every situation. Every result concerned a base-change family of morphisms $X \times_S T \to T$, where $X \to S$ was fixed and $T \to S$ varied. The power of this technique to clarify every situation was certainly not appreciated before him. Finally recognizing an analogy between the role played in topology by the embedding of an open subvariety and by an unramified covering map seems to be a step for which there were no precedents. These were the parts of his work that I saw first hand. I'm not in a position to summarize his later work, crystals and children's drawings, etc. What I did see was enough to make him a unique genius in my lifetime of mathematical friends.

Descent

Carlos T. Simpson

ABSTRACT. The notion of "descent" dates back to medieval cartography, and has taken on a role of increasing importance in geometrically-oriented mathematics ever since then. Numerous techniques, constructions, and definitions are related to this mode of thinking, among them the construction of parameter spaces in algebraic geometry, categorical notions of site and topos together with their corresponding very general notions of sheaves and cohomology, and the notion of stack. One natural direction of development goes toward the introduction of higher categories, leading to higher stacks, derived algebraic geometry, and higher nonabelian cohomology.

1. Introduction

The notion of descent, piecing together a global picture out of local pieces and glueing data, permeates Grothendieck's work. The history of this idea dates to the middle ages with mapmakers drawing an ever more precise picture of the world, as modern terminology of "atlases" and "charts" reminds us. It is crucial to the notion of cohomology, where we first meet higher glueing data.

Descent comes into Grothendieck's philosophy and work in a myriad of forms, starting with his series of papers "Technique de descente et théorèmes d'existence en géométrie algébrique" [**61, 62, 63, 64**]. The technical requirements of this theory incited him to introduce the notion of fibered category. He extended the domain of application of this point of view in a revolutionary way by introducing the notion of "Grothendieck topology", integrating all of Galois theory and giving us etale cohomology. A further transformation occured with the notion of topos. Another incarnation of the idea was cohomological descent used in Deligne's papers on Hodge theory, where simplicial objects enter in a way which differs significantly from their original occurences in algebraic topology.

CNRS, Laboratoire J. A. Dieudonné, UMR 7351, Université Nice Sophia-Antipolis, 06108 Nice, Cedex 2, France. carlos@unice.fr.

Between the 1960's and the 1980's, the notion of descent for objects of a category, slowly gave way to a notion of "higher descent" for objects in generalized categorical situations. Examples include Breen's calculation of etale Ext groups using the cohomology of simplicial Eilenberg-MacLane presheaves, and the theory of twisted complexes of Toledo and Tong. Stasheff and Wirth investigated higher cocycles in the 1960's although Wirth's thesis was only recently made public. In algebraic topology, Quillen's theory of model categories was gradually applied to simplicial diagrams. Illusie introduced the notion of weak equivalence of simplicial presheaves on a Grothendieck site.

Grothendieck came out of isolation with the manuscript *La Poursuite des Champs* [**68**], which at its start refers to a letter from Joyal to Grothendieck developing the model category structure on simplicial sheaves. This led to Jardine's model category structure for simplicial presheaves enhancing Illusie's weak equivalences. We enter into the modern period in which Jardine's model structure and its variants have been used and developed with applications in a wide range of mathematics including Thomason's work in K-theory and then Voevodsky's theory of $\mathbb{A}^1$-homotopy and motives. Algebraic stacks, the first step in the "higher descent" direction, are now used without restraint in all of algebraic geometry. In Grothendieck's vision as set out in "La poursuite des champs", higher descent is just the same as usual descent, but for n-stacks of n-categories over a site. The theory of 2-categories was developed early on by Bénabou, having occured also in the book of Gabriel and Zisman. The theory of strict n-categories was thoroughly investigated by Street and the Australian school, and Brown and Loday introduced other related algebraic objects which could model homotopy types. Grothendieck set out the goal of finding an adequate theory of weak n-categories where composition would be associative only up to a coherent system of higher equivalences. Similar ideas were being developed by Dwyer and Kan in algebraic topology, and Cordier and Porter in category theory. Several definitions of weak n-categories have been proposed, by Baez-Dolan, Tamsamani, Batanin and others.

It is now well understood that the homotopy coherence problems inherent in higher categories, are basically the same as those which were studied by topologists for delooping machines. Segal's simplicial approach and May's operadic approaches play important roles in all of the current definitions. Maltsiniotis points out that Batanin's definition is the closest to Grothendieck's original idea. The topologists, notably Rezk and Bergner, have developed model structures on simplicial categories and simplicial spaces, and Joyal gives a model structure on the restricted Kan complexes originally defined by Boardman and Vogt in the 1960's. Cisinski and Maltsiniotis have built on a somewhat different direction of "La poursuite des champs" which aims to characterize the algebraic models for homotopy theory.

My own work in this area is inspired by the phrase in "La poursuite des champs" where Grothendieck forsees n-stacks as the natural coefficients for higher nonabelian cohomology. With Hirschowitz, we have developed the notion of n-stack based on Tamsamani's definition of n-category, and proven that the association $U \mapsto \{ n\text{-stacks on } U\}$ is an $(n+1)$-stack. The theory of "derived algebraic geometry" originated by Kontsevich, Kapranov and Ciocan-Fontanine is now cast by Toën, Vezzosi and Lurie in a foundational framework which relies on higher categories and higher stacks for glueing. In the future derived geometry should be a key ingredient in Hodge theory for higher nonabelian cohomology, to be compared with Katzarkov, Pantev and Toën's Hodge theory on the schematic homotopy type. The latter is a higher categorical version of Grothendieck's reinterpretation of Galois theory, forseen in "La poursuite des champs", or really its Tannakian counterpart. Grothendieck also mentionned, somewhat cryptically, a potential application to stratified spaces. The respective theses of Treumann and Dupont go in this direction by using exit-path n-categories to classify constructible complexes of sheaves.

Up-to-the-minute developments include Hopkins and Lurie's proof of a part of the Baez-Dolan system of conjectures relating higher categories to topological quantum field theory. And derived algebraic geometry permits us to imagine a local notion of descent as was explained to me by David Ben-Zvi: using the derived non-transverse intersections, Schlessinger-Stasheff-Deligne-Goldman-Millson theory should be viewed as descent for the inclusion of a point into a local formal space, with the neighborhood intersection being the derived loop space of Ben-Zvi and Nadler.

It would be impossible to give a thorough treatment of this vast subject here. We will start by discussing some of the basic and main ideas. This materiel is of course by now very well-established, so our purpose here is to give a survey rather than a detailed treatment. The reader should refer to the numerous available resources for more precise definitions and statements.

One of the main aims of our discussion will be to indicate some of the directions in which the theory of descent is or will be developing, in relation to the theory of higher categories as Grothendieck previewed in *Pursuing Stacks*. So the discussion will be at times rather "chatty" without proofs. Much of what is said can be found in some form in the literature, but some aspects have not yet been written up in detail. In these cases, the statements we make here should be considered as "pre-theorems" in the sense explained by Adams in [**2**]. I hope that this will not detract from the value of future works providing these results in the form of theorems, because everything flows in a logical way from the initial ideas of descent and ∞-categories.

As with all of Grothendieck's ideas, we can look with wonder at how the core notion of descent is leading to a fantastically diverse and ever-widening circle of research touching all areas of mathematics.

Acknowledgments. This paper is an expanded version of what I talked about in the Peyresq and IHES conferences on Grothendieck's legacy. The introduction is the same as my talk abstract for the IHES conference, available on the *nLab* website [**101**]. This work is supported in part by the Agence Nationale de la Recherche grant ANR-09-BLAN-0151-02 (HODAG).

2. Glueing

During the middle ages and at the beginning of the Rennaissance, European explorers were gathering more and more information about the shape of the world, gathered together by cartographers in the form of maps. Modern geometry springs from the observation that different map views have to be "glued together" on the overlapping territory, to obtain a global picture. The sphere S^2 is the first real-life manifold which was explicitly covered by charts.

This idea leads to all kinds of things in geometry, ranging from the definition of a manifold as being obtained by glueing together local charts, to Čech cohomology built up out of open sets and their intersections. In this context, *descent* means the construction of global objects over a space, by glueing together objects over the open sets of a covering.

Suppose X is a topological space (possibly with some other kind of structure such as a differentiable, complex analytic, or algebraic variety). If $X = \bigcup_i U_i$ is a covering by open subsets $U_i \subset X$, then we define $U_{ij} := U_i \cap U_j$ and $U_{ijk} := U_i \cap U_j \cap U_k$. If $Z_i \to U_i$ is a collection of some type of object, such as fibrations, then the additional data needed to construct a global $Z \to X$ with $Z|_{U_i} \cong Z_i$ is a collection of *glueing isomorphisms*

$$g_{ij} : Z_i|_{U_{ij}} \xrightarrow{\cong} Z_j|_{U_{ij}},$$

which should be isomorphisms of objects relative to U_{ij}. These will give rise to a global Z if and only if they satisfy the *cocycle condition*:

$$(g_{jk}|_{U_{ijk}}) \circ (g_{ij}|_{U_{ijk}}) = (g_{ik}|_{U_{ijk}}),$$

as well as the normalization that g_{ii} is the identity of Z_i.

Now the component pieces Z_i can be any of a wide variety of object, such as vector bundles, flat bundles, sheaves, families of spaces or varieties, and so on. This situation is formalized by considering a *family of categories* $\mathscr{C}(U)$ associated to open sets $U \subset X$. The Z_i are objects of $\mathscr{C}(U_i)$, and we search for a global object $Z \in \mathscr{C}(X)$. It is crucial here to have the *restriction functors*

$$r_{U/V} : \mathscr{C}(U) \to \mathscr{C}(V)$$

whenever $V \subset U$. Furthermore, these should satisfy a transitivity condition whenever $W \subset V \subset U$:

$$r_{V/W} \circ r_{U/V} = r_{U/W}.$$

Recall that the family of all categories has a structure of 2-*category*, and it is natural to think of functors $f, g : \mathscr{A} \to \mathscr{B}$ as "the same" whenever there is a natural isomorphism $u : f \to g$ (i.e. a natural transformation or 2-morphism, admitting an inverse $v : g \to f$). So the transitivity condition for our family of categories can naturally be weakened to the requirement of a natural isomorphism

$$\alpha_{U/V/W} : r_{V/W} \circ r_{U/V} \overset{\cong}{\to} r_{U/W}.$$

This should satisfy its own cocycle relation, whenever $Y \subset W \subset V \subset U$, which is that the two natural ways of composing pairs of α's to get natural transformations from $r_{W/Y} \circ r_{V/W} \circ r_{U/V}$ to $r_{U/Y}$, should be the same. This kind of higher relationship shows up all the time, and is generically known as a *coherence relation.*

The above collection of data is called a *prestack* over X. As we'll see below, there is another conceptually simpler presentation in terms of *fibered categories*, in which the restriction functors $r_{U/V}$ are no longer even precisely defined, but in which the coherence relations are taken care of automatically by the categorical formalism.

To get back to glueing, if we are given an abstract family of categories $U \mapsto \mathscr{C}(U)$, organized into a prestack, then we can formulate the glueing property as a question: given objects $Z_i \in \mathscr{C}(U_i)$ and isomorphisms g_{ij} satisfying the cocycle condition, does there exist a global $Z \in \mathscr{C}(X)$ together with isomorphisms $f_i : r_{X/U_i}(Z) \cong Z_i$ such that the glueing data arising from $(Z, \{f_i\})$ is "the same" as the given ones?

Before getting to that question, the same natural question should be posed for morphisms, one level prior: given two global objects $Z, Z' \in \mathscr{C}(X)$ and a localized collection of maps $h_i : Z|_{U_i} \to Z'|_{U_i}$ such that h_i and h_j induce the same maps over U_{ij}, does there exist a global map $h : Z \to Z'$ which induces the h_i on restrictions? Furthermore, given two morphisms h and h' such that $h|_{U_i} = h'|_{U_i}$ does this imply that $h = h'$? In the affirmative, these conditions may be rephrased as saying that the assignment $U \mapsto \mathscr{C}(U)(Z|_U, Z'|_U)$ should be a *sheaf.* The notion of a sheaf of sets is the 0-th order version of the glueing condition we are considering here.

If $U \mapsto \mathscr{C}(U)$ admits a positive answer to all of these gluing questions (and this should hold in fact with X replaced by any of its open subsets $X' \subset X$), then we say that the prestack $\mathscr{C}$ *satisfies effective descent*, or that it is a *stack.*

With this terminology, the glueing properties for various kinds of objects can all be phrased in a uniform way: the prestacks

$$\begin{aligned}
\mathscr{C}(U) &= \text{the category of vector bundles over } U,\\
\mathscr{C}(U) &= \text{the category of flat bundles over } U,\\
\mathscr{C}(U) &= \text{the category of fibrations } Z \to U,\\
\mathscr{C}(U) &= \text{the category of sheaves of sets, groups, abelian groups, } \ldots \text{on } U,\\
&\vdots
\end{aligned}$$

are all stacks.

These kinds of glueing properties were well-known in individual cases; then Grothendieck synthesized them into the notion of stack.

In the descent theory we will be discussing here, we are mostly interested in glueing together objects over a given generalized topological situation. The problem of glueing together abstract pieces to form something like a manifold, is viewed by Grothendieck in terms of glueing these pieces together in the world of sheaves over a bigger site such as Aff the site of affine schemes. A somewhat different abstract point of view to creating global objects, inspired by Thurston's work, has been proposed by Paul Feit [**53**]. The reader is invited to think about how Feit's point of view might mesh with the theory of (higher) stacks.

It is interesting to look, at this stage, at how the notion of stack leads to nonabelian cohomology. Suppose G is a sheaf of groups over X, so for each open set $U \subset X$ we have a group $G(U)$, with restrictions $G(U) \to G(V)$ for $V \subset U$, and sections satisfy the glueing properties for a sheaf in case of an open covering.

Each group $G(U)$ may be considered as a 1-object groupoid denoted

$$B(G(U)) \text{ or } \begin{pmatrix} G \\ * \end{pmatrix}.$$

The group of automorphisms of the single object $*$ is by definition $G(U)$. From the restriction maps for G these groupoids fit together to form a prestack which we denote by $B^{\mathrm{pre}}(G)$ defined by

$$B^{\mathrm{pre}}(G)(U) := B(G(U)).$$

However, this prestack *is not a stack*. Indeed, for an open covering $X = \bigcup_i U_i$, a locally defined object would consist of objects $Z_i \in \mathrm{Ob}(B^{\mathrm{pre}}(G)(U))$, hence in fact $Z_i = *$ for all i; together with glueing isomorphisms $g_{ij} \in G(U_{ij})$ satisfying the cocycle condition. In other words, this data is exactly a Čech 1-cocycle for the sheaf of groups G with respect to the given covering. It isn't too hard to see that two cocycles viewed as collections of glueing data, are isomorphic if and only if the cocycles are related by a coboundary. So, in order to "stackify" $B^{\mathrm{pre}}(G)$ we should introduce new objects which would be classified by 1-cocycles. A *G-torsor* over X is a sheaf of sets F together with a left action $G \times F \to F$ of the sheaf of groups G, such that F is locally isomorphic to G with its left action. More concretely this condition says that there should exist an open covering $X = \bigcup_i U_i$ and sections $f_i \in F(U_i)$ such that

$$(G|_{U_i}) \times \{f_i\} \overset{\cong}{\to} F|_{U_i}.$$

A *morphism* of G-torsors is a morphism $F \to F'$ compatible with the action of G; any such morphism is necessarily an isomorphism, as may be seen on the local trivializations. The category of G-torsors over X is therefore a groupoid. Define

$$B(G)(U) := \{(G|_U)\text{-torsors over } U\}.$$

This is a stack. The effective descent condition follows from the fact that sheaves (such as G-torsors F in this case) satisfy a glueing property which exactly says that the assignment

$$U \mapsto \{\text{sheaves of sets over } U\}$$

is a stack.

The automorphism group of the trivial torsor (i.e. G itself) is naturally identified with the group G acting from the right. This gives a morphism of prestacks

$$B^{\mathrm{pre}}(G) \xrightarrow{i} B(G).$$

And i is the stackification, a property which may be expressed by a 2-universal property: if H is any stack with a morphism

$$B^{\mathrm{pre}}(G) \xrightarrow{u} H$$

then there exists a morphism $v : B(G) \to H$ and a 2-isomorphism $\alpha : v \circ i \cong u$. And the pair (v, α) is unique up to unique 2-isomorphism.

The groupoid of global sections is the *nonabelian cohomology groupoid*

$$\mathbf{H}^1(X, G) := B(G)(X).$$

The set of isomorphism classes of objects of $\mathbf{H}^1(X, G)$ is the set, denoted by $H^1(X, G)$, of isomorphism classes of G-torsors on X. This cohomological notation is justified by the consideration of Čech cocycles considered above. One can obtain the stackification $B(G)$ by iterating three times a process of adding in objects and morphisms locally defined over various open coverings, and identifying morphisms which are locally equal.

If G is a sheaf of *abelian* groups, then $H^1(X, G)$ has a structure of abelian group object in the category of groupoids; the set $H^1(X, G)$ of isomorphism classes of torsors is just the usual first cohomology group of X with coefficients in G, and the group of automorphisms of any object is $H^0(X, G)$.

The theory of stacks serves to understand the glueing of objects with automorphisms. However, stacks themselves are one of the first examples of objects which have "higher automorphisms": since a stack is basically a family of categories, between two morphisms of stacks we can have 2-*morphisms*. Think of what BG would mean when G is a *stack with group structure*. Another example is the case of complexes: between two morphisms of complexes we might have *homotopies*, and this goes higher—between two homotopies there might be a 2-homotopy, and so on.

The question of glueing objects with higher automorphisms, leads to the notion of *higher stack*. For instance, in terms of the previous notations, suppose we are trying to glue together a family of stacks Z_i over open sets U_i. The glueing cocycle is again a family of equivalences of stacks g_{ij} over the U_{ij}; but the cocycle relation is no longer expressible as an equality, rather one should introduce 2-isomorphisms

$$\alpha_{ijk} : (g_{jk}|_{U_{ijk}}) \circ (g_{ij}|_{U_{ijk}}) \xrightarrow{\cong} (g_{ik}|_{U_{ijk}}),$$

and these are subject to a 2-cocycle relation over the quadruple intersections U_{ijkl}. This picture was discussed by Breen in [**26**], and constitutes one of the main motivations for "La poursuite des champs". It turns out to be quite difficult to deal with all of the higher coherence relations generated when we try to do it for n-stacks. Many recent authors are working in this direction, particularly in differential geometry. An abstract higher-categorical approach, taking advantage of Grothendieck's categorical definition of the topologies occuring in algebraic geometry, and integrating simplicial objects, model categories and other tools from homotopy theory, has proven to be fruitful.

3. From spaces to categories

The first main aspect of Grothendieck's approach to the theory of descent, was his reinterpretation of topology by the introduction of the notion of a *Grothendieck topology* promoting a category to a *site*. A brief introduction to this notion may be obtained by considering the notion of sheaf on a usual topological space in a somewhat more categorical way.

Suppose $(X, \mathscr{T})$ is a topological space. Then we can consider the category $\mathbf{Op}(X, \mathscr{T})$ whose objects are the open sets of X (that is, the elements of $\mathscr{T}$), and whose morphisms are the inclusions of open sets. A useful variant is $\mathbf{Op}^{\sqcup}(X, \mathscr{T})$, the category whose objects are formal disjoint unions of elements of $\mathscr{T}$ and whose morphisms are collections of inclusions. Formally, an object of $\mathbf{Op}^{\sqcup}(X, \mathscr{T})$ is a pair (I, U) where I is a set and $U : I \to \mathscr{T}$ is a function denoted $i \mapsto U_i$. A morphism

$$f : (I, U) \to (J, V)$$

consists of a morphism $f : I \to J$ such that $U_i \subset V_{f(i)}$ for any $i \in I$.

A presheaf on $(X, \mathscr{T})$ with values in a category $\mathscr{C}$ is by definition a functor $F : \mathbf{Op}(X, \mathscr{T})^o \to \mathscr{C}$. For $\mathscr{C} = \text{Set}$, $\mathscr{C} = \text{Ab}$ or $\mathscr{C} = \text{Ab}$ we recover the usual notions of presheaves of sets, of groups, or of abelian groups respectively.

All of these categories $\mathscr{C}$ *admit limits.* If Φ is a category and $F : \Phi^o \to \mathscr{C}$ is a functor, then we may define $\Gamma(\Phi, F) := \lim_{\leftarrow} F$ to be the limit. In the usual cases, where an object of $\mathscr{C}$ is a set with extra structure and the limits are calculated on the underlying sets, then $\Gamma(\Phi, F)$ is an object of $\mathscr{C}$ whose underlying set is the set of families $\{x_a \in F(a)\}_{a \in \Phi}$ such that along any arrow $a \to b$ in Φ, x_b pulls back to x_a. The notion of limit allows us to consider the *locally defined sections* of a presheaf, with respect to a covering.

Suppose $U \in \mathbf{Op}(X, \mathscr{T})$. A *covering* of U is a collection of open sets $U_i \subset U$, which is to say objects of the category $\mathbf{Op}(X, \mathscr{T})/U$, such that their union is equal to U. Given such a covering, we define the *associated sieve* to be the subcategory $\mathscr{B} \subset \mathbf{Op}(X, \mathscr{T})/U$ defined as the subcategory of all $V \subset U$ such that V is contained in some U_i. In categorical terms, $\mathscr{B}$ is the full subcategory of all objects $V \to U$ which admit a factorization through some U_i. It is a "sieve", in that $V \in \mathscr{B}$ and $V' \to V$ implies $V' \in \mathscr{B}$.

DEFINITION 3.1. *A* covering sieve *of the open set U is any full subcategory $\mathscr{B} \subset \mathbf{Op}(X, \mathscr{T})/U$ satisfying the sieve property, and such that the following equivalent conditions hold:*

- *$\mathscr{B}$ contains a sieve associated to a covering as considered above;*
- *$\mathscr{B}$ is a sieve associated to a covering; or*
- *the elements of $\mathscr{B}$ cover U.*

Suppose $\mathscr{C}$ is a category admitting limits, and $F : \mathbf{Op}(X, \mathscr{T})^o \to \mathscr{C}$ is a $\mathscr{C}$-valued presheaf. The category $\mathbf{Op}(X, \mathscr{T})/U$ has U as its coinitial object, which means that the natural restriction functor

$$\Gamma(\mathbf{Op}(X, \mathscr{T})/U, F_{\mathbf{Op}(X,\mathscr{T})/U}) \to F(U)$$

is an isomorphism. For a sieve $\mathscr{B} \subset \mathbf{Op}(X, \mathscr{T})/U$ the diagram

$$F(U) \stackrel{\cong}{\leftarrow} \Gamma(\mathbf{Op}(X, \mathscr{T})/U, F_{\mathbf{Op}(X,\mathscr{T})/U}) \to \Gamma(\mathscr{B}, F|_{\mathscr{B}})$$

therefore yields a natural map

$$F(U) \to \Gamma(\mathscr{B}, F|_{\mathscr{B}}). \tag{3.1}$$

We say that F *satisfies the descent condition* for a sieve $\mathscr{B} \subset \mathbf{Op}(X, \mathscr{T})/U$, if (3.1) is an isomorphism. The elements of $\Gamma(\mathscr{B}, F|_{\mathscr{B}})$ should be thought of as the *sections of F locally defined on the covering $\mathscr{B}$*. The "descent condition" thus requires that the object (set, group, etc.) of sections $F(U)$ is the same as the object of sections locally defined over the sieve $\mathscr{B}$.

We say that F is a *sheaf* if it satisfies the descent condition for any open set $U \in \mathbf{Op}(X, \mathscr{T})$ and any covering sieve $\mathscr{B} \subset \mathbf{Op}(X, \mathscr{T})/U$.

EXERCISE 3.2. *Suppose $\mathscr{C}$ is a category of sets with extra structure, and $F : \mathbf{Op}(X, \mathscr{T})^o \to \mathscr{C}$ is a presheaf.*

Suppose $\mathscr{B}$ is the covering sieve associated to a covering $U = \bigcup U_i$ of an open set in a topological space. Let $U_{ij} := U_i \cap U_j$. Then the object of locally-defined sections $\Gamma(\mathscr{B}, F|_{\mathscr{B}})$ is equal to the object of $\mathscr{C}$ whose underlying set is the set of collections of $x_i \in F(U_i)$ satisfying the classical glueing condition $x_i|_{U_{ij}} = x_j|_{U_{ij}}$.

Thus, the presheaf F is a sheaf if and only if it satisfies the classical sheaf conditions.

We may similarly endow the extended category $\mathbf{Op}^{\sqcup}(X, \mathscr{T})$ with a notion of covering sieve.

PROPOSITION 3.3. *Suppose $F : \mathbf{Op}^{\sqcup}(X, \mathscr{T})^o \to \mathscr{C}$ is a presheaf. Then its restriction to the subcategory $\mathbf{Op}(X, \mathscr{T})^o \to \mathscr{C}$ is also a presheaf. If F is a sheaf then its restriction is a sheaf, and this restriction establishes an equivalence between the category of sheaves on $\mathbf{Op}^{\sqcup}(X, \mathscr{T})$ and the category of sheaves on $\mathbf{Op}(X, \mathscr{T})$. A sheaf on $\mathbf{Op}^{\sqcup}(X, \mathscr{T})$ will have the property that*

$$F(\coprod_i U_i) = \prod_i F(U_i)$$

so its values on formal sums of open sets are determined by its values on the original open sets of $\mathbf{Op}(X, \mathscr{T})$.

The notion of *Grothendieck topology* on a category Φ just means a family of covering sieves $\mathscr{B} \subset \Phi/U$ for any $U \in \Phi$, satisfying some axioms, so that we can directly generalize the above discussion. Recall that a full subcategory $\mathscr{B} \subset \Phi/U$ is said to be a sieve if, whenever $V \to U$ is in $\mathscr{B}$, then for any morphism $W \to V$ the composition $W \to U$ is also in $\mathscr{B}$.The terminology "sieve" means that we are trying to specify a certain smallness of open sets, and anything which is even smaller should also fall through.

AXIOMS 3.4. *The axioms for a Grothendieck topology are as follows (see* [**101**] *for a lot of good discussion):*

- *the full* Φ/U *is a covering sieve of* U*;*
- *if* $\mathscr{B} \subset \mathscr{B}' \subset \Phi/U$ *are sieves, and* $\mathscr{B}$ *is a covering sieve, then so is* $\mathscr{B}'$*;*
- *if* $\mathscr{B}, \mathscr{B}' \subset \Phi/U$ *are covering sieves, then their intersection* $\mathscr{B} \cap \mathscr{B}'$ *is a covering sieve;*
- *if* $V \to U$ *is a morphism and* $\mathscr{B} \subset \Phi/U$ *is a covering sieve, then* $\mathscr{B}|_V$*, defined as the subcategory of all morphisms* $W \to V$ *whose composition* $W \to U$ *is in* $\mathscr{B}$*, is a covering sieve;*
- *and the* locality condition *which says that if* $\mathscr{B} \subset \Phi/U$ *is a covering sieve, and* $\mathscr{B}' \subset \Phi/U$ *is a sieve such that for each* $V \to U$ *in* $\mathscr{B}$*, the restriction* $\mathscr{B}'|_V$ *is a covering sieve of* V*, then* $\mathscr{B}'$ *is a covering sieve.*

Using fiber products in Φ, if the appropriate ones exist, one may also give a definition of topology in terms of "covering families" rather than covering sieves. See [**154**] for example.

A *site* is a pair consisting of a category Φ provided with a Grothendieck topology. A presheaf $F : \Phi \to \mathscr{C}$ is a *sheaf* if, for any covering sieve $\mathscr{B} \subset \Phi/U$, the natural map $F(U) \to \Gamma(\mathscr{B}, F)$ is an isomorphism. This reworking of the notion of sheaf, and introduction of the more general notion of "site", allowed Grothendieck to vastly expand the scope of situations to which the basic topological techniques of sheaf theory could apply.

Getting back to our original example of a topological space, let us remark that the extended site $\mathbf{Op}^{\sqcup}(X, \mathscr{T})$ has the following additional nice property: any covering sieve has a refinement (i.e. a sub-sieve which is also covering) generated by a single object. This is to say, that for any $U \in \mathbf{Op}^{\sqcup}(X, \mathscr{T})$ we may restrict our attention to coverings of the form $V \to U$. Indeed, given a covering $\{V_i \to U\}$, form the object $V := \coprod V_i$ which is again in $\mathbf{Op}^{\sqcup}(X, \mathscr{T})$.

This property is somewhat usual in the sites which come up in algebraic geometry, so we may often consider a *descent situation* as being just a morphism $V \to U$. In this case (and assuming that our site Φ admits fiber products), the simplicial object

$$V. := \cdots V \times_U V \times_U V \substack{\rightarrow\\ \rightarrow\\ \rightarrow} V \times_U V \rightrightarrows V$$

together with its augmentation map $V. \to U$, plays a crucial role. When $V = \coprod V_i$ is a disjoint union of the open sets in a covering of a topological space, then $V \times_U V$ is the disjoint union of the double intersections V_{ij}, $V \times_U V \times_U V$ is the disjoint union of the triple intersections, and so forth. Another way of thinking of the "locally defined sections" with respect to the covering V, is to take $\Gamma(\Delta, F \circ \nu)$ where $\nu : \Delta^o \to \mathscr{A}$ is the functor corresponding to the simplicial object $V.$. This encloses the totality of the *Čech cocycle data* involved in the descent data. Although for sheaves and 1-stacks it suffices to look at the first few stages, it becomes necessary, of course, to consider the full simplicial object when we want to do higher descent.

4. Some Grothendieck topologies in algebraic geometry

For completeness, we record here some of the main examples of Grothendieck topologies applied to various areas of algebraic geometry. The first and perhaps most famous, is the *etale topology.* An *etale morphism* is a map $V \to U$ of schemes, which is smooth of relative dimension 0.

Pictorially, and working in a category of smooth manifolds rather than algebraic varieties, we should think of V as a kind of locally topologically trivial covering space of U. An example of an etale covering of the circle could thus be the $2:1$ map

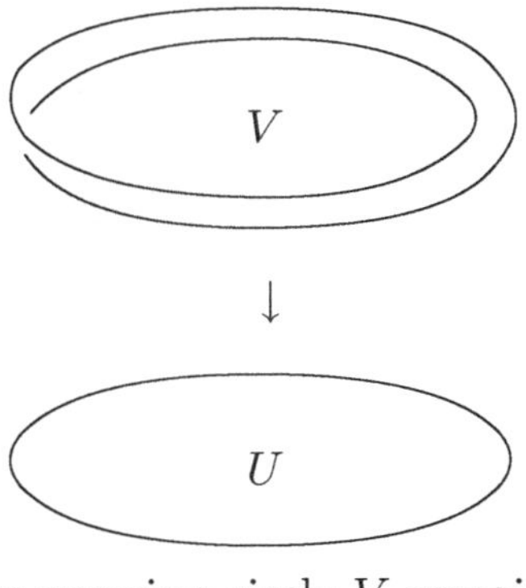

where, notably, the upper covering circle V remains connected, so that no uniform choice of section is possible. This is geometrically quite different from the idea of a disjoint union of open sets—allowing new coverings such as these was Grothendieck's conceptual leap.

EXERCISE 4.1. *Understand the simplicial object $V.$ associated to the above $2:1$ covering $V \to U$ of the circle by itself.*

Returning to the general algebro-geometric situation, in order to specify a site there is first a choice of underlying category. The *big etale site* is the category of all schemes, probably cut down by some kind of finiteness properties such as being noetherian or of finite type over a field or adequate base ring. The *small etale site* of a scheme X consists only of the schemes which are etale over X. The distinction between these two things is of a rather technical nature; much of the original theory was developed in the context of the small site, whereas more recent works tend to prefer the big site because it is conceptually more flexible (for example one can pass between different topologies). It often turns out that the invariants which can be calculated, are obtained using the simplicial object corresponding to a covering $V. \to U$, so they will be the same for the two cases. It would go beyond our present scope to delve into the technical hypotheses of this type of statement.

Cohomology of sheaves on the etale sites, developed by Grothendieck and his co-workers in the *Seminaire de Géométrie Algébrique* volumes, played a spectacular role in a wide range of results on arithmetic algebraic geometry, such as Deligne's proof of the Weil conjectures [**42**].

The *Zariski topology* goes back to the original example of $\mathbf{Op}(X, \mathscr{T})$ where X is a scheme and $\mathscr{T}$ is the topology of Zariski open sets, giving the small Zariski site. The *big Zarisk site* (which comes up in [**69**] for example) remains an example of Grothendieck's new notion, concerning the underlying category of all schemes satisfying an appropriate finiteness condition.

The *Nisnevich topology* lies halfway in between the Zariski and etale topologies. A *Nisnevich covering* is an etale covering, which admits sections over each piece of a Zariski locally-closed stratification of the base scheme. This has played an important role in the work of Voevodsky and Morel on $\mathbb{A}^1$-homotopy theory [**99**]. Algebraic K-theory turns out to be quite sensitive to the choice of underlying topology, a theme present already in Thomason's [**136**], and working with the Nisnevich topology allowed Voevodsky to prove the Milnor conjecture [**155**].

Some other rather famous Grothendieck topologies on the big category of schemes, are the *fppf* (fidèlement plate de présentation finie), and *fpqc* (fidèlement plate quasi-compacte) topologies. In these, a covering map $Y \to X$ is expected to be *faithfully flat*, which means flat and surjective, and to satisfy a finiteness condition such as finite presentation or (weaker) quasicompactness.

A key result is *faithfully flat descent.* Suppose $V \to U$ is a faithfully flat morphism of schemes, and form the simplicial object $V.$ as above. Suppose $Y. \to V.$ is a morphism of schemes such that for any map $n \to m$ in Δ, the square

$$\begin{array}{ccc} Y_m & \to & Y_n \\ \downarrow & & \downarrow \\ V_m & \to & V_n \end{array}$$

is cartesian. Then there exists a map of schemes $Z \to U$ and a compatible system of isomorphisms $Y_n \cong V_n \times_U Z$. The analogous descent result holds for sections, morphisms to schemes and stacks, and so forth.

It is particularly interesting to notice that for these flat topologies, the covering objects V start to have only the most tenuous of geometric relationships with the underlying open set U which is being covered. For example, the dimension of V will generally be bigger than that of U, and furthermore V can have arbitrarily bad singularities even over very nice points of U. Cohomology groups for these topologies are notoriously hard to calculate, and indeed one of the first (somewhat hidden) appearances of higher stacks was in Breen's deep work on this problem [**25**].

We may also cite the crystalline site, in which the rather simpler Zariski topology, or pretty much equivalently its etale or fppf brethren, play a supporting role to the fundamental properties of the underlying category of infinitesimal thickenings which build in the relationship with de Rham cohomology. In this case, the covering object can sort of "overflow", going beyond the edges of the U which is being covered. One may view the whole de Rham picture as a case of *infinitesimal descent*, we'll touch on a modern version of this briefly at the end of the paper.

An interesting topology, which seems to be well-known in a folkoric sense but which was explained to me by Constantin Teleman, is the *proper-etale topology*. This is the smallest topology, i.e. the one with the smallest sets of covering sieves, such that proper surjective maps are coverings, and etale surjective maps are coverings. Descent with respect to this topology corresponds to what was called *cohomological descent* in SGA [**121**], and was notably used in Deligne's Hodge theory for singular spaces [**43**].

The literature concerning the applications of algebraic stacks over these various Grothendieck topologies, to classical problems in algebraic geometry, is by now vast and we have been able to include only a small sampling [**1, 4, 12, 22, 24, 29, 31, 34, 45, 46, 49, 50, 51, 52, 55, 57, 59, 74, 77, 85, 86, 87, 88, 89, 96, 104, 105, 106, 112, 119, 150, 151, 153**]. A good foundational reference is the *stacks project* [**40**], a cooperative book aiming to provide a complete groundwork for the theory of stacks in algebraic geometry.

This has been a very rapid overview. The reader may appreciate the vast implications of Grothendieck's approach to topology and sheaf theory, through the extensive and long-term utilisation of these concepts in wide swaths of the literature. For our present purposes, we will be interested in how Grothendieck saw the development of the *formal* aspects of these ideas, which gradually leads toward the ideas of stacks, higher categories, and higher stacks.

5. A philosophical interlude

Before getting on with how the idea of glueing leads naturally to the notion of higher categories, let's take a break to discuss what kind of philosophical connections might be drawn from this circle of ideas. This was a significant aspect of the discussion at Peyresq.

We can clearly trace the notion of glueing together local pieces to form a manifold, to the ancient practice of cartography. Indeed, humanity's first meeting with a nontrivial manifold was the spherical surface of the earth itself. The rediscovery of this idea was at the heart of the renaissance. The role of cartography in one of the main movements leading into that period, figures nicely in the narrative [**158**].

If you look at the back cover of a typical road atlas, you get a graphic depiction of exactly what it means for a manifold to be covered by open sets, each map outlined and numbered by page. Then on each page you have four or more arrows directing to the relevant pages for the nearby maps. These arrows represent the glueing isomorphisms on the overlaps between the charts.

Using the word "descent" to describe such a glueing procedure, somehow implies that the local pictures are to be found *above* the "ground level" of that which we would like to represent. In Grothendieck's vision, this "upper level" could take on a wide variety of new forms: rather than just being a collection of open subsets of the ground-level space, the upper level could be an etale covering, or even more exotic things such as a smooth, flat, or even just proper surjective covering scheme. The *glueing data* therefore take on a more elaborate texture.

The passage from the theory of sheaves or stacks on a Grothendieck site, to topos theory, pushes this philosophy to a new level. The different pieces making up a space don't have to share a common existence in a single place. The glueing data, abstracted into the categorical structure of the topos of sheaves, provide the links which form a virtual reality from which the geometric object emerges. The original "ground level" fades from view, replaced by the abstract collection of glueing data itself as the only true reality. It is as if the captain of the boat, in the chartroom with all his maps, no longer needs to go out on deck to look at the ocean upon which he is navigating. If that sounds like it might be a bit dangerous in a 15th-century context, consider a hyper-modern example: the *internet* is based on the same philosophy—even though the servers are physically located in many different places, they link up together to create a virtual space which, to the user, seems like a single continuum.

What does this tell us about *ideas*? I would like to claim that such a cartographic approach, and particularly Grothendieck's new reworking of this ancient idea, tells us that there is *no single prefered point of view* to any given question. In Grothendieck's etale, and more specially fppf and fpqc topologies, the covering spaces are no longer even viewed as being composed of little pieces of the real world, but just arbitrary spaces mapping in a sufficiently surjective way to the real world. Heretofore, a given point on the ground level is represented by many different points on the chart level. So, wherever you are, there might be some viewpoints which are better adapted than others, but there is not a single unique best viewpoint, and the various different views work together to generate a picture of the whole.

If we start trying to identify a "best approach", then inevitably the best approach for one problem will not work quite as well for the next one. Think of trying to map the earth's curved surface by a flat map. Somehow toward the corners of each chart it doesn't work quite as well, and for the next page you will change the projection. So, the notion that there should be a single best way of approaching any given problem, leads to the division of the realm into little zones where each zone is covered by a certain approach. The joining together of the different zones becomes a more abstract and far-off question best left to "the experts". This hierarchichal segmentation is wrong: it limits the possibilities for communication between different sectors, blocking the harmonious flourishing of the whole.

When we think of the complicated "ground level" as being a reality that is best apprehended (and might only really exist) through a covering representing a wide variety of different approaches and viewpoints, the place where things are really happening, and what we should concentrate on understanding, is the glueing data which explain how to pass between the various different points of view and how they are bound together.

This philosophy applies notably to mathematics: we shouldn't try to hope that a single theory will be the best approach, but rather we should work on a wide diversity of different theories and try to understand how they are interrelated. We'll see this in action up ahead.

6. Toward higher categories

In our discussion so far, we have skipped the somewhat technical work of defining sheaf cohomology, even though that was, of course, the main application of the theory of sheaves on a site or the more advanced topos theory. One could expound at length the passage between the notion of Čech glueing data, to Čech cocycle, and the relationship with the topos of sheaves, on the one hand, and the simplicial objects $V_\cdot$ corresponding to coverings $V \to U$, on the other.

Rather than doing that, we would like to continue on our way by noticing that one might well wonder how to reconcile two seemingly disjoint observations: on the one hand, the descent condition for sheaves really only involves the very start of the simplicial object

$$V \times_U V \rightrightarrows V$$

whereas, on the other hand, the Čech calculation of cohomology eventually ends up using the full simplicial object $V_\cdot$. It seems natural to look for a *generalization of sheaf theory* which would also use the full simplicial object.

The theory of 1-stacks, which we talked about in the beginning of the paper, is a step in this direction. The descent condition for a stack involves a cocycle relation of the form $g_{ik} = g_{jk} \circ g_{ij}$ over triple intersections U_{ijk}, and this is indeed exactly the condition needed to obtain a 1-cocycle and hence a degree-1 cohomology class. In terms of the simplicial object $V_\cdot$, this condition therefore uses the first three terms V_2, V_1 and V_0.

Formally speaking, one therefore wants to go toward a notion of *higher stack*. In the 1960's and 1970's, this flow came up against a number of technical as well as sociological obstacles. However, there were several important developments which constituted a very essential start.

For example, Verdier's notion of *hypercovering* [**6**] [**7**] was cast in the language of *simplicial objects* of a topos. Since the topos is viewed as the category of sheaves on a site, this means that a hypercovering is a simplicial sheaf. The purpose is to enlarge the family of *covering sieves* in a way that takes into account the idea of starting with some open sets U_i, but then furthermore choosing a covering of the intersections U_{ij}, and continuing in this way. Among other things, it resolved the technical issue that Čech cohomology doesn't coincide with sheaf cohomology in the general setting.

Then came the notion of "Illusie weak equivalence". Here, we are more resolutely in the situation of simplicial sheaves, or even simplicial presheaves (the naturalness of the latter having been pointed out only later by Jardine [**79**]). Given a simplicial presheaf T over a site $\mathscr{X}$, and if necessary a section $t \in T(U)$, we may form its *homotopy group presheaves* $\pi_i(T|_{\mathscr{X}/U}, t)$. These are presheaves on $\mathscr{X}/U$. For $i = 0$ the section t is not needed. Illusie then said, a morphism of simplicial presheaves $T \to T'$ is a *weak equivalence* if it induces isomorphisms on the *sheafifications* of the homotopy group sheaves. We can see that this is a very natural notion: in the case when the site $\mathscr{X}$ has enough points, which covers virtually all of the relevant examples, then we may define the *stalk* T_x of a simplicial presheaf T at a point x, which is a simplicial set and therefore viewed as a space; a map $T \to T'$ is an Illusie weak equivalence if and only if it induces weak equivalences of simplicial sets $T_x \xrightarrow{\sim} T'_x$ at all points.

With this definition in hand, it became possible to consider the *homotopy theory* of simplicial presheaves, obtained by localizing the category of simplicial presheaves along the weak equivalences. This was our first approach to the homotopy theory of higher stacks.

Cohomology then has a nice interpretation analogous to "Brown representability". If G is a sheaf of groups, then we can form its classifying simplicial presheaf $BG = K(G, 1)$. If it is abelian, this may furthermore be iterated to obtain the simplicial presheaf $K(G, n)$, and for any sheaf F member of the topos of sheaves F, we have that $H^n(F, G)$ is just the mapping set $[F, K(G, n)]$ in the homotopy theory generated by Illusie weak equivalences.

Breen used this theory to great effect in his calculation of some of the most complicated sheaf cohomology groups [**25**].

The effort of formalising things was, by this point, lagging fairly seriously behind. It was then in "La poursuite des champs" that Grothendieck explained how we really needed a full-fledged theory of *higher categories.*

On a personal note, the Princeton math library in Fine Hall had a unique copy, available for consultation in a back room. Between 1987 and 1990, I made numerous trips down to the library, just to read snippets and try to think about how to understand what Grothendieck was trying to say.

A very first piece of Grothendieck's text spurred the response by Joyal, in which Joyal explained that one could develop a model-category-theoretical approach to the homotopy theory of simplicial sheaves. This was then fully worked out, in the much more natural context of simplicial presheaves, by Jardine [**79**]; these works formed a narrative concurrent to the formation of "La poursuite" as it now exists.

Thomason applied Jardine's work to K-theory in his groundbreaking paper [**136**], and starting from there, it has become clear that Jardine's paper pretty much answers the question of what is a higher stack of ∞-groupoids even though that language wasn't yet in vogue.

We have now seen that the quest for a formal groundwork underlying descent theory, leads naturally to the notion of higher stack. With the notion of higher stack, should also come the notion of higher category. (Abstractly one would expect that the notion of higher category should come first, but that wasn't the way things happened.) We shall next start in on a somewhat more specific discussion of n-categories, as the theory was imagined by Grothendieck and as it has come to be understood today.

7. ∞-groupoids

The notion of strict n-category is defined by induction on n: a strict n-category is just a category enriched over strict $(n-1)$-categories. If one writes out explicitly what this means, the definition may be easily extended to the case $n = \infty$, yielding the notion of strict ∞-category. In a strict ∞-category, the various compositions which can be defined, of i-morphisms with respect to j-morphisms for $j < i$, are strictly associative. Also the units or identity morphisms satisfy the unit axioms strictly. As Grothendieck explained [**68**], the strict ∞-groupoids unfortunately don't model homotopy types in the way we would like (see the discussion of this issue and references in [**129**]), and one should expect a more complicated theory of *weak n-category* (including the case $n = \infty$) in which the associativity and perhaps unit axioms would only hold up to natural equivalences the next higher level up.

In the past years, a number of different approaches to the notions of weak n-categories have been developed. The reader may refer to Leinster's compendium [**90**], to the references in my book [**129**], to the *nLab* site [**101**], or to a number of other places, some of which will be discussed more in detail below.

In all these approaches, one can furthermore define a notion of *weak n-groupoid*, that being a weak n-category such that all of the i-morphisms are *invertible up to equivalence*. In this context invertibility of $f : u \to v$ up to equivalence (where f is an i-morphism between u and v which are $(i-1)$-morphisms themselves sharing the same source and target) just means that there exists an i-morphism $g : v \to u$ together with $(i+1)$-morphisms $a : fg \to 1_v$ and $b : gf \to 1_u$. This is because the hypothesis applied to a and b says that they will be equivalences too; in the context of a general ∞-category, an equivalence f is required to have a quasi-inverse (g, a, b) such that the a and b are themselves equivalences[1].

One basic desired property of the theory of ∞-categories is that the ∞-groupoids should model homotopy types, in the sense that there should exist adjoint functors

$$\Pi_\infty : \text{spaces} \longrightarrow \infty\text{-groupoids}$$

$$|\ | : \infty\text{-groupoids} \longrightarrow \text{spaces}$$

such that the adjunction maps are equivalences on both sides:

$$G \xrightarrow{\sim} \Pi_\infty(|G|),$$

$$|\Pi_\infty(X)| \xrightarrow{\sim} X.$$

At finite values of n, the same should hold but for the category of *n-truncated spaces* X with $\pi_i(X) = 0$ for $i > n$. This property is known for some theories, see Tamsamani [**133**] for example, and expected to hold in all cases.

As Maltsiniotis pointed out [**98**], Grothendieck actually specified a theory of ∞-groupoids in [**68**], although it is not altogether easy to extract the definition. For this definition, the full statement of the equivalence with homotopy types is not yet known, although Berger [**17**] and Ara [**5**] have strong partial results saying notably that any homotopy type is realized.

The upshot of this whole discussion is that in order to understand ∞-groupoids, we don't really need to understand the details of any specific theory of ∞-categories: the notion of ∞-groupoid can be just replaced by that of topological spaces up to homotopy. Then, given any theory of ∞-groupoids, whatever we may say using spaces is transfered to ∞-groupoids using the functors Π_∞ and $|\ |$.

It is usual to replace the notion of topological space by the notion of simplicial set. And indeed, for the theory of quasi-categories (this is a theory of $(\infty, 1)$-categories, see below), the groupoids are exactly the Kan simplicial sets.

So we adopt the working definition that an ∞-groupoid is the same thing as a simplicial set; a morphism of ∞-groupoids is a morphism of simplicial sets, and such a morphism is an equivalence if it induces an isomorphism on π_0 and all the homotopy groups.

This brings us to a remark coming from notion of model category: when looking at the set of morphisms from A to B one should suppose that B is

---

[1]The notion of equivalence needs to be known recursively for $(i+1)$-morphisms, a process which stops if n is finite; for $n = \infty$ the collection of morphisms which are said to be "equivalences" needs to be specified as extra data.

fibrant, which for simplicial sets is just the Kan horn-extension condition. In this case, we get furthermore a simplicial set

$$\underline{Hom}(A, B)$$

of maps from A to B, defined by adjunction with the direct product in the sense that a map $E \to \underline{Hom}(A, B)$ is the same thing as a map $E \times A \to B$.

8. $(\infty, 1)$-categories

An $(\infty, 1)$-category means an ∞-category A such that for every pair of objects x, y, the ∞-category $A(x, y)$ is an ∞-groupoid. We can therefore speak of $(\infty, 1)$-categories without knowing about a full theory of ∞-categories: rather we just need to have a theory of ∞-groupoids, plus a theory of categories enriched over these things. Adopting the point of view that an ∞-groupoid is the same thing as a homotopy type of simplicial set, we come upon the notion of $(\infty, 1)$-category as being a category enriched over simplicial sets. In the strict version of this theory, $\mathrm{Ob}(A)$ is a set, and for each $x, y \in \mathrm{Ob}(A)$ we have a simplicial set $A(x, y)$. The composition map is a map of simplicial sets

$$A(y, z) \times A(x, y) \to A(x, z)$$

which is strictly associative in the sense that the diagram

$$\begin{array}{ccc} A(z, w) \times A(y, z) \times A(x, y) & \to & A(z, w) \times A(x, z) \\ \downarrow & & \downarrow \\ A(y, w) \times A(x, y) & \to & A(x, w) \end{array}$$

commutes. The identity maps are elements $1_x \in A(x, x)_0$ which may also be viewed as maps $* \to A(x, x)$ of simplicial sets such that the resulting composed maps

$$A(x, y) \to A(x, y) \times A(x, x) \to A(x, y)$$

and

$$A(x, y) \to A(y, y) \times A(x, y) \to A(x, y)$$

are the identities.

Various weaker versions of this notion have been proposed. For example, a *Segal category* is a bisimplicial set, viewed as consisting of a set $\mathrm{Ob}(A)$ together with simplicial sets $A(x_0, \ldots, x_n)$ for any sequence $x_0, \ldots, x_n \in \mathrm{Ob}(A)$ and simplicial transition maps between these, satisfying the Segal condition that

$$A(x_0, \ldots, x_n) \to A(x_0, x_1) \times A(x_1, x_2) \times \cdots \times A(x_{n-1}, x_n)$$

is a weak equivalence of simplicial sets. The composition is then only weakly defined, being given by the map $A(x_0, x_1, x_2) \to A(x_0, x_2)$.

A *quasicategory* is a simplicial set A satisfying the "restricted Kan condition" saying that certain horns have fillers. These horns are those which are canonically filled in the nerve of a category. The set of objects is $\mathrm{Ob}(A) = A_0$. If $x, y \in A_0$ then the simplicial set of maps from x to y is a little bit more involved, given by

$$A(x,y) := \underline{Hom}^{x,y}(\Delta[1], A)$$

where the superscript indicates maps from the simplicial interval $\Delta[1]$ to A sending the two endpoints to x and y respectively.

A *Rezk category* or *complete Segal space* is a bisimplicial set A such that the Segal maps

$$A_n \to A_1 \times_{A_0} A_1 \times_{A_0} \cdots \times_{A_0} A_1$$

are weak equivalences, also satisfying conditions insuring that the two maps $A_1 \to A_0$ are fibrations, and satisfying a further condition which basically says that the simplicial set A_0 is equivalent to the "interior" of A, a notion to be discussed below. In this theory, the "set of objects" is really only defined up to homotopy, although for a given model A we can take $\mathrm{Ob}(A) := A_{0,0}$ to be the set of vertices of A_0. The philosophy behind having a simplicial set A_0 is that in a usual category, one should really only think of the set of objects as being defined up to equivalence.

Following Tamsamani [**133**], the notion of Segal category may be iterated to give the notion of Segal n-category [**70**], which is one model for the theory of (∞, n)-categories. Similarly, Rezk's definition [**117**] may be iterated, as is being treated in current work of Bergner and Rezk [**21**].

Higher categories may be viewed as an example of *enriched category theory*: an n-category should be thought of as a category enriched over $(n-1)$-categories, even though applying the strict definition of enrichment leads to strict rather than weak n-categories. Other kinds of enriched categories come up in closely related areas, such as *dg*-categories whose weak versions, A_∞-categories, play an important role in symplectic geometry and mirror symmetry. We don't yet have a comprehensive theory of weak enrichment. Some works in this direction include Stanculescu's model category for comonoids [**130**], and the development of Leinster's idea for extending the Segal formalism to weak enrichment with coefficients in monoidal model categories, going in several different ways by Bacard [**8**].

There are now numerous references for all of these notions, for the corresponding model category structures, and for the interconnections between them. Without pretending to give an exhaustive list, here are a few places to start:

- The *nLab* website `ncatlab.org` [**101**] is evolving into a very complete web reference.
- The comparison papers of Bergner [**19**], Toën [**139**], and Barwick and Schommer-Pries [**11**] will clearly provide good introductions and overviews.
- Tom Leinster has made numerous contributions, and his papers provide good points of entry into the subject. He brought together 10 very distinct different approaches in [**90**], discussing the different textures of each one. In [**91**], he developed a general machinery of multi-sorted operads and multicategories, designed to clarify

what was going on with Batanin's definition, leading to what is now known as the Batanin-Leinster approach. Cheng has further developed this formalism [**32**].
- I would like to refer the reader to the references in my book [**129**] for a more complete list than we could include here.

As was forseen by Grothendieck in a famous passage of [**68**], this plethora of different definitions and approaches opens up the main question of comparison. For the first basic case of $(\infty, 1)$-categories, the comparison question has been very well understood thanks to work of Bergner [**19**] and Toën [**139**], also Tabuada for *dg*-categories [**132**]. Porter's notes [**111**] give a succinct discussion of the comparison issues for $(\infty, 1)$-categories.

For n-categories or (∞, n)-categories, Barwick and Schommer-Pries have axiomatized the theory, shown that many of the main approaches satisfy their axioms, and proven that any two theories which satisfy their axioms are canonically equivalent (up to inverting directions of arrows) [**11**]. This is a very satisfactory answer, but it leaves open some questions. For example, in many of the approaches, there are one or more underlying Quillen model categories. It would be good to have natural concrete Quillen equivalences between these models, with progress being made by Bergner and Rezk [**21**]. Furthermore, as may already be seen in the passage between quasicategories and simplicial or Segal categories, there might well be several natural functors between two theories; one would then like to have good natural equivalences between these functors. There remain some theories such as Baez-Dolan's and Batanin's, where we don't yet know how to show the axioms of Barwick and Schommer-Pries. This could enclose some substantial combinatorial problems. Eugenia Cheng has nevertheless obtained a comparison between Batanin's and Trimble's definitions [**33**].

In using $(\infty, 1)$-categories, the basic idea is to proceed as for usual 1-categories, but bearing in mind that the $A(x, y)$ are simplicial sets, to be considered up to homotopy, rather than sets. An *arrow* between x and y is a vertex of the simplicial set $A(x, y)$, and two arrows are "equivalent" if they are connected by a series of edges. One usually assumes that $A(x, y)$ also satisfy the Kan condition, in which case two arrows are equivalent if and only if they are connected by a single edge.

An arrow $x \xrightarrow{u} y$ is an *inner equivalence* (or just "equivalence" if no confusion is feared) if there exists $y \xrightarrow{v} x$ such that $u \circ v$ is equivalent to 1_y and $v \circ u$ is equivalent to 1_x.

Given an $(\infty, 1)$-category A, we get a usual 1-category denoted $\tau_{\leq 1}(A)$ sharing the same set of objects, by setting

$$\tau_{\leq 1}(A)(x, y) := \pi_0(A(x, y)).$$

The arrows in $\tau_{\leq 1}(A)$ are therefore the equivalence classes of arrows in A. This 1-category is called the *1-truncation of* A. There is a map of $(\infty, 1)$-categories $A \to \tau_{\leq 1}(A)$ universal among maps from A to usual 1-categories.

From the previous definition, an equivalence class of inner equivalences from x to y is the same thing as an isomorphism from x to y in $\tau_{\leq 1}(A)$. Thus, x is inner-equivalent to y if and only if they are isomorphic objects of $\tau_{\leq 1}(A)$.

We can now define the notion of *equivalence of* $(\infty, 1)$*-categories*, which could be called "external equivalence", often called *Dwyer-Kan equivalence* in the literature. A morphism $F : A \to B$ between two $(\infty, 1)$-categories will mean a morphism in any one of our chosen categories of structures such as strict simplicially enriched categories, Segal categories, quasicategories or complete Segal spaces. Such a morphism is said to be *essentially surjective* if any object $y \in \mathrm{Ob}(B)$ is inner-equivalent, within B, to the image $F(x)$ of some object of A. A morphism F is said to be *fully faithful* if, for every pair of objects $x, y \in \mathrm{Ob}(A)$ the map $F : A(x, y) \to B(x, y)$ is a weak equivalence of simplicial sets. Putting these together, F is an *equivalence* if it is fully faithful and essentially surjective.

This notion of equivalence between $(\infty, 1)$-categories determines the homotopy theory of these objects. For example, in the model categories where usually at least the fibrant objects satisfy the required conditions (such as the Segal condition) to be $(\infty, 1)$-categories, a morphism between fibrant objects is a weak equivalence if and only if it is an equivalence in the above sense.

One of the most usual ways in which an $(\infty, 1)$-category arises in practice is via *Dwyer-Kan localization.* If C is a 1-category provided with a subcategory W, then Dwyer and Kan construct a simplicial category $L_{DK}(C, W)$ provided with a functor

$$C \to L_{DK}(C, W)$$

which sends arrows of W to inner equivalences in $L_{DK}(C, W)$. The $(\infty, 1)$-category $L_{DK}(C, W)$ may be characterized, up to canonical equivalence of $(\infty, 1)$-categories, as the universal $(\infty, 1)$-category with functor from C sending the arrows of W to inner equivalences.

Barwick and Kan have expanded further upon this idea to propose a model category for $(\infty, 1)$-categories whose objects are pairs (C, W) [**10**]. Using this theory the localization becomes tautological: $L_{DK}(C, W)$ is just the pair (C, W) itself. The point of their theory is to show that the full homotopy theory of $(\infty, 1)$-categories is obtained by looking at pairs (C, W).

9. The interior ∞-groupoid

The notion of "interior" of an $(\infty, 1)$-category provides an important link with the notion of ∞-groupoid. Let's start with the interior of a usual 1-category. If A is a 1-category, for any $x, y \in \mathrm{Ob}(A)$ let

$$A^{\mathrm{int}}(x, y) := \mathrm{Iso}_A(x, y) \subset A(x, y)$$

be the subset of arrows $x \xrightarrow{u} y$ such that u is an isomorphism. These are closed under composition and contain the identity arrows, so they fit together to form a subcategory

$$A^{\rm int} \subset A$$

called the *interior of* A. By construction all the arrows of $A^{\rm int}$ are invertible, so $A^{\rm int}$ is a groupoid. In fact it is easy to see that it is the *biggest groupoid contained in* A and this property is an alternative definition.

Turning now to the case when A is an $(\infty,1)$-category, we can similarly define a sub-$(\infty,1)$-category $A^{\rm int} \subset A$ with $\mathrm{Ob}(A^{\rm int}) = \mathrm{Ob}(A)$ defined by the property that for each $x, y \in \mathrm{Ob}(A)$, $A^{\rm int}(x,y) \subset A(x,y)$ is the union of connected components of $A(x,y)$ corresponding to arrows which are inner equivalences. Since inner equivalences are exactly those arrows which map to isomorphisms in $\tau_{\leq 1}(A)$, another way of saying the definition is by the pullback diagram

$$\begin{array}{ccc} A^{\rm int} & \rightarrow & A \\ \downarrow & & \downarrow \\ \tau_{\leq 1}(A)^{\rm int} & \rightarrow & \tau_{\leq 1}(A) \end{array}$$

with the interior of the 1-truncation.

An $(\infty,1)$-category is said to be an $(\infty,0)$-category or an ∞-groupoid if all the arrows are inner equivalences, or equivalently if its 1-truncation is a groupoid. With this notation, $A^{\rm int}$ is the biggest sub-∞-groupoid of A.

We now have two conflicting meanings of the terminology "∞-groupoid" which need to be reconciled. For this discussion, in order to distinguish the two notions, we'll say "simplicial set" for an ∞-groupoid defined as previously, and "$(\infty,0)$-category" for an $(\infty,1)$-category where all the arrows are inner equivalences. We construct from an $(\infty,0)$-category A a simplicial set denoted $|A|$. If A is given in the Segal point of view by a bisimplicial set, then $|A|_m := A_{m,m}$ is just the diagonal realization. If A is a strict simplicial category, apply the previous sentence to its simplicial nerve which is a Segal category. If A is a quasicategory, then $|A| = A$ is just the same simplicial set.

Going in the other direction, suppose X is a simplicial set which we assume satisfies the Kan condition. We can define a Segal category $\Pi_{Se}(X)$ whose set of objects is $\mathrm{Ob}(\Pi_{Se}(X) := X_0$ the set of vertices of X, by setting $\Pi_{Se}(X)_m(x_0, \ldots, x_m)$ equal to the simplicial mapping space from the standard m-simplex to X sending the vertices to $x_0, \ldots, x_m$. Let $\Pi_{Se}(X)_m$ be the disjoint union of the $\Pi_{Se}(X)_m(x_0, \ldots, x_m)$ over all sequences of objects.

These two constructions are adjoint and set up an equivalence between the $(\infty,0)$-categories and the Kan simplicial sets. In view of this equivalence, we can call either type of object an "∞-groupoid".

One could also try following the Grothendieck-Maltsiniotis definition of ∞-groupoid. Most basic elements of the equivalence between that theory and the ones used previously are known by work of Berger [**17**], Cisinski and Maltsiniotis [**37**] and Ara [**5**]. However as Georges Maltsiniotis pointed out, the full statement of the equivalence is not yet known.

10. Homotopy groups

For a simplicial set X with basepoint $x \in X_0$ the *homotopy groups* $\pi_i(X,x)$ are defined. Say that X is *n-truncated* if for any basepoint, $\pi_i(X,x) = 0$ for $i > n$. For an $(\infty,0)$-category A the homotopy groups are defined as follows: given $x \in \mathrm{Ob}(A)$ we have a basepoint $1_x \in A(x,x)$ and put

$$\pi_i(A,x) := \pi_{i-1}(A(x,x),1_x).$$

If $i = 1$ so $i-1 = 0$ the basepoint is not needed on the right side. If A corresponds to a simplicial set X in the sense that $X \sim |A|$ and $A \sim \Pi_{Se}(X)$ then objects of A correspond to vertices of X and $\pi_i(A,x) \cong \pi_i(X,x)$. We say that A is *n-truncated* if $\pi_i(A,x) = 0$ for any basepoint $x \in \mathrm{Ob}(A)$ and any $i > n$. This is equivalent to requiring that each $A(x,y)$ be an $(n-1)$-truncated simplicial set, making the convention that a (-1)-truncated simplicial set is a 0-truncated one which is either empty or connected (if connected it is consequently contractible). To see the equivalence note that by the hypothesis that A is an ∞-groupoid, given any nonempty $A(x,y)$ with an arrow $f \in A(x,y)$, composition with f gives an equivalence of based spaces

$$(A(x,x),1_x) \stackrel{\sim}{\to} (A(x,y),f),$$

so vanishing of the homotopy groups of all $(A(x,x),1_x)$ is equivalent to vanishing of the homotopy groups of all $(A(x,y),f)$.

More generally, for $n \geq 1$ we can say that an arbitrary $(\infty,1)$-category is *n-truncated* if $A(x,y)$ is an $(n-1)$-truncated simplicial set for each pair of objects $x,y \in \mathrm{Ob}(A)$. In keeping with Lurie's terminology, an n-truncated $(\infty,1)$-category can be called an $(n,1)$-category, whereas an n-truncated $(\infty,0)$-category can be called an $(n,0)$-category. In the language of n-categories, the $(n,1)$-categories (resp. $(n,0)$-categories) are the (weakly associative) n-categories such that the i-morphisms are invertible (in the weak sense i.e. up to equivalence) for all $i > 1$ (resp. $i > 0$).

If we consider a 0-truncated simplicial set to be the same thing as its corresponding discrete set obtained by applying π_0, then the 1-truncated $(\infty,1)$-categories are the same as usual 1-categories. More precisely, if A is a 1-category then it is also naturally an $(\infty,1)$-category which is 1-truncated, indeed the $A(x,y)$ are discrete sets so they are 0-truncated when viewed as simplicial sets. In the other direction if A is a $(1,1)$-category then the truncation $\tau_{\leq 1}(A)$ is a usual 1-category and the functor $A \to \tau_{\leq 1}(A)$ is an equivalence. With this equivalence, the $(1,0)$-categories are the same thing as 1-groupoids.

On the other hand the $(\infty,0)$-categories correspond to simplicial sets. A simplicial set X corresponds to a $(1,0)$-category if and only if it is 1-truncated, which in turn is equivalent to requiring that X be a disjoint union of $K(\pi,1)$-spaces. For a 1-truncated X (which we assume, by replacement if necessary, to satisfy the Kan condition) the $(\infty,0)$-category $\Pi_{Se}(X)$ is

1-truncated hence there is an equivalence

$$\Pi_{Se}(X) \xrightarrow{\sim} \tau_{\leq 1}(\Pi_{Se}(X)).$$

The term on the right is the usual Poincaré groupoid $\Pi_1(X)$, and indeed the equality

$$\tau_{\leq 1}(\Pi_{Se}(X)) = \Pi_1(X)$$

holds for any Kan simplicial set X.

One of Grothendieck's stated aims in *La poursuite des champs* was to investigate algebraic models of homotopy types. The initial idea was that the notion of n-groupoid would provide a good combinatorial model for n-truncated homotopy types. Such an application has not yet been achieved, for several reasons.

First of all, it turns out that algebraic topologists have really been working with a number of different algebraic models for homotopy types, ever since the inception of the theory. The most basic of these is the notion of Kan simplicial complex. Then, the various flavors of operadic delooping machines provide algebraic models, which have been used most particularly for the study of stable homotopy theory and spectra. Quillen introduced the notion of *simplicial group*, and this leads to localization and completion theories, to the Curtis spectral sequence, and other tools which can be used to calculate homotopy-theoretic invariants. The popularly held idea that "we don't know how to calculate the homotopy groups of spheres" turns out to be a myth: there are actually a number of different algorithms answering this question, they just have pretty bad runtime estimates.

Answering somewhat more directly the problem posed by Grothendieck, Brown and Loday constructed an algebraic model using Cat^n-groups, roughly speaking n-cubes of groups. This comes a lot closer to the weak n-groupoid model which Grothendieck was looking for, and indeed Paoli and Pronk have recently shown how the two ideas might well link up [**108**]. In all, the algebraic topologists have no real need for yet another model.

It next turns out that some of the most economical ways of approaching the theory of higher categories, are by using exactly the topologists' categorical tools such as simplicial objects. So these theories don't really answer the query for a "new" algebraic model of homotopy theory. The homotopy hypothesis in, say, Tamsamani's theory rather serves to verify that the n-groupoids, modeling n-truncated homotopy types, do indeed play their expected role within the larger theory of n-categories.

Maltsiniotis has pointed out that one can extract a combinatorial definition of n-groupoid from [**68**], a definition which is similar in spirit to Batanin's higher-operadic method [**98**]. As Maltsiniotis says, the question of the "homotopy hypothesis" for this particular combinatorial model remains an open question. An answer to which, would constitute an answer to Grothendieck's question in [**68**]. As soon as we can show that Batanin's model admits the right kind of limits and colimits and has a small generation property, such as in the axiomatic system of Barwick and Schommer-Pries

[**11**], the homotopy hypothesis for these combinatorial models might well follow formally. At the present time, such a proof or even a more direct proof remain somewhat far off.

Finally, the notion of a *calculatory approach* allowing us to reduce homotopy-theoretical questions to calculations, has recently taken a wider turn, with the arrival of *homotopy type theory* [**156**]. Voevodsky's univalence axiom leads to explicit calculations in homotopy theory, which are based on the structure of logical λ-calculus, which is quite different from the combinatorial structures envisioned in the theory of n-groupoids.

The idea of obtaining a great new way of calculating homotopy-theoretical invariants out of higher categories, has therefore not (yet) come to fruition. The reason why the influence of [**68**] has begun to permeate a wide range of mathematical currents, is rather that the higher-categorical framework provides a conceptual way of understanding the phenomena of localization and glueing, in which homotopy-theory turns out to play a central role.

Perhaps sensing that n-groupoids were not actually best directed toward the question of algebraic models, and out of curiosity about why the simplicial category Δ should play such an important role, Grothendieck changed directions in the middle of [**68**]. He turned to look at the question of what properties of Δ would make its presheaf category into a good model for homotopy type. His investigations led to the notion of *test category* and *localizer.* These ideas have now been worked out in great detail by Maltsiniotis [**97**] and Cisinski [**35**]. These ideas, toward which Grothendieck makes a clear transition in [**68**], go beyond the scope of the present article. At the end, [**68**] makes yet another transition, as Grothendieck heads toward the consideration of "Dessins d'enfants"....

11. Topos theory

One of the original ways of thinking about a sheaf $\mathscr{F}$ over a topological space X, was in terms of its *espace étalé* $\mathrm{Esp}(\mathscr{F}) \to X$. This is a space obtained by taking the union of one copy of each open set $U \subset X$ for each element of $\mathscr{F}(U)$, then dividing by the natural identifications. The set of sections $U \to \mathrm{Esp}(\mathscr{F})$ over an open set U, is naturally identified with $\mathscr{F}(U)$. In this way of looking at a sheaf, the sheaf becomes identified with a space lying over X.

Grothendieck introduced the notion of *topos*, consisting of an abstract characterization of the properties held by the category $\mathrm{Sh}(\Phi)$ of sheaves over a site Φ. The topos $\mathscr{T} := \mathrm{Sh}(\Phi)$ has its own *canonical topology*, and the category of sheaves on $\mathscr{T}$ with its canonical topology, is naturally equivalent to $\mathscr{T}$ the category of sheaves on Φ. This equivalence comes from the natural inclusion $\Phi \to \mathscr{T}$ sending $U \in \Phi$ to the sheaf it represents. Thus, the site Φ is enlarged to $\mathscr{T}$ by adding in all the sheaves which weren't representable by elements of Φ. Intuitively, we think of adding to Φ all of the "espaces étalés" corresponding to sheaves. This gives a canonical representation of $\mathscr{T}$ as the category of sheaves on a site.

Breaking free from this geometrical intuition, it turns out that we can very productively work *entirely within* the topos $\mathscr{T}$ just starting from its axiomatic properties, without actually identifying the objects $F \in \mathscr{T}$ with sheaves on a particular site.

It is natural to ask whether a similar advance could be made for the theory of higher stacks. This seemed a long way off, posing seemingly insurmountable problems, when Jacob Lurie came along with his book *Higher Topos Theory* [**94**]. He solved all of the technical difficulties in one gigantic swoop, starting out by using the theory of *quasicategories* to represent $(\infty, 1)$-categories, and adding in many techniques to solve the specific problems of Quillen model categories which therefore arise.

This advance has led to numerous works in many different directions, allowing mathematicians from diverse fields to use $(\infty, 1)$-categories right where they most immediately occur rather than passing through various more classical work-arounds. Although it would go beyond our present scope to cite all of these, a notable example is the work of Barwick [**9**] on the universal characterization of K-theory, generalizing work of Tabuada [**131**] who worked with stable ∞-categories using dg-categories and derivators in place of quasicategories.

As Eduard Balzin has explained to me, Rezk [**118**] introduced a characterization of $(\infty, 1)$-topoi based on a "descent" property explaining how pushouts and pullbacks interact, see also [**94**]. This leads to a whole new direction of generalization of the notion of descent, which goes beyond our scope here.

One of the latest developments is the notion of *homotopy type theory*, explained in the "HoTT book" [**156**]. This combines formal logic and computer theorem-proving, with ∞-topos theory, in a strikingly simple way. The basic idea is to just pretend we are working with sets, but a notation X for a set really means a homotopy type (or more generally an object of an ∞-topos [**56**]). If $x, y \in X$ then the "set" (i.e. *type*) of proofs of the statement $x = y$ is considered to be the space of paths[2] between x and y in X. This works really well, thanks to Voevodsky's *univalence axiom* which says that the space of equalities between types $X = Y$ is "equal to" (which here means "weak equivalent to") the space of isomorphisms $X \xrightarrow{\cong} Y$. We refer to [**156**] for all the fun things one can do from here. Schreiber and Shulman [**123**]

[2]The idea that one might be able to do such a thing, had actually been mentioned to me by Jacques Sauloy in Toulouse in the very early 1990's, just as Zouhair Tamsamani and I were getting interested in the question of how to define higher categories. Sauloy suggested that the notion of identity type in formal type theory, could lead to a notion of how to organise the higher arrows in an n-category. At the time, I didn't know anything about type theory and we never pursued this path. My later interest in type theory came about during poolside discussions with André Hirschowitz about the possibility of computer-checked theorem proving.

explain how higher cocycle theory and homotopy type theory combine in a foundational way for physics. In the long term, we can hope that it would give a *calculatory method* for dealing with the homotopy-coherence questions that arise in higher stack theory.

12. Fibered categories

One-half of the first volume of SGA introduced the theory of fibered categories. Reading this part of SGA 1 was my first direct contact with Grothendieck's writing. As a young graduate student, I did a handwritten translation of it, as a way to learn French and algebraic geometry at the same time.

The main goal is to be able to discuss families of categories indexed by a base category Φ. If we introduce the notation CAT for the 2-category whose objects are categories, whose morphisms are functors and whose 2-morphisms are natural transformations, then we would like to consider 2-functors

$$\mathscr{F} : \Phi \to CAT.$$

This notion involves, on the one hand, additional homotopy-coherence structure: given a composable pair of arrows $a \xrightarrow{f} b \xrightarrow{g} c$ in Φ, there would be a homotopy $\alpha_{f,g}$ between $\mathscr{F}(gf)$ and $\mathscr{F}(g) \circ \mathscr{F}(f)$, subject to compatibility conditions. The theory of 2-categories as developed by Bénabou [**15**] allows us to make this definition precise. Another kind of problem also shows up, that of universes: in order to consider a family of large categories we would need to define CAT as the 2-category of large categories; this requires introducing a next-higher universe level. Again this poses no fundamental technical problem if, as Grothendieck did, we adopt Tarski's axiom that every set is contained in some universe. It is nevertheless natural to look for an alternative solution.

Instead of considering a functor from Φ to a parametrizing 2-category, Grothendieck looked at "fibrations"

$$P : \mathscr{F} \to \Phi$$

whose fibers $\mathscr{F}_x := P^{-1}$ correspond to the categories being parametrized by $x \in \mathrm{Ob}(\Phi)$. Some additional properties are required before we can say that this really corresponds to a family of categories indexed by Φ, specially to define the pushforward maps $u_* : \mathscr{F}_x \to \mathscr{F}_y$ whenever $x \xrightarrow{u} y$ is an arrow in Φ.

Given a functor $P : \mathscr{F} \to \Phi$, suppose $a \in \mathscr{F}_x$ and $b \in \mathscr{F}_y$. If $b \xrightarrow{f} a$ is an arrow in $\mathscr{F}$, it induces an arrow $u := P(f)$ from y to x in Φ. We say that f is *cartesian* if, for every arrow $c \xrightarrow{g} a$ with $P(g) = u$, there exists a unique arrow $c \xrightarrow{h} b$ in $\mathscr{F}_y$ (meaning that $P(h) = 1_y$) such that $g = f \circ h$.

A functor $P : \mathscr{F} \to \Phi$ is *a fibered category* if:

(FC1)—for every arrow $y \xrightarrow{u} x$ in Φ and every object $a \in \mathscr{F}_x$ there exists a

cartesian arrow $b \xrightarrow{f} a$ such that $P(f) = u$; and
(FC2)—the composition of cartesian arrows is again cartesian.

The combination of these two conditions can be expressed in a unified way by introducing the notion of *strongly cartesian arrow*. An arrow $b \xrightarrow{f} a$ with $P(f) = u$ in $\mathscr{F}$ is *strongly cartesian* if, for any object $c \in \mathrm{Ob}(\mathscr{F})$ and any composable sequence

$$P(c) \xrightarrow{v} P(b) \xrightarrow{v} P(a)$$

in Φ, composition with f induces an isomorphism between the set of arrows $c \to b$ mapping to v, and the set of arrows $c \to a$ mapping to $v \circ u$. Note that the cartesian condition is the same but only when v is the identity of $P(b)$ (in particular strongly cartesian implies cartesian). The conjunction of conditions (FC1) and (FC2) is equivalent to:
(FC)—for every arrow $y \xrightarrow{u} x$ in Φ and every object $a \in \mathscr{F}_x$ there exists a strongly cartesian arrow $b \xrightarrow{f} a$ such that $P(f) = u$.

In a fibered category, cartesian arrows are strongly cartesian.

If these axioms hold, we can make the following construction for an arrow $y \xrightarrow{u} x$ in Φ: choose for each $a \in \mathscr{F}_x$ a cartesian arrow denoted $u^*(a) \xrightarrow{f_a} a$ with $P(f_a) = u$. For any arrow $b \xrightarrow{g} a$ in $\mathscr{F}_x$, by hypothesis (FC1) there exists a unique arrow denoted

$$u^*(b) \xrightarrow{u^*(g)} u^*(a)$$

in $\mathscr{F}_y$ i.e. with $P(u^*(g)) = 1_y$, such that

$$g \circ f_b = f_a \circ u^*(g).$$

The uniqueness part of (FC1) implies that the construction u^* is compatible with composition in the sense that $u^*(gg') = u^*(g)u^*(g')$, and similarly it is compatible with identities. Therefore we get a functor denoted

$$u^* : \mathscr{F}_x \to \mathscr{F}_y.$$

It should be stressed that this functor is not uniquely defined: it depends on the choices of liftings of u to cartesian arrows f_a starting from each object $a \in \mathscr{F}_x$. However, any two different sets of choices will lead to functors which are naturally isomorphic by a uniquely determined isomorphism. If we make choices of cartesian liftings for all arrows and all starting objects, then we obtain a family of functors u^* depending on arrows u of Φ, and these are compatible with composition by axiom (FC2), up to unique isomorphism. This means that if $z \xrightarrow{v} y \xrightarrow{u} x$ is a composable sequence in Φ, then there is a natural isomorphism of functors from $\mathscr{F}_x$ to $\mathscr{F}_z$,

$$\alpha_{u,v} : v^* \circ u^* \xrightarrow{\cong} (vu)^*.$$

Furthermore these isomorphisms in turn satisfy a cocycle condition so they define a weak 2-functor $\Phi^o \to CAT$, that is to say a weak 2-functor contravariant on Φ. Conversely, given such a weak 2-functor there is a fibered category and choice of cartesian liftings which corresponds to it.

The cartesian family itself may be considered as a replacement for the functor $\Phi^o \to CAT$, indeed the functors and the families define notions which, while not identical, are equivalent up to natural equivalences on both sides.

Notice that $\mathscr{F}$ can be a large category in this definition, leading to a notion of family of large categories which only requires the notion of "class" rather than a full higher universe.

There is a dual notion of cofibered category over Φ which corresponds to a functor $\Phi \to CAT$. A morphism $a \stackrel{f}{\to} b$ projecting to $x \stackrel{u}{\to} y$ is *cocartesian* if for any object $c \in \mathscr{F}_y$ and any $a \stackrel{g}{\to} c$ with $P(g) = u$, there is a unique $b \stackrel{h}{\to} c$ in $\mathscr{F}_y$ with $h \circ f = g$. A functor P is a *cofibered category* if cocartesian liftings exist and the composition of cocartesian morphisms is cocartesian. Again after making a choice of cocartesian liftings, we get a functor $u_* : \mathscr{F}_x \to \mathscr{F}_y$, compatible with compositions up to canonical isomorphism, yielding a 2-functor $\Phi \to CAT$.

If $P : \mathscr{F} \to \Phi$ is a fibered category corresponding to a functor $S : \Phi^o \to CAT$ then $\mathscr{F}^o \to \Phi^o$ is a cofibered category, corresponding to the functor $S^o : \Phi^o \to CAT$ with $S^o = (\text{op}) \circ S$ where

$$(\text{op}) : CAT \to CAT$$

is the 2-functor taking a category to its opposite.

Here is an example to show that condition (FC2) is needed in the definition (thanks to J. Bénabou for pointing out this phenomenon during the Peyresq meeting). The category Φ will consist of objects $0, 1, 2$ with arrows $0 \to 1$ and $1 \to 2$ composing to $0 \to 2$, plus the identities. The category $\mathscr{F}$ consists of pairs (i, a) where $i \in \text{Ob}(\Phi)$ and $a \in \mathbb{N}$. There is at most one arrow

$$(i, a) \to (j, b)$$

between any pair of objects in $\mathscr{F}$, and such an arrow exists whenever $i = j$ and $b \leq a$, or $i < j$ and $b + 1 \leq a$. The cartesian arrows are of the form

$$(0, a+1) \to (1, a), \quad (1, a+1) \to (2, a)$$

and

$$(0, a+1) \to (2, a),$$

so the composition of cartesian arrows lying over $0 \to 1$ and $1 \to 2$ is not a cartesian arrow lying over $0 \to 2$.

13. Sections and strictification

If $P : \mathscr{F} \to \Phi$ is a fibered category, there might not exist a choice of splittings compatible with composition, so it might not come from a 1-functor $\Phi^o \to \text{Cat}$ where Cat denotes the 1-category of categories and functors rather than the 2-category CAT which includes natural transformations. A fibered category comes from a 1-functor to Cat if and only if it admits a choice of cartesian liftings which is strictly compatible with composition. Such a choice might not exist. However, we can find an equivalent fibered category such that there is one, as will now be shown. This provides an important first illustration of the phenomenon of "strictification", and also serves to explain how to glue together descent data, an application we'll discuss afterward.

If $P : \mathscr{F} \to \Phi$ is a functor between categories, define the category $\Gamma(\Phi, \mathscr{F})$ of *sections* by the formula

$$\Gamma(\Phi, \mathscr{F}) := \underline{Hom}(\Phi, \mathscr{F}) \times_{\underline{Hom}(\Phi,\Phi)} \{\mathrm{Id}_\Phi\}.$$

A section is therefore a functor $s : \Phi \to \mathscr{F}$ such that $P \circ s = \mathrm{Id}_\Phi$, and a morphism between sections is a natural transformation of functors $\eta : s \to s'$ such that $P(\eta)$ is the identity transformation of Id_Φ.

Suppose now that P is a fibered category. A section is said to be *cartesian* if $s(u)$ is a cartesian arrow for any arrow u of Φ. Denote by $\Gamma^{\mathrm{cart}}(\Phi, \mathscr{F})$ the full subcategory of cartesian sections.

Suppose P is a fibered category and Φ has an coinitial object x_0. Evaluation at x_0 provides a functor

$$\Gamma^{\mathrm{cart}}(\Phi, \mathscr{F}) \to \mathscr{F}_{x_0}, \quad s \mapsto s(x_0).$$

We claim that this functor is an equivalence of categories. Note first that it is essentially surjective and in fact surjective: suppose $a \in \mathrm{Ob}(\mathscr{F}_{x_0})$ and construct a cartesian section s with $s(x_0) = a$. For any $y \in \Phi$ there is a unique arrow $y \xrightarrow{u_y} x_0$. Choose a cartesian arrow $s(y) \xrightarrow{f_y} a$ with $P(f_y) = u_y$. If $z \xrightarrow{v} y$ is an arrow of Φ then $u_y \circ v = u_z$. Choose a cartesian arrow $b \xrightarrow{h} s(y)$ with $P(h) = v$; then $f_y \circ h$ is a cartesian arrow over u_z by axiom (FC2), so by uniqueness of cartesian lifts up to isomorphism, there is a unique isomorphism $k : s(z) \cong b$ in $\mathscr{F}_z$ such that $f_y \circ h \circ k = f_z$. Define $s(v) := h \circ k$. Thus $s(v)$ is a cartesian arrow lying over v and $f_y \circ s(v) = f_z$. Uniqueness of factorizations through cartesian arrows implies that $s(v) \circ s(v') = s(v \circ v')$ and $s(1_y) = 1_{s(y)}$, so s defines a cartesian section. In the construction, we could start by choosing f_{x_0} to be the identity of a, so $s(x_0) = a$.

To prove that the evaluation at x_0 is fully faithful, suppose s and t are cartesian sections. Let $a := s(x_0)$ and $b := t(x_0)$. If $a \xrightarrow{f} b$ is an arrow in $\mathscr{F}_{x_0}$ then for any object $y \in \mathrm{Ob}(\Phi)$ we have the unique map u_y from y to x_0 hence cartesian arrows $s(u_y)$ from $s(y)$ to a and $t(u_y)$ from $t(y)$ to b respectively. The cartesian property of $t(u_y)$ applied to $f \circ s(u_y)$ provides a unique map η_y from $s(y)$ to $t(y)$ such that $t(u_y) \circ \eta_y = f \circ s(u_y)$. Uniqueness allows us to see that the η_y fit together to form a natural transformation from s to t, and also that this is the only possible one with $\eta_{x_0} = f$. Thus evaluation at x_0 is fully faithful, and we have now shown that it is an equivalence of categories.

Suppose now that $P : \mathscr{F} \to \Phi$ is a fibered category over any small base category Φ. For any $x \in \mathrm{Ob}(\Phi)$ consider the category Φ/x of arrows $y \to x$.

The restriction $\mathscr{F}|_{\Phi/x}$, defined to be the pullback of P along the functor $(\Phi/x) \to \Phi$, is a fibered category whose base has a coinitial object (denoted also by x). From the previous paragraphs we obtain that the functor of evaluation at x is an equivalence of categories

$$\mathscr{G}(x) := \Gamma^{\text{cart}}(\Phi/x, \mathscr{F}|_{\Phi/x}) \stackrel{\sim}{\to} \mathscr{F}_x.$$

However, the $\mathscr{G}(x)$ vary functorially in x, which is to say that we have a strict 1-functor

$$\mathscr{G} : \Phi^o \to \text{Cat}.$$

Furthermore, starting from such a functor one defines the *Grothendieck integral* which is a fibered category

$$\int_\Phi \mathscr{G} \to \Phi$$

and the evaluation functors $\mathscr{G}(x) \to \mathscr{F}_x$ fit together into a morphism of fibered categories

$$\begin{array}{ccc} \int_\Phi \mathscr{G} & \to & \mathscr{F} \\ \downarrow & & \downarrow \\ \Phi & = & \Phi. \end{array}$$

The construction of this morphism is similar to the arguments done above and is left to the reader. It is an equivalence of categories, indeed any morphism between fibered categories which induces equivalences on the fibers, is an equivalence. This shows that any fibered category is equivalent to one which comes from a strict 1-functor, i.e. one which admits a choice of cartesian liftings compatible with composition.

This strictification result is one of the main outcomes of that part of SGA1 concerning fibered categories. It means that we don't need to use a definition using the "homotopy coherencies" involved in the definition of a weak 2-functor There are a number of different "strictification" or "rectification" results in the algebra of homotopy-coherent objects. For complicated structures, what generally happens is that we can strictify some of the homotopy-coherent structures, but usually not all at once. In the case of a single functor $F : \Phi^o \to CAT$, Grothendieck's result allows us to strictify completely.

In the higher-categorical situation, the same argument using sections of a "fibered category" allows us to strictify functors which correspond to families of categories over a base 1-category. This played an important role in Tamsamani's approach to iterating the Segal delooping machinery to obtain a definition of n-categories. His idea was that, instead of considering some kind of weak family of $(n-1)$-categories indexed by Δ, which would have been complicated to say the least, we could instead just consider strict functors $\Delta^o \to (n-1)\text{Cat}$. Iterating leads directly to multisimplicial sets [**133**].

14. Stacks

Suppose given a fibered category $\mathscr{F} \to \Phi$ corresponding to a 2-functor $F : \Phi \to CAT$. The category of cartesian sections $\Gamma^{\text{cart}}(\Phi, \mathscr{F})$ is the 2-limit of the diagram F. For this reason, the cartesian section construction is crucial in the theory of descent.

Suppose $P : \mathscr{F} \to \Phi$ is a fibered category and Φ has a Grothendieck topology $\mathscr{T}$ so it is a site. The topology $\mathscr{T}$ consists of specifying which are the covering sieves $\mathscr{B} \subset \Phi/x$ for any object x. For any such sieve, we can consider the diagram

$$\mathscr{F}_x \stackrel{\sim}{\leftarrow} \Gamma^{\text{cart}}(\Phi/x, \mathscr{F}|_{(\Phi/x)^o}) \to \Gamma^{\text{cart}}(\mathscr{B}, \mathscr{F}|_{\mathscr{B}^o}).$$

We think of the category of cartesian sections over $\mathscr{B}$ as representing the "sections of $\mathscr{F}$ at x, locally defined with respect to the sieve $\mathscr{B}$". It is the 2-limit of the corresponding diagram $\mathscr{B}^o \to CAT$. Up to inverting the leftward equivalence of categories, the above diagram corresponds to the restriction functor from $\mathscr{F}_x$ to the category of locally defined sections.

The fibered category P is said to have *effective descent* if, for any $x \in \Phi$ and any covering sieve $\mathscr{B} \subset (\Phi/x)$, the arrow

$$\Gamma^{\text{cart}}(\Phi/x, \mathscr{F}|_{(\Phi/x)^o}) \to \Gamma^{\text{cart}}(\mathscr{B}, \mathscr{F}|_{\mathscr{B}^o})$$

is an equivalence of categories. Since the left term is equivalent to $\mathscr{F}_x$, this means that $\mathscr{F}_x$ is equivalent to the category of sections locally defined over $\mathscr{B}$.

A 1-*stack* over the site Φ is a fibered category $P : \mathscr{F} \to \Phi$ which has effective descent. In keeping with this terminology, a 1-*prestack* on a site Φ is simply a different notation for a fibered category over Φ.

A category C is said to be 0-*truncated* if it is a groupoid, and if between each pair of objects there is at most one morphism. For any groupoid C, the set $\pi_0(C)$ is defined to be the set of isomorphism classes of objects. A set X may be considered as a category in the following way: the set of objects is X itself, and the set of morphisms between two objects is empty if they are distinct, and contains just the identity if they are the same. In this way the set $\pi_0(C)$ may be considered as a category, and there is a natural functor $C \to \pi_0(C)$. A groupoid C is 0-truncated if and only if this natural functor is an equivalence of categories.

A 0-*stack* (resp. 0-*prestack*) is a 1-stack (resp. 1-*prestack*) $\mathscr{F} \to \Phi$ such that each $\mathscr{F}_x$ is 0-truncated. The pullback operations then provide well-defined maps $u^* : \pi_0(\mathscr{F}_x) \to \pi_0(\mathscr{F}_y)$ whenever $y \xrightarrow{u} x$ is an arrow in Φ. This yields a presheaf of sets denoted $\pi_0(\mathscr{F}/\Phi)$.

It is an exercise to show that a 0-prestack is a 0-stack if and only if this presheaf of sets is a sheaf. The effective descent condition is equivalent to the sheaf condition in this case.

Going in the other direction, given a presheaf of sets $F : \Phi^o \rightarrow$ Sets we obtain the fibered category

$$\int_{\Phi} F \rightarrow \Phi^o.$$

If $\mathscr{F}$ is a 0-prestack on Φ, there is a natural equivalence of fibered categories

$$\mathscr{F} \rightarrow \int_{\Phi} \pi_0(\mathscr{F}/\Phi)$$

and similarly a natural isomorphism of presheaves of sets

$$\pi_0\left(\int_{\Phi} F\right) \cong F.$$

Therefore, a presheaf F is a sheaf if and only if $\int_{\Phi} F$ is a 0-stack, i.e. satisfies effective descent.

15. Fibered $(\infty, 1)$-categories

Lurie **[94]** **[95]** tells us that theory of fibered categories can be generalized to the context of $(\infty, 1)$-categories. Our description will be heuristic; the reader is refered to **[94]** **[95]** where quasicategories facilitate a complete approach.

Keep a usual 1-category Φ as base. A functor of $(\infty, 1)$-categories

$$P : \mathscr{F} \rightarrow \Phi$$

consists of specifying the map P on sets of objects, and for each sequence of objects $a_0, \ldots, a_m$ in $\mathrm{Ob}(\mathscr{F})$, a map from the simplicial set $\mathscr{F}(a_0, \ldots, a_m)$ to the discrete set $\Phi(P(a_0), \ldots, P(a_m))$. In particular this data is the same as specifying a map of sets

$$\pi_0(\mathscr{F}(a_0, \ldots, a_m)) \rightarrow \Phi(P(a_0), \ldots, P(a_m))$$

so giving P is equivalent to giving a functor $\tau_{\leq 1}(\mathscr{F}) \rightarrow P$ of 1-categories.

Therefore if $x \stackrel{u}{\rightarrow} y$ is an arrow in Φ, and if $P(a) = x$ and $P(b) = y$, we can consider the sub-simplicial set $\mathscr{F}(a, b)^u$ which is the union of connected components of $\mathscr{F}(a, b)$ which map to u by P.

We adapt the alternative form of the definition of fibered category. Say that an arrow $f \in \mathscr{F}(b, a)_0$ is *very cartesian* if, for every $c \in \mathrm{Ob}(\mathscr{F})$ and every diagram

$$P(c) \stackrel{u}{\rightarrow} P(b) \stackrel{P(f)}{\rightarrow} P(a)$$

in Φ, the map of composition with f

$$\mathscr{F}(c, b)^u \rightarrow \mathscr{F}(c, a)^{P(f) \circ u}$$

is a weak equivalence of simplicial sets. Notice that the homotopy class of this map is well defined even in the weaker versions of $(\infty, 1)$ categories which don't have a strictly well-defined composition operation; thus the condition that the map be a weak equivalence is well-defined. The very cartesian condition is invariant under homotopy: if f is very cartesian then so is any other arrow in the same connected component of $\mathscr{F}(b,a)$. A functor of $(\infty,1)$-categories $P : \mathscr{F} \to \Phi$ is a *fibered* $(\infty,1)$*-category* if, for every $a \in \mathrm{Ob}(\mathscr{F})$ and every arrow $y \xrightarrow{u} P(a)$ in Φ, there exists a very cartesian arrow $f \in \mathscr{F}(b,a)_0$ with $P(f) = u$.

The same constructions as were done previously carry over to this context. However, before looking at morphisms into an $(\infty,1)$-category, the target should be replaced by a fibrant object in the model category structure, and the source by a cofibrant object. The latter condition leads to combinatorial complications if we use the model category structure of Dwyer-Kan-Bergner [**18**] for strict simplicial categories: our 1-category Φ is not generally a cofibrant object so it would need a cofibrant replacement. This can be done functorially in Φ, so it works but requires some additional notation which is left to the reader. On the other hand if we use the model category structures for Segal categories, Rezk categories or quasicategories, then all objects are cofibrant. In these cases we still need to suppose that the functor $P : \mathscr{F} \to \Phi$ is a fibration.

If $P : \mathscr{F} \to \Phi$ is a fibration which is a fibered $(\infty,1)$-category, then we can consider the $(\infty,1)$-category of sections $\Gamma(\Phi,\mathscr{F})$ and the sub-$(\infty,1)$-category $\Gamma^{\mathrm{cart}}(\Phi,\mathscr{F})$ of *cartesian sections*, meaning those sections such that the image of any arrow in Φ is very cartesian.

The same result as previously holds here: if x is a coinitial object of Φ, the evaluation map

$$\Gamma^{\mathrm{cart}}(\Phi,\mathscr{F}) \to \mathscr{F}_x$$

is an equivalence of $(\infty,1)$-categories, and more generally for any object x,

$$\Gamma^{\mathrm{cart}}(\Phi/x, \mathscr{F}|_{\Phi/x}) \to \mathscr{F}_x$$

is an equivalence. Note that the restriction $\mathscr{F}|_{\Phi/x}$ remains a fibration over (Φ/x). As before, this allows us to *strictify* $\mathscr{F}$, making it equivalent to a fibered category which comes from a strict 1-functor from Φ to $\mathrm{Cat}^{(\infty,1)}$ denoting whichever model category of $(\infty,1)$-categories we're using. Given any fibered $(\infty,1)$-category we can first replace it by an equivalent one which is fibrant over Φ and then use the above procedure to strictify.

An $(\infty,1)$*-prestack* over a category Φ is a fibered $(\infty,1)$-category $\mathscr{F} \to \Phi$. By the strictification procedure, this may also be viewed as a functor $\Phi \to \mathrm{Cat}^{(\infty,1)}$. If the functor takes values in, or equivalently if the $\mathscr{F}_x$ are $(\infty,0)$-categories (resp. $(n,1)$-categories, resp. $(n,0)$-categories) then we say that $\mathscr{F}$ is an $(\infty,0)$*-prestack* (resp. an $(n,1)$-prestack, resp. an $(n,0)$-prestack). Denote by

$$\begin{array}{ccc} \mathrm{Prestack}^{(n,0)}(\Phi) & \subset & \mathrm{Prestack}^{(\infty,0)}(\Phi) \\ \cap & & \cap \\ \mathrm{Prestack}^{(n,1)}(\Phi) & \subset & \mathrm{Prestack}^{(\infty,1)}(\Phi) \end{array}$$

the various categories of prestacks when viewed as strict functors from Φ^0 to $\mathrm{Cat}^{(\infty,1)}$ (resp. $\mathrm{Cat}^{(n,1)}$ etc.). The categories $\mathrm{Prestack}^{(\infty,0)}(\Phi)$ and

$\text{Prestack}^{(n,0)}(\Phi)$ are equivalent respectively to the categories of simplicial presheaves and presheaves of $(\infty, 1)$-categories over Φ, and $\text{Prestack}^{(n,0)}(\Phi)$ is equivalent to the category of n-truncated simplicial presheaves.

On the other hand, we have $(\infty, 2)$-categories of fibered $(\infty, 1)$-categories which may be denoted by

$$\begin{array}{ccc} \text{PRESTACK}^{(n,0)}(\Phi) & \subset & \text{PRESTACK}^{(\infty,0)}(\Phi) \\ \cap & & \cap \\ \text{PRESTACK}^{(n,1)}(\Phi) & \subset & \text{PRESTACK}^{(\infty,1)}(\Phi). \end{array}$$

The terms are respectively $(n+1,1)$, $(\infty,1)$, $(n+1,2)$ and $(\infty,2)$-categories. There are (several different) model structures for the category $\text{Prestack}^{(\infty,1)}(\Phi)$ of Φ^o-diagrams in $\text{Cat}^{(\infty,1)}$ The Dwyer-Kan localizations of these model categories provide $(\infty, 1)$-categories which are equivalent to the $(\infty, 1)$-interior of the $(\infty, 2)$-category $\text{PRESTACK}^{(\infty,1)}(\Phi)$. Similarly there are (again several different) classical model structures on the category $\text{Prestack}^{(\infty,0)}(\Phi)$ of Φ^o-diagrams in $\text{Cat}^{(\infty,0)} \equiv$ (simplicial sets), which is just the category of simplicial presheaves over Φ. In this case since $\text{PRESTACK}^{(\infty,0)}(\Phi)$ is already an $(\infty, 1)$-category, it is equivalent to the Dwyer-Kan localization of any of the relevant model structures.

The structure of $(\infty, 2)$-category on $\text{PRESTACK}^{(\infty,1)}(\Phi)$ can be recovered from the model category $\text{Prestack}^{(\infty,1)}(\Phi)$ using the *internal Hom* if we take care to use a cartesian model category structure. In fact this gives a structure of category enriched over $(\infty, 1)$-prestacks, yielding a category enriched over $(\infty, 1)$-categories by application of the global section construction.

Higher stacks may thus be seen to fit into the same formal framework as sheaves or 1-stacks. Historically, this wasn't the first approach. Jardine's paper [**79**] takes a different tack: he constructs a Quillen model category whose objects are all presheaves of simplicial sets over the category Φ underlying our site. Such a category has various *diagram category* model structures where the weak equivalences are the maps which are objectwise weak equivalences between simplicial sets. The *projective model structure* retains the objectwise fibrations and modifies the cofibrations accordingly, whereas the *injective model structure* retains the objectwise cofibrations, and modifies the fibrations accordingly. These model categories have equivalent Dwyer-Kan localizations, the $(\infty, 1)$-category of $(\infty, 1)$-functors from Φ to spaces. Jardine's approach is then to proceed by *left Bousfield localization*, increasing the class of weak equivalences from the objectwise ones, to the Illusie weak equivalences. This yields a new model category structure, corresponding to the theory of higher stacks, in which the fibrant objects satisfy the descent condition. We are able to get here without studying the descent condition itself beforehand. The basic question, in any case, is how to calculate the space of maps $Hom(A, B). = Hom_{\text{STACK}}(A, B).$ between two objects. In Jardine's formalism, this mapping space is intrinsic to the model category, and is calculated by making replacements up to weak equivalence, so that A should be cofibrant and B should be fibrant, then take the standard simplicial mapping space $k \mapsto Hom(A \times \Delta[k], B)$ (it is best to use the injective model structure here, otherwise one would need to make a projective cofibrant replacement of $A \times \Delta[k]$). One may even note that the mapping-space information was already contained in the notion of Illusie weak equivalence: we can take the category of all simplicial presheaves and Dwyer-Kan localize by the Illusie weak equivalences, then look at the mapping space in the resulting simplicial category. However, stated in that way without an accompanying model structure, it would have been essentially impossible to work with such a definition.

The role of the descent condition may be envisioned in the following way. Suppose that we already know how to calculate the simplicial mapping space $Hom_{\mathrm{diag}}(A, B)$. in the diagram category, i.e. the $(\infty, 1)$-category of functors from Φ to spaces, equivalently the Dwyer-Kan localization of the category of simplicial presheaves by the objectwise weak equivalences. Now, if B also satisfies the descent condition for being a higher stack, then $Hom_{\mathrm{diag}}(A, B). \sim Hom_{\mathrm{STACK}}(A, B)$. will also be the correct simplicial mapping space for the Jardine model structure. Technically speaking, if we are working with non-truncated ∞-groupoids here, then B should be required to satisfy *hyperdescent*, i.e. descent for all hypercoverings rather than just for all covering sieves. This distinction is played out in [**94, 95**]. For n-prestacks of groupoids, either the notion of Illusie weak equivalence, or the descent condition equivalently serve to define the homotopy theory of n-stacks.

16. Notions of effective descent for $(\infty, 1)$-prestacks

Suppose Φ is a site with Grothendieck topology $\mathscr{T}$. There are two distinct notions of descent for $(\infty, 1)$-prestacks. These are also distinct for $(\infty, 0)$-prestacks, however they become the same if we impose a finite level of n-truncation. The distinction turns on whether we require descent for hypercoverings, or only for usual covering sieves.

Recall the notion of "Illusie weak equivalence" [**76**] [**81**] [**79**]. If F is a simplicial presheaf or $(\infty, 0)$-stack over a site Φ, the *homotopy group presheaves* are just the most basic families of homotopy groups which can be defined. Thus, $\pi_0^{\mathrm{pre}}(F)$ is the presheaf $Y \mapsto \pi_0(F(Y))$. For $i \geq 1$ basepoints are needed. So, consider $X \in \Phi$ and suppose $x \in F(X)$ is a vertex (or object in the $(\infty, 0)$-category viewpoint). We get a presheaf of groups $\pi_i^{\mathrm{pre}}(F, x)$ on Φ/X by saying that for any $Y \to X$ in Φ/X,

$$\pi_i^{\mathrm{pre}}(F, x)(Y) := \pi_i(F(Y), x|_Y).$$

The homotopy group presheaves don't take into account the topology on Φ. Define the *homotopy group sheaves* $\pi_0(F)$ and $\pi_i(F, x)$ by sheafifying the above presheaves with respect to the topology.

Now, a morphism of simplicial presheaves $f : F \to G$ over a site Φ is an *Illusie weak equivalence* if the induced maps of sheaves

$$\pi_0(F) \to \pi_0(G),$$

and, for any $i \geq 1$ and any choice of $X \in \Phi$ and $x \in F(X)$,

$$\pi_i(F, x) \to \pi_i(G, f(x)),$$

are isomorphisms.

If Φ has enough points, one can form the "stalk" of a simplicial presheaf at a point P, and $F \to G$ is an Illusie weak equivalence if and only if it induces weak equivalences of simplicial sets $F_P \to G_P$ for all points P [**79**].

Suppose Φ is a site and $X \in \Phi$. A simplicial presheaf H on Φ/X is a *weak hypercovering* (of X) if the map $H \to *_{\Phi/X}$ is an Illusie weak equivalence; and it is a *hypercovering* if for any simplicial level $n \in \Delta$, the map $H_n \to \mathrm{csk}_n(H)_{[n]}$ is a surjection of sheaves (i.e. a "local epimorphism" [**101**]). Being a hypercovering implies being a weak hypercovering, and the nerve of a usual Čech covering is a hypercovering; similarly the characteristic presheaf of a covering sieve $\mathcal{B} \subset \Phi/X$ is a hypercovering. We call either of these last two examples of hypercoverings, just *usual coverings*.

Suppose H is at least a weak hypercovering of $X \in \Phi$. Let Φ/H denote the integral of H, which is a fibered category lying over Φ/X. Let $\mathcal{F}$ be a fibered $(\infty, 1)$-category over Φ. We get a map

$$\Gamma^{\mathrm{cart}}(\Phi/X, \mathcal{F}) \to \Gamma^{\mathrm{cart}}(\Phi/H, \mathcal{F}).$$

Say that $\mathcal{F}$ *satisfies effective descent with respect to* H if this map is an equivalence of $(\infty, 1)$-categories.

Say that $\mathcal{F}$ is an $(\infty, 1)$*-stack* over Φ, if it satisfies effective descent with respect to all usual coverings; and that it is an $(\infty, 1)$*-hyperstack* if it satisfies effective descent with respect to all hypercoverings. The latter condition is the same whether we use all weak hypercoverings, or just hypercoverings themselves.

These conditions are invariant under morphisms which are levelwise weak equivalences. They apply to presheaves of $(\infty, 1)$-categories or of $(\infty, 0)$-categories or simplicial presheaves since all of these give rise to split fibered categories.

In the case of $(\infty, 0)$-stacks identified with simplicial presheaves, the category of simplicial presheaves has the *Jardine model structure* [**79**] in which the weak equivalences are Illusie weak equivalences, and the cofibrations are all injections (i.e. monomorphisms). If F is a simplicial presheaf, then F is a hyperstack if and only if for one or any Illusie weak equivalence $F \to F'$ to a fibrant object in the Jardine model structure, $F(Y) \sim F'(Y)$ for all $Y \in \Phi$ (i.e. the fibrant replacement map is a levelwise weak equivalence).

Lurie points out [**94**] that there is similarly a model structure which has the same property with respect to stacks. It is characterized by being a simplicial model category in which the fibrant objects are the ones which are stacks and which are fibrant for the injective model structure with levelwise weak equivalences (call this the "injective prestack structure"). A map of simplicial presheaves $F \to G$ is a *Lurie weak equivalence* if, for any stack H fibrant in the injective prestack structure, the induced morphism on mapping spaces $Hom(G, H) \to Hom(F, H)$ is a weak equivalence. The cofibrations are again injections, and this gives the Lurie model structure. Lurie weak equivalences between stacks are levelwise weak equivalences, so the characterization given above for hyperstacks using the Jardine structure, transposes to stacks using the Lurie structure.

These things can be extended to $(\infty, 1)$ or indeed (∞, n)-prestacks, following a discussion such as in [**70**]. Note however that we were unaware of the stack/hyperstack distinction at the time of writing [**70**]. It was pointed out by Dan Dugger as a mistake in the first version, although Constantin Teleman had made a number of valiant efforts in that direction too.

One of the reasons why this distinction didn't seem immediately obvious was that it disappears as soon as the objects in question are n-truncated for any finite n. A space or $(\infty, 0)$-category is said to be n-truncated if its homotopy groups vanish in degrees $i > n$. An $(\infty, 1)$-category A is *n-truncated* if, for any pair x, y the $A(x, y)$ is an $(n-1)$-truncated space[3]. Terminologically speaking, an n-truncated $(\infty, 1)$-category is the same thing as an n-category such that the morphism $(n-1)$-categories are $(n-1)$-groupoids; and an n-truncated $(\infty, 0)$-category is the same thing as an n-groupoid.

A prestack is said to be n-truncated, or an n-prestack, if its values are n-truncated. For an n-truncated $(\infty, 1)$-prestack (or n-prestack or n-truncated $(\infty, 0)$-prestack or simplicial presheaf), being a stack and being a hyperstack are equivalent.

In the above discussion, we are using implicitly statements saying things of the form, an $(\infty, 0)$-prestack is a stack, or hyperstack, if and only if it is one when considered as an $(\infty, 1)$-prestack. Of course these kinds of statements would need to be made explicit for a more technical presentation.

17. Descent for n-stacks

The $(n+1)$-prestack of n-stacks, is actually an $(n+1)$-stack. This is perhaps the most fundamental type of descent statement. It was the subject of my work with André Hirschowitz [**70**], which started out with his encouragement that we should work toward a theory of higher stacks in order to be able to discuss the ∞-stack **Perf** of perfect complexes.

Recall a classical fact about sheaves: the 1-prestack of sheaves is a 1-stack. In concrete terms, this says that we can *glue together* sheaves as discussed in the section on glueing above: if $\{U_i \to X\}$ is an open covering

[3]Here, for $n = 0$ one must introduce the terminology of a -1-truncated space as one which is either $\emptyset$ or $*$; that can be continued to saying that -2-truncated means being exactly $*$

of a topological space, given sheaves $\mathcal{F}_i$ on U_i, and isomorphisms ϕ_{ij} : $\mathcal{F}_i|_{U_{ij}} \xrightarrow{\cong} \mathcal{F}_j|_{U_{ij}}$ satisfying for all i, j, k the cocycle condition $\phi_{jk}\phi_{ij} = \phi_{ik}$ when everything is restricted to the triple intersection U_{ijk}, then there is a global sheaf $\mathcal{F}$ on X and isomorphisms $\mathcal{F}|_{U_i} \cong \mathcal{F}_i$ compatible with the ϕ_{ij}. And there is a similar descent statement for morphisms between sheaves, expressed abstractly as saying that the mapping presheaf is a sheaf[4].

The collection of descent conditions defining the notion of 1-stack is just an abstraction of these descent properties of sheaves.

Once we introduce the generalization of 1-stacks to n-stacks, it is completely natural to expect that the corresponding glueing property holds. Perhaps the original instance of this statement in the context of higher stacks was Breen's remark in [**26**] that 2-stacks satisfy descent to form a 3-stack. And the general statement is indeed the case: with Hirschowitz in [**70**] we show that the $(n+1)$-prestack of n-stacks $nSTACK$ is an $(n+1)$-stack. Given a site $(\Phi, \mathscr{T})$, the association

$$\text{nStack} : x \mapsto \{n\text{-stacks on } \Phi/x\}$$

is a strict functor from Φ^o to the category $(n+1)Cat$ of fibrant $(n+1)$-categories, so we can in particular think of it as a weak functor from Φ^o to $(n+1)CAT$, i.e. an $(n+1)$-prestack.

THEOREM 17.1 ([**70**]). *The functor $x \mapsto \{n\text{-stacks on } \Phi/x\}$ is in fact an $(n+1)$-stack.*

The proof is pretty much analogous to the proof for sheaves, or to the proof that one could give that we get a 2-stack of 1-stacks.

This leaves open the question of descent for (∞, n)-stacks and (∞, n)-hyperstacks. This becomes more delicate, because of the distinction between stacks and hyperstacks. I won't venture beyond what is written in the current version of [**70**], and the reader is asked to refer there.

We have a similar descent result for complexes: the ∞-prestack of complexes of sheaves, is an ∞-stack. Some care may be required when considering unbounded complexes on spaces not of finite cohomological dimension; this case is not covered by our paper [**70**], and we will not discuss boundedness hypotheses here.

The case of complexes serves to illustrate why it is important to introduce ∞-categories. Recall that the usual *derived category* $D(X)$ is the homotopy category of the category of complexes of sheaves of abelian groups over a space X. It is the 1-categorical Gabriel-Zisman localization of the category of complexes, by inverting the quasiisomorphisms. This by existence of a model category structure on the category of unbounded complexes [**75**]; in the classical literature one usually passes first to a subcategory such as

[4]The reader may notice that the glueing property also holds in the category of presheaves; however, maps between presheaves don't satisfy a glueing property, so it is actually the descent conditions for morphisms that set apart the stack of sheaves.

the injective complexes, in order to obtain a good calculus of fractions for the localization.

The derived category may be seen as the 1-truncation of the ∞-category $CP(X)$ obtained by Dwyer-Kan simplicial localization instead of 1-categorical localization:

$$D(X) = \tau_{\leq 1}(CP(X)).$$

Suppose we are in the simplest situation of a covering of a space X by two open sets $X = U \cup V$. In general, the diagram

$$\begin{array}{ccc} D(X) & \rightarrow & D(U) \\ \downarrow & & \downarrow \\ D(V) & \rightarrow & D(U \cap V) \end{array}$$

is not a homotopy-cartesian diagram of categories. This says that the association $U \mapsto D(U)$ is not a 1-stack. On the other hand, by going to the ∞-categorical level, the diagram

$$\begin{array}{ccc} CP(X) & \rightarrow & CP(U) \\ \downarrow & & \downarrow \\ CP(V) & \rightarrow & CP(U \cap V) \end{array}$$

is a cartesian diagram of ∞-categories: the association $U \mapsto CP(U)$ is an ∞-stack. The truncation operation $\tau_{\leq 1}$ doesn't preserve limits.

A very recent reference is Paulin's paper [**109**], using this higher categorical glueing property to do the theory $\mathcal{D}$-modules on smooth algebraic stacks.

18. Artin n-stacks

Let Φ be the site of schemes of finite type over a ground ring k with the etale topology. A sieve $\mathscr{B} \subset \Phi/X$ is covering, if there exists a covering of X by a collection of etale maps $U_i \rightarrow X$ each contained in $\mathscr{B}$.

Artin's representability theorem led him to formulate a property of stacks which he called "algebraic stacks". This terminology has sometimes taken on various different meanings, and it is now more usual to call them *Artin stacks*. The basic idea is that an Artin stack $\mathscr{F}$ is one which has a chart

$$Z \rightarrow \mathscr{F}$$

from a scheme Z of finite type. Here a scheme $Z \in \Phi$ represents a functor which is a sheaf or 0-stack on Φ, which can then be considered also as a 1-stack of discrete groupoids.

To give the definition one then needs to say what it means for a morphism to be a "chart". The definition involves the *fiber product*: if $\mathscr{F} \rightarrow \mathscr{G} \leftarrow \mathscr{H}$ are morphisms of prestacks of groupoids, then the fiber product denoted

$$\mathscr{F} \times_{\mathscr{G}} \mathscr{H}$$

is the prestack of groupoids whose fiber over an $X \in \Phi$ is the "homotopy fiber product" $\mathscr{F}_X \times^h_{\mathscr{G}_X} \mathscr{H}_X$. In turn the homotopy fiber product of groupoids is the groupoid whose objects are triples (a, b, γ) where $a \in \mathscr{F}_X$, $b \in \mathscr{H}_X$ and γ is an isomorphism in $\mathscr{G}$ between the images of a and b.

When $\mathscr{F}$, $\mathscr{G}$ and $\mathscr{H}$ are stacks, then so is the fiber product. This generalizes the usual property that limits of sheaves (such as the kernel of a map between sheaves) remain sheaves. The reason is that the fiber product is a 2-limit, the stack property may be expressed in terms of 2-limits, and 2-limits commute with other 2-limits.

With the notion of fiber product, say that

$$\mathscr{F} \to \mathscr{G}$$

is a *very representable morphism* (resp. *representable morphism*) between stacks of groupoids, if for any $Y \in \Phi$ and any map $Y \to \mathscr{G}$ (equivalent to an object of $\mathscr{G}_Y$), the fiber product

$$\mathscr{F} \times_{\mathscr{G}} Y$$

is represented by a scheme of finite type (resp. an algebraic space). The projection to Y is then a morphism of schemes (resp. algebraic spaces). A representable morphism is *smooth* if for any $Y \to \mathscr{G}$ the resulting morphism of schemes

$$\mathscr{F} \times_{\mathscr{G}} Y \to Y$$

is smooth. A *smooth chart* for a stack $\mathscr{F}$ is a representable smooth morphism $Z \to \mathscr{F}$ such that Z itself is the 0-stack represented by a scheme.

A 1-stack of groupoids $\mathscr{F}$ over Φ is an *Artin 1-stack locally of finite type* (resp. *of finite type*) if there exists a family (resp. finite family) of smooth charts $Z_i \to \mathscr{F}$ which covers $\mathscr{F}$ (i.e. such that any map from a scheme $Y \to \mathscr{F}$ admits, locally on Y in the etale topology, liftings to the Z_i).

Artin's definition generalizes readily to the case of $(n, 0)$-stacks, i.e. stacks of n-groupoids. This was first suggested to me by Charles Walter for $n = 2$, and that was quite sufficient for understanding the general case [**127**].

For this iteration, we need to know how to take the homotopy fiber product of $(n, 0)$-stacks. One way of doing this is by using the model category of simplicial presheaves: the homotopy fiber product is obtained by replacing at least one of the maps by a fibration and then taking the usual fiber product. More abstractly we can consider it as just a limit in the $(n+1, 1)$-category of $(n, 0)$-stacks.

An $(n, 0)$-stack is -1-*Artin* if it is in fact a 0-stack (i.e. a sheaf of sets) represented by a scheme of finite type. We know what it means for a morphism between these to be smooth. Starting from there, make the following definitions for $(n, 0)$-stacks by induction on m:

- a morphism $\mathscr{F} \to \mathscr{G}$ is *m-representable* if for any scheme $Y \in \Phi$ and map $Y \to \mathscr{G}$, the fiber product $\mathscr{F} \times_{\mathscr{G}} Y$ is m-Artin;
- an m-representable morphism is *smooth* if, with the same notations, for any $Y \to \mathscr{G}$ the projection $\mathscr{F} \times_{\mathscr{G}} Y \to Y$ is a smooth morphism of m-Artin stacks;

- an $(n,0)$-stack $\mathscr{F}$ is *m-Artin* if there exists a scheme Z and a map $Z \to \mathscr{F}$ which is $(m-1)$-representable and smooth;
- a morphism $\mathscr{F} \to Y$ from an m-Artin $(n,0)$-stack to a scheme, is *smooth* if for the smooth covering $Z \to \mathscr{F}$ considered previously, the composed map $Z \to Y$ is a smooth map of schemes.

The above mutually inductive schema of definitions is recursive and serves to define all of the terms. An $(n,0)$-stack is Artin if it is m-Artin for some m (and m may be taken as at most $n+2$). Any fiber product of Artin $(n,0)$-stacks is again Artin, in particular, any morphism between Artin $(n,0)$-stacks is $n+2$-representable.

The above collection of definitions is made for the case of Artin n-stacks of finite type; they can be modified to obtain a definition of Artin n-stack locally of finite type. The covering chart $Z \to \mathscr{F}$ is replaced by an arbitrary family of $X_i \to \mathscr{F}$.

Toën has shown in [**146**], following Laumon and Moret-Bailly [**89**] for the case $n=1$, that if we make the same inductive definition as above, but using flat rather than smooth morphisms, then we obtain the same notion of Artin n-stack. In other words, having flat charts implies having smooth charts. This shows that the definition is quite naturally since it is stable in a strong sense.

Pridham has recently shown that the notion of Artin n-stack can be described in a different way [**115**]. Namely, an n-stack is Artin if and only if there exists a presentation by a simplicial scheme satisfying a certain smoothness condition, expressed briefly as saying that the maps which are required to be surjective in Kan's condition for simplicial sets, should actually be smooth and surjective. This condition was, in fact, suggested by Grothendieck in [**68**]. Benzeghli has constructed a canonical model of this form for the stack of perfect complexes and similar examples [**16**]. See also Hollander's papers [**72**], [**73**], and Wolfson [**160**].

The relationship between stacks and complexes was pointed out in [**89**], taken up more recently for length 3 complexes in [**134**]. The total space of a perfect complex of amplitude n is an interesting first example of an Artin n-stack.

19. Higher nonabelian cohomology

One of the main propositions advanced by Grothendieck in [**68**], was the notion the n-stacks over a site should form the most natural systems of coefficients for *higher nonabelian cohomology*. To understand the geometric reasoning behind this idea, let's go back to the "representability" point of view on cohomology. For any $n \geq 0$, if A is an (abelian for $n \geq 2$) group, then we can form a pointed space $(K(A,n),o)$, which by the way admits a good explicit description as a simplicial set, such that $\pi_i(K(A,n),o)=0$ for $i \neq n$ and $\pi_n(K(A,n),o)=A$. This space represents cohomology: for any

CW-complex X, there is a natural isomorphism

$$H^n(X, A) \cong [X, K(A, n)]$$

between the n-th cohomology of X with coefficients in A, and the set of homotopy classes of maps from X to $K(A, n)$.

More general "cohomology theories" are obtained by looking at $[X, S]$ where S is a *spectrum*, which is some kind of a limit of a tower of spaces, in which the composition of loops is as abelian as possible. Roughly speaking, a spectrum is a "stable homotopy type", in which the homotopy-theoretical phenomena occuring in the stable range of homotopy groups persist, but the phenomena of the unstable range are pushed away.

Starting from here, we can conceptualize the notion of *nonabelian cohomology* as something represented by an arbitrary space T rather than a spectrum; the nonabelian cohomology with coefficients in T is just defined to be the functor $X \mapsto [X, T]$. This may be further redefined in terms of ∞-groupoids: if T is an ∞-groupoid then we obtain the ∞-groupoid $\underline{Hom}(X, T)$ whose π_0 is $[X, T]$.

Although the previous paragraph should be considered as entirely tautological, it nonetheless paves the way for an important extension. Suppose Φ is a Grothendieck site. If $\mathscr{F}$ is an ∞-stack of ∞-groupoids over Φ, then $\Gamma(\Phi, \mathscr{F})$ is an ∞-groupoid which should be thought of as the *nonabelian cohomology of Φ with coefficients in $\mathscr{F}$*. By making the coefficient space vary in an ∞-stack over our site, we obtain a nontrivial notion of "nonabelian cohomology". More generally, if $\mathscr{X}$ is a sheaf, a 1-stack or even another ∞-stack on Φ, then the ∞-category $Hom(\mathscr{X}, \mathscr{T})$ or its internalized version the ∞-stack $\underline{Hom}(\mathscr{X}, \mathscr{T})$, may be called the *nonabelian cohomology* of $\mathscr{X}$ with coefficients in $\mathscr{T}$.

The case when our ∞-stack is a simplicial presheaf whose values are actually spectra already played an important role in Thomason's work [**136**], and had been investigated by Joshua [**80**].

It is instructive to consider how this notion of nonabelian cohomology links up with Giraud's nonabelian 2-cohomology [**58**]. Consider $1STACK$, the 2-stack classifying 1-stacks on the site Φ. This associates to each $U \in \Phi$ the 2-category of 1-stacks (say, of groupoids) on the relative site Φ/U. The prestack defined in this way satisfies effective descent, so it is a 2-stack [**26**].

Consider now the truncation $\tau_{\leq 0}(1STACK)$ to a sheaf of sets. A section $f \in \tau_{\leq 0}(1STACK)(U)$ is a specification of a compatible family of *equivalence classes* of stacks, locally over U. That is to say, f corresponds to giving an open covering $U_i \to U$ and for each U_i, an equivalence class of stacks $f(U_i)$ on Φ/U, up to *local equivalence* (two stacks being locally equivalent if there is a covering over which they are equivalent), such that the restrictions of the equivalence classes $f(U_i)$ and $f(U_j)$ are locally equivalent on U_{ij}. A global section f of the sheaf $\tau_{\leq 0}(1STACK)$ corresponds to what Giraud called a *band*. A stack F on Φ is *bound* by the band f, if for each U the restriction $F|_{\Phi/U}$ is locally equivalent to $f(U)$.

If we let $1STACK^{\text{int}}$ denote the *interior stack of groupoids*, then there is a morphism $1STACK^{\text{int}} \to \tau_{\leq 0}(1STACK)$. One should be careful that no such morphism exists starting from $1STACK$ itself, due to the presence of morphisms which are not equivalences, which would have nowhere to go in $\tau_{\leq 0}(1STACK)$.

Fix a band f, and consider the full sub-2-stack of groupoids $1STACK^f \subset 1STACK^{\text{int}}$ classifying stacks bound by f. It fits into the cartesian commutative diagram

$$\begin{array}{ccc} 1STACK^f & \to & 1STACK^{\text{int}} \\ \downarrow & & \downarrow \\ * & \xrightarrow{f} & \tau_{\leq 0}(1STACK). \end{array}$$

In the situation of [**58**], we are usually furthermore considering stacks F which are *connected*, i.e. $\pi_0(F) = 0$. This is a local property stable under equivalence, so it corresponds to a subsheaf of $\tau_{\leq 0}(1STACK)$, whose pullback is the full sub-2-stack $GERB \subset 1STACK^{\text{int}}$ of *gerbs*, i.e. connected 1-stacks of groupoids. A gerb F is determined, up to local equivalence, by its sheaf of groups $\pi_1(F)$. To state this more precisely, let GRP denote the 1-stack of groupoids of sheaves of groups, i.e. $GRP(U)$ is the category of sheaves of groups on Φ/U and morphisms the isomorphisms. Its truncation $\tau_{\leq 0}(GRP)$ is the sheaf whose sections over U are the local equivalence classes of sheaves of groups. We have

$$\tau_{\leq 0}(GERB) \cong \tau_{\leq 0}(GRP),$$

because locally a gerb is determined by its isomorphism class of groups. These are subsheaves of $\tau_{\leq 0}(1STACK)$, so we can say that a band f is a *band for gerbs* if it is in this subsheaf. The sheaf of bands for gerbs is isomorphic to the sheaf of local equivalence classes of sheaves of groups $\tau_{\leq 0}(GRP)$. Again define the full sub-2-stack $GERB^f$ consisting of the sections which project to f, and it is equivalent to $1STACK^f$. Again we have a cartesian square

$$\begin{array}{ccc} GERB^f & \to & GERB \\ \downarrow & & \downarrow \\ * & \xrightarrow{f} & \tau_{\leq 0}(GRP). \end{array}$$

In particular, $GERB^f$ (and more generally any $1STACK^f$) is *connected*, that is to say

$$\tau_{\leq 0}(GERB^f) = *.$$

What does the higher structure of $GERB^f$ look like? To simplify things, we assume that there exists a section of GRP corresponding to f, that is to say, assume that there exists a sheaf of groups G representing the equivalence class f. Then, the classifying space construction BG yields a section $\sigma \in \Gamma(\Phi, GERB^f)$. As $GERB^f$ is a 2-stack, it has only two nontrivial homotopy group sheaves, and they are given by

$$\pi_1(GERB^f, \sigma) = \text{Out}(G),$$

$$\pi_2(GERB^f, \sigma) = Z(G).$$

These notations are defined as follows. Let $\mathrm{Aut}(G)(U)$ be the group of automorphisms of $G|_{\Phi/U}$; these form a sheaf of groups $\mathrm{Aut}(G)$, which by the way is $\pi_1(GRP^f, G)$. Inner automorphisms give a map $G \to \mathrm{Aut}(G)$. The *sheaf of groups of outer automorphisms* $\mathrm{Out}(G)$ is the quotient sheaf $\mathrm{Out}(G) := \mathrm{Aut}(G)/G$, and the *center* is the kernel $Z(G) := \ker(G \to \mathrm{Aut}(G)$. If we think of the morphism $G \mapsto BG$ as a fibration between 2-stacks of groupoids $GRP^f \to GERB^f$, the homotopy fiber over σ is BG itself, and the long exact sequence of homotopy group sheaves comes from the above map:

$$\begin{array}{ccccccccc} \pi_2(GRP^f, G) & \to & \pi_2(GERB^f, \sigma) & \to & \pi_1(BG, o) & \to & \pi_1(GRP^f, G) & \twoheadrightarrow & \pi_1(GERB^f, \sigma) \\ \| & & \| & & \| & & \| & & \| \\ 0 & \to & Z(G) & \to & G & \to & \mathrm{Aut}(G) & \twoheadrightarrow & \mathrm{Out}(G). \end{array}$$

The precise 2-stack structure of $GERB^f$ may be seen as corresponding to the classical *crossed module* structure on the map $G \to \mathrm{Aut}(G)$.

The group of automorphisms $\mathrm{Aut}(BG)$ of the stack BG is itself a stack, with a group structure. We can therefore take its classifying 2-stack $B(\mathrm{Aut}(BG))$.

PROPOSITION 19.1. *Assuming that f corresponds to a global sheaf of groups, $GERB^f \cong B(\mathrm{Aut}(BG))$.*

Giraud's *nonabelian 2-cohomology* classifying the gerbs bound by f, may be seen as the 2-*groupoid of sections* of the 2-stack in groupoids $GERB^f$. This extends to the classification of gerbs over any object of the topos corresponding to Φ.

PROPOSITION 19.2. *If X is a sheaf or even a stack on Φ, we obtain the site Φ/X. The 2-groupoid of gerbs over Φ/X bound by $f|_{\Phi/X}$ is equivalent to the nonabelian cohomology 2-groupoid $Hom(X, GERB^f)$.*

The nonabelian cohomology 2-groupoid may be approached by looking at the Postnikov tower for $GERB^f$. Maintaining the assumption of a global sheaf of groups G, the Postnikov tower is the fibration sequence

$$\begin{array}{ccc} K(Z(G), 2) & \to & GERB^f \\ & & \downarrow \\ & & K(\mathrm{Out}(G), 1). \end{array}$$

We get a map $Hom(X, GERB^f) \to H^1(X, \mathrm{Out}(G))$, the target being seen as a 1-groupoid. The fiber over the trivial section is $H^2(X, G)$; over a nontrivial section we get a twisted 2-cohomology.

To motivate the extension of nonabelian cohomology to higher stacks, it is worth mentioning that higher stacks give us a way of thinking about generalized spaces of sections. Suppose X is a topological space. If $Y \to X$ is a fibration, then we may consider the following ∞-stack $\underline{\mathrm{Sect}}(Y/X)$ on X: it takes an open set $U \subset X$ to the space of sections $U \to Y$ over X. It turns out that this satisfies the homotopy glueing property so it is an ∞-stack. If the fibers of Y/X are n-truncated then the spaces $\underline{\mathrm{Sect}}(Y/X)$ are n-truncated so we get an n-stack of n-groupoids. We may roughly speaking think of any n-stack as being something like a stack of sections of something. Our $\underline{\mathrm{Sect}}(Y/X)$ has the following additional property: its homotopy group sheaves are locally constant. The study of higher stacks with this property, in particular proving the descent property for the space of sections, has been done by Toën [**145**], Shulman [**125**], Cisinki [**36**], Lurie [**95**], and others.

"Cohomological descent" is the subject of a significant literature, starting from [**121**]. We can use the notion of higher stack, from the viewpoint of nonabelian cohomology, to give a very natural statement of this idea. Given an n-truncated space T, let $\underline{T}$ be the n-prestack on the category $\mathrm{Sch}_{\mathbb{C}}$ of schemes of finite type over $\mathbb{C}$, which to each such scheme X associates the space of maps from X^{top} to T. Then we say that a Grothendieck topology on $\mathrm{Sch}_{\mathbb{C}}$ satisfies *nonabelian cohomological descent* if for any T the n-prestack $\underline{T}$ is an n-stack, i.e. satisfies descent for that topology.

THEOREM 19.3. *The Zariski, etale (hence Nisnevich), fppf or equivalently fpqc, and specially the proper-etale topologies, satisfy nonabelian cohomological descent.*

Taking $T = K(A, n)$ yields the usual notion of cohomological descent. This formulation explains, if nothing else, the origin of the terminology "cohomological descent". It indicates that somehow, the higher-stack way of thinking was already present from the beginning in the SGA.

The reader may extend this to various other types of coefficients such as locally constant or constructible n-stacks. For $n = \infty$ we may run up against questions about the distinction between descent and hyperdescent, upon which I am not qualified to comment.

Grothendieck mentions, in just a single intriguing half-phrase in [**68**], the possible utility of the notion of *constructible stacks.* These generalize locally constant stacks just as constructible sheaves generalize locally constant sheaves. Very roughly speaking, they serve to classify families of objects which are not really fibrations in a homotopical sense, but which may have jumps along a stratification. The serious development of this theory has now begun with work of Treumann [**149**], and also in Dupont's thesis [**49**]. Treumann introduces the notion of *exit-path categories* as a way of categorifying the topological information contained in a stratification. An important direction for the future will be to integrate this notion into the theory of nonabelian cohomology, and in particular to see what kind of Hodge-theoretic information should go along.

Arriving in Toulouse, I had the great fortune to make contact with some people from the Ehresmann school: Jean Pradines and Joseph Tapia, together with many members of the Toulouse department interested in various aspects of foliation theory. We are concentrating here on Grothendieck's point of view on higher stacks and categories, but Ehresmann and his students were also going toward a very similar picture, and the eventual communication between these streams leads to great progress.

Among other things, Ehresmann's viewpoint is much more resolutely differential-geometric. The association

$$\{\text{smooth algebraic varieties}/\mathbb{C}\} \to \{\mathscr{C}^\infty \text{ manifolds}\}$$

means that higher stacks in algebraic geometry will lead to higher differential stacks. Pradines was one of the first to study higher differential stacks, and Tapia cast these into a topos-theoretic framework. There are several present currents of thought directed toward the study of "higher monodromy" for topological or differential stacks. See for example Breen and Messing [**28**], Polesello and Waschkies [**110**], Aldrovandi and Noohi [**3**] [**102**], the work of Aaron Smith and Jonathan Block [**23**] on higher connections and differential cocycles, together with Schreiber and Waldorf [**124**] in the same spirit (as well as several other papers by Schreiber and various co-authors which the reader may look up), and Chenchang Zhu's work [**161**] on stacky groupoids. Their introduction of higher cocycles might be viewed, to some extent, as generalizing the 2 and 3-cocycles considered by Breen [**26**]. No formal comparison has yet been made with the model-category approach, but Wirth's 1965 thesis [**159**] represents a contribution which undoubtedly would have initiated a lot of progress relating higher cocycles and homotopy theory, had it been better known.

In the world of algebraic geometry, the interplay between the homotopical directions, represented by the level of higher arrows, and the geometrical directions, comes about through the choice of underlying site. When we envision the study of nonabelian cohomology, an addition choice of the *kind of coefficient stacks* becomes important. This may be seen at the level of sheaves of groups G and their classifying stacks BG. If we choose to look at locally constant finite sheaves of groups over the etale site, then $Hom(X, BG)$ is the *etale cohomology of* X *with coefficients in* G, classifying finite G-torsors over X. Taken as a functor in the variable G, this data determines the etale fundamental group.

On the other hand, we might choose to let G be an affine algebraic group, considered as a representable sheaf over one of the big sites. In this case, $Hom(X, BG)$ classifies principal G-bundles over X; for example when $G = GL(r)$ it classifies vector bundles of rank r.

The notion of Artin n-stack provides one of the important kinds of coefficient stacks T to consider, generalizing the BG for affine algebraic groups G. For good domain objects $\mathscr{X}$, and possibly with further restrictions on T, the nonabelian cohomology stack $\underline{Hom}(\mathscr{X}, T)$ will again be an Artin n-stack. This allows us to envision doing geometry with these entities. Considered as a functor in the variable T, the nonabelian cohomology defines a theory of the *shape* of the domain stack $\mathscr{X}$. The *de Rham shape* of a smooth variety X is obtained by considering

$$\mathscr{X} := X_{\mathrm{dR}} : Y \mapsto X(Y^{\mathrm{red}}).$$

For $T = BG$ with G a linear algebraic group, the nonabelian cohomology $\underline{Hom}(X_{\mathrm{dR}}, T)$ is the classical moduli stack of principal G-bundles with flat connection. For certain more general Artin n-stacks T, the higher stack $\underline{Hom}(X_{\mathrm{dR}}, T)$ should retain a geometrical structure and we can hope that it is again an Artin n-stack. This was shown for $T = \mathbf{Perf}$ in [**128**].

One of the recurrent themes of "La poursuite des champs" is the *schematization of homotopy types*. Modeled on *rational homotopy theory*, the aim is to import the notion of homotopy type into the world of schemes. Very roughly speaking, a schematic homotopy type is a higher stack which would have something like an underlying scheme for π_0, then whose π_1 would be something like a group scheme over π_0, and whose higher π_i would be abelian group schemes with action of π_1. Breen and Ekedahl made a start in this direction [**27**].

Higher Artin stacks provide a fairly good notion of schematic homotopy type. However, as may already be seen from the case of Artin 1-stacks, the different homotopy levels are connected together. In particular, when T is an Artin stack, $\pi_0(T)$ will not be a scheme or algebraic space, as the truncation destroys the interconnections between the levels. A fully splittable theory, stable under truncation and Postnikov towers, may also be constructed [**69**] [**126**].

When $\pi_0 = *$ the splitting question doesn't arise, and there is a single natural notion of schematic type. It is often useful to impose some further conditions, that π_1 be an affine group scheme and the higher π_i be unipotent abelian group schemes. Grothendieck asked whether we could "schematize" a classical homotopy type. Toën defines a functor denoted $X \mapsto X \otimes k$ from spaces to schematic homotopy types over a field k, with the universal property that if T is any other k-schematic homotopy type, then the maps from X to T are the same as the maps from $X \otimes k$ to T. As shown in Pridham's approach [**113**], the schematization generalizes the process of taking a rational homotopy type, representing it by a dga $A^{\cdot}$ over $\mathbb{Q}$, and then taking $A^{\cdot} \otimes_{\mathbb{Q}} k$.

The schematization process goes to the heart of why one might be interested in higher stack theory for complex algebraic geometry. The point is that Hodge-theoretical invariants are attached to *complexified* topological invariants of X. For example

$$H^k(X^{\mathrm{top}}, \mathbb{Q}) \otimes_{\mathbb{Q}} \mathbb{C} = \bigoplus_{p+q=k} H^{p,q}(X)$$

when X is a smooth projective complex variety. In order to go toward Hodge theory for nonabelian topological invariants, we need to be able to cast these nonabelian invariants into a complexified world, that is to say we need to be able to apply $- \otimes_{\mathbb{Q}} \mathbb{C}$. Katzarkov, Pantev and Toën give a nonabelian Hodge structure on $X \otimes_{\mathbb{Q}} \mathbb{C}$ [**84**], generalizing the mixed Hodge structure on rational homotopy types. Pridham investigates further the resulting structures in [**114**].

Schematizing the homotopy type itself corresponds to acting on the homological side. A different way is to consider nonabelian cohomology with coefficients in a complexified schematic homotopy type T. A conjectural picture of nonabelian mixed Hodge structures on higher stacks of the form $\underline{Hom}(X^{\mathrm{top}}, T)$ is developed in [**83**].

In either picture, the passage from homotopy types to higher stacks over the site $\mathrm{Sch}_{\mathbb{C}}$ is crucial. On the level of 1-stacks, an affine algebraic group G gives rise to a stack BG, and de Rham cohomology of X with coefficients in G is the mapping stack $\underline{Hom}(X_{\mathrm{dR}}, BG)$ where $X_{\mathrm{dR}} : Y \mapsto X(Y^{\mathrm{red}})$. As we said above, if T is a higher stack satisfying some kind of algebraicity condition, such as an Artin n-stack, then $\underline{Hom}(X_{\mathrm{dR}}, T)$ gives a good notion of de Rham cohomology with coefficients in T.

The relationship between the schematization and nonabelian cohomology, is given by a *universal property* of the schematic homotopy type: if X is a topological space, let $\underline{X}$ denote the constant presheaf of ∞-groupoids whose values are the ∞-groupoid corresponding to X. We have a morphism of ∞-stacks

$$\underline{X} \to X \otimes k,$$

which is universal for maps from X to schematic homotopy types T over k. That is to say, it induces an equivalence of mapping spaces

$$Hom(X \otimes k, T) \overset{\cong}{\to} Hom(\underline{X}, T).$$

Unfortunately, if X is not simply connected then this doesn't extend to an equivalence of mapping stacks: the schematization $X \otimes k$ doesn't "see" the geometric structure of the *space of representations* of $\pi_1(X)$ into a linear algebraic group G, or rather more precisely its corresponding Artin 1-stack $\underline{Hom}(\underline{X}, BG)$, which is the stack version of the famous *character variety.*

When X is simply connected, the above equivalence does extend to mapping stacks. The $\mathbb{Q}$-schematization $X \otimes \mathbb{Q}$ is essentially the same information as the *rational homotopy type* of X, and this functorial point of view serves to show how the rational homotopy type varies as a function of X, complete with all higher-level homotopy coherencies. It also gives a geometric interpretation to the "extension of scalars" going from $X \otimes \mathbb{Q}$ to $X \otimes \mathbb{C}$. Hodge theory for the schematization coincides with the previously known Hodge theory for the rational homotopy type [**84**].

The schematization of homotopy may also be thought of in terms of *higher Tannaka duality.* The schematic homotopy type $X \otimes k$ may be thought of as obtained by Tannaka duality from the ∞-$\otimes$-category of perfect complexes over X. We don't have room to discuss this in detail here, but it is worth mentioning that descent theory already played an important role in the original Tannakian theory [**120**] [**44**]. The higher categorical versions of

Lurie [**95**], Fukuyama and Iwanari [**54, 78**], Moriya [**100**], and Wallbridge [**157**] make essential use of the ∞-categorical and derived settings.

20. Derived stacks

The introduction of stacks meant that we could consider a new "circle", it is the stack $B\mathbb{Z} = K(\mathbb{Z}, 1)$, i.e. the stack associated to the constant prestack which associates to any object of our site, the space S^1. We can then consider the *free loop space*

$$\mathscr{K}X := \underline{Hom}(B\mathbb{Z}, X) = X \times_{X\times X} X.$$

If X is a usual scheme, we get back X itself. When X is a stack, this gives further information which is often known as the *inertia stack* of X.

Recent work of Ben-Zvi and Nadler [**13**] [**14**], Toën and Vezzosi [**147**], and others, has shown that this construction becomes really useful when it is done in the world of *derived stacks.*

In the derived point of view, originally introduced by Kontsevich and Kapranov, and developed by Toën-Vezzosi [**140, 141**] and Lurie [**95**], even if X is a smooth scheme, will yield an $\mathscr{K}X$ which, as a non-transverse fiber product, becomes a nontrivial derived scheme. A good introduction is Vezzosi's "What is..." [**152**].

Recall that our circle of ideas, as it relates to algebraic geometry over a field k, starts of with sheaves of sets which are functors

$$\mathrm{Alg}_k^{\mathrm{comm}} \to \mathrm{Sets}$$

from commutative k-algebras to sets. Going to stacks means we look at (possibly weak, but strictifiable) functors

$$\mathrm{Alg}_k^{\mathrm{comm}} \to \mathrm{Gpd}$$

now taking values in groupoids. Going to higher stacks means that we look at functors

$$\mathrm{Alg}_k^{\mathrm{comm}} \to (n,0)\mathrm{Gpd}$$

or (with appropriate care taken for the infinite range of homotopy groups)

$$\mathrm{Alg}_k^{\mathrm{comm}} \to (\infty,0)\mathrm{Gpd} \equiv \mathrm{Spaces} \equiv \mathrm{Sets}^{\Delta^o}.$$

In all of these steps, we have been including homotopy theory on the right side or target side of the functors. The basic idea of *Derived Algebraic Geometry* is to include homotopy theory also on the left or source side of the functors. Notice that a first basic requirement is that the representable functors should be included. If we include higher homotopies in the source category, then it is basically necessary to keep the highest level of generality for the target. So we are led to consider various kinds of functors of the form

$$\{\text{spectral algebras}\} \to \{\infty\text{-groupoids}\},$$

where "spectral algebras" mean commutative algebras in some kind of category of "spectra", with the meaning of this word being understood in the wide sense as any kind of stable homotopy object. Here there are many possible flavours, starting with the easiest case where "spectra" are just complexes. In that case, the source category is $CDGA$, the category of graded-commutative differential graded algebras. Other possible choices include cosimplicial algebras, simplicial-cosimplicial algebras, E_∞ ring spectra, and more general possibilities considered, in the most complete treatment, by Lurie in [**95**].

One of the main advantages of derived algebraic geometry is that it renders automatic the consideration of virtual intersections which replace non-transversal intersections. Thus, for example, when faced with a fiber product of schemes, if we take the fibered product in the world of derived algebraic geometry, it will give a "good answer" whatever the transversality properties of the intersection.

The theory of Artin n-stacks extends naturally into the world of derived algebraic geometry, as a way of explaining how to glue together representable objects into global ones. This leads to the notion of *geometric derived stack* [**141**]. The previous remark extends to this case: fibered products of geometric derived stacks are again geometric derived stacks which are supposed to have the best possible structure.

It is therefore most natural to use derived geometry to look at nonabelian cohomology: if X is a CW-complex with finitely many cells and T is a geometric derived stack, then the nonabelian cohomology stack $Hom(X,T)$ is again a geometric derived stack.

This is interesting even in the first basic case $X = S^1$: it leads to the "derived loop stack" $Hom(S^1,T)$ for any geometric derived stack T, studied by Ben-Zvi, Nadler, Francis, and others, and used by Toën and Vezzosi to study the Chern character.

We finish by describing a beautiful idea described to me by David Ben-Zvi. One of the main benefits of derived stacks over regular stacks, is that they allow for a good computation of non-transverse fiber products. Now, one of the most non-transverse fiber products possible is

$$N_1(X,x) := \{x\} \times_X \{x\}$$

where X is, say, a variety and $x \in X$ is a closed point. In usual geometry, of course, $N_1(X,x)$ would just be $\{x\}$ itself. However, if we define $N_1(X,x)$ to be the derived fiber product, then it is a derived scheme which retains information about the local structure of X infinitesimally near to the point x. We can continue to form a *simplicial derived scheme* $N.(X,x)$ with levels

$$N_k(X,x) := \{x\} \times_X \{x\} \times_X \cdots \times_X \{x\}$$

with $k+1$ factors in the fiber product. This is exactly the same kind of object as the usual nerve of a covering; but here we are "covering" X by the point x. The idea explained by Ben-Zvi is that the above simplicial object allows us to recover the formal completion $\widehat{X}^x$ of X at x by some kind of derived descent theory. Such a theory of descent for the "covering" $x \to \widehat{X}^x$ would constitute a fantastic generalization of Grothendieck's original conception.

Nobody seems yet to have fully exposed the theory according to this philosophical point of view. The discussion with Ben-Zvi was the first time I heard it expressed so naturally, although Paul Bressler had also previously said something similar. This idea is closely related to the folkloric observation exploited by Kapranov, Caldararu and others [**82**] [**30**], that $T_X[1]$ has a structure of Lie-algebra object in the derived category, and on a technical level it should hook up with the more classical idea of describing deformation theory by differential graded algebras and the like, discussed in ∞-categorical terms by Lurie in DAG-X.

The ramifications of this theory should, in the future, form key contributions to our understanding of the local structure of higher nonabelian cohomology stacks, and thus be crucial to the development of nonabelian mixed Hodge structures.

References

[1] D. Abramovich, T. Graber, A. Vistoli. Gromov-Witten theory of Deligne-Mumford stacks. *Amer. J. Math.* **130** (2008), 1337-1398.

[2] J. Adams. *Infinite Loop Spaces*, ANN. OF MATH. STUDIES **90**, Princeton University Press (1978).

[3] E. Aldrovandi, B. Noohi. Butterflies, I. Morphisms of 2-group stacks. *Adv. Math.* **221** (2009), 687-773.

[4] J. Alper. A guide to the literature on algebraic stacks. Preprint `http://www.math.columbia.edu/~jarod/stacks_guide.pdf` (2009).

[5] D. Ara. On the homotopy theory of Grothendieck8-groupoids. *J. Pure Appl. Alg.* **217** (2013), 1237-1278.

[6] M. Artin, A. Grothendieck, J.L. Verdier. Théorie des topos et cohomologie étale des schémas. Séminaire de Géométrie Algébrique du Bois-Marie 1963-1964 (SGA 4). Avec la collaboration de N. Bourbaki, P. Deligne et B. Saint-Donat. LECTURE NOTES IN MATH., Springer: Tome 1—Vol. **269** (1972); Tome 2—Vol. **270** (1972); Tome 3—Vol. **305** (1973).

[7] M. Artin, B. Mazur. *Etale Homotopy*, LECTURE NOTES IN MATH. **100**, Springer (1972).

[8] H. Bacard. Catégories faiblement enrichies sur une catégorie monoïdale symétrique. Doctoral dissertation, Nice (2012).

[9] C. Barwick. On the algebraic K-theory of higher categories. Preprint arXiv:1204.3607 (2012).

[10] C. Barwick, D. Kan. Relative categories: Another model for the homotopy theory of homotopy theories. *Indagationes Mathematicae* **23** (2012), 42-68.

[11] C. Barwick, C. Schommer-Pries. On the unicity of the homotopy theory of higher categories. Preprint arXiv:1112.0040 (2011).

[12] K. Behrend. On the de Rham cohomology of differential and algebraic stacks. *Adv. Math.* **198** (2005), 583-622.

[13] D. Ben-Zvi, D. Nadler. Loop spaces and connections. *J. Topology* **5** (2012), 377-430.

[14] D. Ben-Zvi, D. Nadler. Loop spaces and representations. *Duke Math. J.* **162** (2013), 1587-1619.

[15] J. Bénabou. *Introduction to Bicategories* LECT. NOTES IN MATH. **47**, Springer-Verlag (1967).

[16] B. Benzeghli. Etude explicite de quelques n-champs géométriques. Doctoral dissertation, Nice (2013).

[17] C. Berger. A cellular nerve for higher categories. *Adv. Math.* **169** (2002), 118-175.

[18] J. Bergner. A model category structure on the category of simplicial categories. *Trans. Amer. Math. Soc.* **359** (2007) 2043-2058.

[19] J. Bergner. Three models for the homotopy theory of homotopy theories. *Topology* **46** (2007), 397-436.

[20] J. Bergner. A characterization of fibrant Segal categories. *Proc. Amer. Math. Soc.* **135** (2007), 4031-4037.

[21] J. Bergner, C. Rezk. Comparison of models for (∞, n)-categories, I. Preprint arXiv:1204.2013 (2012).

[22] I. Biswas. Parabolic bundles as orbifold bundles. *Duke Math. J.*, **88** (1997), 305-325.

[23] J. Block, A. Smith. A Riemann-Hilbert correspondence for infinity local systems. Preprint arXiv:0908.2843 (2009).

[24] N. Borne. Fibrés paraboliques et champ des racines. *I.M.R.N.* **16** (2007).

[25] L. Breen. Extensions du groupe additif. *Publ. Math. I.H.E.S.* **48** (1978), 39-125.

[26] L. Breen. *On the classification of* 2-*gerbes and* 2-*stacks.* ASTÉRISQUE **255**, S.M.F. (1994).

[27] L. Breen, T. Ekedahl. Construction et propriétés de la sphère schématique. *C. R. Acad. Sci. Paris Sér. I Math.* **300** (1985), 665-668.

[28] L. Breen, W. Messing. Differential geometry of gerbes. *Adv. Math.* **198** (2005), 732-846.

[29] C. Cadman. Using stacks to impose tangency conditions on curves. *Amer. J. Math.* **129** (2007), 405-427.

[30] D. Calaque, A. Caldararu, J. Tu. On the Lie algebroid of a derived self-intersection. Preprint arXiv:1306.5260 (2013).

[31] F. Campana. Orbifolds, special varieties and classification theory. *Ann. Inst. Fourier* **54** (2004), 499-630.

[32] E. Cheng. Monad interleaving: a construction of the operad for Leinster's weak ω-categories. *J. Pure Appl. Alg.* **214** (2010), 809-821.

[33] E. Cheng. Comparing operadic theories of n-category. *Homology Homotopy Appl.* **13** (2011), 217-248.

[34] A. Chiodo. Stable twisted curves and their r-spin structures. *Ann. Inst. Fourier* **58** (2008), 1635-1689.

[35] D.-C. Cisinski. *Les préfaisceaux comme modèles des types d'homotopie.* ASTÉRISQUE **308**, S.M.F. (2006).

[36] D.-C. Cisinski. Locally constant functors. *Math. Proc. Cambridge Phil. Soc.* **147** (2009), 593-614.

[37] D.-C. Cisinski, G. Maltsiniotis. La catégorie Θ de Joyal est une catégorie test. *J. Pure Appl. Alg.* **215** (2011), 962-982.

[38] D.-C. Cisinski, I. Moerdijk. Dendroidal Segal spaces and infinity-operads. Preprint arXiv:1010.4956 (2010).

[39] J.-M. Cordier, J. Porter. Homotopy coherent category theory. *Trans. Amer. Math. Soc.* **349** (1997), 1-54.

[40] A. J. de Jong *et al. The stacks project,* `http://math.columbia.edu/algebraic_geometry/stacks-git/`.

[41] P. Deligne. *Cohomologie Etale. Séminaire de Géométrie Algébrique du Bois-Marie SGA* $4\frac{1}{2}$. Avec la collaboration de J. F. Boutot, A. Grothendieck, L. Illusie et J. L. Verdier. LECTURE NOTES IN MATH. **569**, Springer (1977).

[42] P. Deligne. La conjecture de Weil II. *Publ. Math. I.H.E.S.* **52** (1980, 137-252.

[43] P. Deligne. Théorie de Hodge III. *Publ. Math. I.H.E.S.* **44** (1974), 5-77.

[44] P. Deligne, J. Milne, A. Ogus, K.-y Shih. *Hodge cycles, motives, and Shimura varieties, Lecture Notes in Math.* **900**, Springer (1982).

[45] P. Deligne, D. Mumford. The irreducibility of the space of curves of given genus. *Publ.Math.I.H.E.S.* **36** (1969), 75-109.

[46] A. Dhillon. Hodge theory for algebraic stacks and the cohomology of the moduli space of stable bundles. Ph.D. thesis, Purdue University (2003).
[47] D. Dugger, S. Hollander, D. Isaksen. Hypercovers and simplicial presheaves. *Math. Proc. Cambridge Phil. Soc.* **136** (2004), 9-51.
[48] D. Dugger, D. Isaksen. Topological hypercovers and $\mathbb{A}^1$-realizations *Mathematische Zeitschrift* **246** (2004), 667-689.
[49] D. Dupont. Stacks on stratified space. Preprint arXiv:1003.4236 (2010).
[50] J. Ebert. The homotopy type of a topological stack. Preprint `arXiv:0901.3295` (2009).
[51] B. Fantechi. Stacks for everybody. *European Congress of Mathematics*, Birkhäuser Basel (2001), 349-359.
[52] B. Fantechi, E. Mann, F. Nironi. Smooth toric Deligne-Mumford stacks. *J. reine ang. Math. (Crelle)* **648** (2010), 201-244.
[53] P. Feit. *Axiomatisation of passage from "local" structure to "global" object.* MEM. A.M.S. **485** (1993).
[54] H. Fukuyama, I. Iwanari. Monoidal infinity category of complexes from Tannakian viewpoint. *Math. Ann.* **356** (2010), 519-553.
[55] D. Gepner, A. Henriques. Homotopy theory of orbispaces. Preprint `arXiv:0701916` (2007).
[56] D. Gepner, J. Kock. Univalence in locally cartesian closed infinity-categories. Preprint arXiv:1208.1749 (2012).
[57] H. Gillet. Arithmetic intersection theory on Deligne-Mumford stacks. *Motives and Algebraic Cycles: a celebration in honor of Spencer Bloch* (R. de Jeu, J. Lewis eds), Fields Institute Communications **56** (2009), 93-110.
[58] J. Giraud. *Cohomologie non abélienne.* GRUND. MATH. WISSENSCHAFTEN **179**, Springer-Verlag (1971).
[59] T. Gomez. Algebraic stacks. *Proc. Indian Acad. Sci. (Math. Sci.)*, **111** (2001), 1-31.
[60] A. Grothendieck. Sur quelques points d'algèbre homologique. **Tôhoku Math. J. 9** (1957), 119–221.
[61] A. Grothendieck. Techniques de construction et théorèmes d'existence en géométrie algébrique. III. Préschemas quotients. Séminaire Bourbaki, Vol. 6, Exp. No. 212, 99–118, Soc. Math. France, Paris, 1995.
[62] A. Grothendieck. Techniques de construction et théorèmes d'existence en géométrie algébrique. IV. Les schémas de Hilbert. Séminaire Bourbaki, Vol. 6, Exp. No. 221, 249–276, Soc. Math. France, Paris, 1995.
[63] A. Grothendieck. Technique de descente et théorèmes d'existence en géométrie algébrique. V. Les schémas de Picard: théorèmes d'existence. Séminaire Bourbaki, Vol. 7, Exp. No. 232, 143–161, Soc. Math. France, Paris, 1995.
[64] A. Grothendieck. Technique de descente et théorèmes d'existence en géométrie algébrique. VI. Les schémas de Picard: propriétés générales. Séminaire Bourbaki, Vol. 7, Exp. No. 236, 221–243, Soc. Math. France, Paris, 1995.
[65] A. Grothendieck. Revêtements étales et groupe fondamental. Fasc. I: Exposés 1 à 5. Séminaire de Géométrie Algébrique, 1960/61. Troisième édition, corrigée Institut des Hautes Etudes Scientifiques, Paris (1963).
[66] A. Grothendieck. Revêtements étales et groupe fondamental. Fasc. II: Exposés 6, 8 à 11. Séminaire de Géométrie Algébrique, 1960/61. Troisième édition, corrigée Institut des Hautes Etudes Scientifiques, Paris (1963).
[67] A. Grothendieck. *Catégories cofibrées additives et complexe cotangent relatif.* Lecture Notes in Mathematics **79**, Springer (1968).
[68] A. Grothendieck. *La Poursuite des Champs.*
[69] A. Hirschowitz. Cohérence et dualité sur le gros site de Zariski. *Algebraic curves and projective geometry (Trento, 1988)*, LECT. NOTES IN MATH. **1389**, Springer (1989), 91-102.

[70] A. Hirschowitz, C. Simpson. Descente pour les n-champs. Preprint math.AG/9807049 (1998).
[71] S. Hollander. A homotopy theory for stacks. *Israel J. Math.* **163** (2008), 93-124.
[72] S. Hollander. Characterizing algebraic stacks. *Proc. Amer. Math. Soc.* **136** (2008), 1465-1476.
[73] S. Hollander. Characterizing Artin stacks. *Math. Z.* (2010), DOI 10.1007/s00209-010-0746-x.
[74] S. Hollander. Descent for quasi-coherent sheaves on stacks. *Algebr. Geom. Topol.* **7** (2007), 411-437.
[75] M. Hovey. Model category structures on chain complexes of sheaves. *Trans. Amer. Math. Soc.* **353** (2001), 2441-2457.
[76] L. Illusie. *Complexe Cotangente et Déformations II.* LECTURE NOTES IN MATH. **283**, Springer-Verlag (1973).
[77] I. Iwanari. Stable points on algebraic stacks. *Adv. in Math.* **223** (2010), 257-299.
[78] I. Iwanari. Tannakization in derived algebraic geometry. Preprint arXiv:1112.1761 (2011).
[79] J.-F. Jardine. Simplicial presheaves. *J. Pure Appl. Alg.* **47** (1987), 35-87.
[80] R. Joshua. Generalised Verdier duality for presheaves of spectra. I. *J. Pure Appl. Alg.* **70** (1991), 273-289.
[81] A. Joyal. Letter to A. Grothendieck.
[82] M. Kapranov. Rozansky-Witten invariants via Atiyah classes. *Compositio Math.* **115** (1999), 71-113.
[83] L. Katzarkov, T. Pantev, C. Simpson. Nonabelian mixed Hodge structures. Preprint math/0006213 (2000).
[84] L. Katzarkov, T. Pantev, B. Toën. Schematic homotopy types and non-abelian Hodge theory. *Compositio* **144** (2008), 582-632.
[85] S. Keel, S. Mori. Quotients by groupoids. *Ann. of Math.* **145** (1997), 193-213.
[86] D. Knutson. *Algebraic spaces.* M.I.T. thesis, Springer LECTURES NOTES IN MATH. **203** (1971).
[87] A. Kresch. On the geometry of Deligne-Mumford stacks. *Algebraic geometry (Seattle 2005).* AMS PROC. SYMPOS. PURE MATH. **80** Part 1 (2009), 259-271.
[88] A. Kresch, A. Vistoli. On coverings of Deligne-Mumford stacks and surjectivity of the Brauer map. *Bull. London Math. Soc.* **36** (2004), 188-192.
[89] G. Laumon, L. Moret-Bailly. *Champs algébriques.* ERGEBNISSE DER MATHEMATIK UND IHRER GRENZGEBIETE **39** Springer-Verlag (2000).
[90] T. Leinster. A survey of definitions of n-category. *Theory Appl. Categories* **10** (2002), 1-70.
[91] T. Leinster. *Higher operads, higher categories,* LONDON MATH. SOC. LECTURE NOTES SERIES **298**, Cambridge University Press (2004).
[92] T. Leinster. An informal introduction to topos theory. Preprint arXiv:1012.5647 (2010).
[93] J. Lurie. Tannaka duality for geometric stacks. arXiv preprint math/0412266 (2004).
[94] J. Lurie. *Higher Topos Theory,* ANNALS OF MATH. STUDIES **170**, Princeton University Press (2009).
[95] J. Lurie. *Higher Algebra.* Book, currently in preprint form (2011).
[96] K. Maehara. Algebraic geometry from the point of view of groupoids. Report `http://www.t-kougei.ac.jp/research/pdf/vol1-25-11.pdf` (2002).
[97] G. Maltsiniotis. *La théorie de l'homotopie de Grothendieck* ASTÉRISQUE **301**, S.M.F. (2005).
[98] G. Maltsiniotis. Infini-groupoides d'après Grothendieck. Preprint, Institite de Math. de Jussieu (2007).
[99] F. Morel, V. Voevodsky. $\mathbb{A}^1$-homotopy theory of schemes. *Publ. Math. I.H.E.S.* **90** (1999), 45-143.

[100] S. Moriya. The de Rham homotopy theory and differential graded category. *Math. Z.* **271** (2012), 961-1010.
[101] The *nLab* website, `http://ncatlab.org/nlab/`.
[102] B. Noohi. Mapping stacks of topological stacks. *J. Reine Angew. Math.* **646** (2010), 117-133.
[103] N. O'Brian, D. Toledo, Y. Tong. Hirzebruch-Riemann-Roch for coherent sheaves. *Amer. J. Math.* **103** (1981), 253-271.
[104] M. Olsson. On proper coverings of Artin stacks. *Adv. in Math.* **198** (2005), 93-106.
[105] M. Olsson. *Crystalline cohomology of algebraic stacks and Hyodo-Kato cohomology.* Astérisque **316**, S.M.F. (2007).
[106] M. Olsson, J. Starr. Quot functors for Deligne-Mumford stacks. *Comm. Alg.* **31** (2003), 4069-4096.
[107] S. Paoli. Weakly globular cat^n-groups and Tamsamani's model. *Adv. Math.* **222** (2009), 621-727.
[108] S. Paoli, D. Pronk. Bicategories and weakly globular double categories. Preprint arXiv:1210.4056 (2012).
[109] A. Paulin. The Riemann-Hilbert correspondence for algebraic stacks. Preprint arXiv:1308.5890 (2013).
[110] P. Polesello, I. Waschkies. Higher monodromy. *Homology, Homotopy and Applications* **7** (2005), 109-150.
[111] T. Porter. $\mathcal{S}$-categories, $\mathcal{S}$-groupoids, Segal categories and quasicategories. Preprint arXiv:math/0401274 (2004).
[112] E. Pribble. Algebraic stacks for stable homotopy theory and the algebraic chromatic convergence theorem. Ph.D. thesis, Northwestern University (2004).
[113] J. Pridham. Pro-algebraic homotopy types. *Proc. London Math. Soc.* **97** (2008), 273-338.
[114] J. Pridham. Formality and splitting of real non-abelian mixed Hodge structures. Preprint arXiv:0902.0770 (2009).
[115] J. Pridham. Presenting higher stacks as simplical schemes. *Adv. Math.* **238** (2013), 184-245.
[116] D. Quillen. *Homotopical algebra.* Lecture Notes in Math. **43**, Springer (1967).
[117] C. Rezk. A model for the homotopy theory of homotopy theory. *Trans. Amer. Math. Soc.* **353** (2001), 973-1007.
[118] C. Rezk. Toposes and homotopy toposes. Lecture notes (2005) `http://www.math.uiuc.edu/~rezk/homotopy-topos-sketch.pdf`
[119] D. Rydh. Submersions and effective descent of etale morphisms. Preprint `arXiv:0710.2488v2`.
[120] N. Saavedra Rivano. *Catégories tannakiennes*, Lecture Notes in Math. **265**, Springer (1972).
[121] B. Saint-Donat. Techniques de descente cohomologique. In SGA 4 [**6**] Tome 3.
[122] I. Satake. On a generalization of the notion of manifold. *Proc. Nat. Acad. Sci. U.S.A.* **42** (1956), 359-363.
[123] U. Schreiber, M. Shulman. Quantum gauge field theory in cohesive homotopy type theory. *Proceedings of Quantum Physics and Logic* `ncatlab.org` (2012).
[124] U. Schreiber, K. Waldorf. Parallel transport and functors. *J. Homotopy Rel. Struct.* **4** (2009), 187-244.
[125] M. Shulman. Parametrized spaces model locally constant homotopy sheaves. *Topology and its Applications* **155** (2008), 412-432.
[126] C. Simpson. A relative notion algebraic Lie group and applications to n-stacks. Preprint alg-geom/9607002 (1996).
[127] C. Simpson. Algebraic (geometric) n-stacks. Preprint alg-geom/9609014 (1996).
[128] C. Simpson. Geometricity of the Hodge filtration on the ∞-stack of perfect complexes over X_{DR}. *Mosc. Math. J* **9** (209), 665-721.

[129] C. Simpson. *Homotopy Theory of Higher Categories.* NEW MATH. MONOGRAPHS **19**, Cambridge University Press (2011).
[130] A. Stanculescu. A Quillen model category structure on some categories of comonoids. *J. Pure Appl. Alg.* **214** (2010), 629-633.
[131] G. Tabuada. Higher K-theory via universal invariants. *Duke Math. J.* **145** (2008), 121-206.
[132] G. Tabuada. Differential graded versus simplicial categories. *Topology and its Applications* **157** (2012), 563-593.
[133] Z. Tamsamani. Sur des notions de n-categorie et n-groupoide non strictes via des ensembles multi-simpliciaux *K-Theory* **16** (1999), 51-99.
[134] A. Tatar. Length 3 complexes of abelian sheaves and Picard 2-stacks. *Adv. in Math.* **226** (2011), 62-110.
[135] C. Teleman, C. Simpson. De Rham's theorem for stacks. Preprint.
[136] R. Thomason. Algebraic K-theory and étale cohomology. *Ann. Sci. E.N.S.* **18** (1985), 437-552.
[137] B. Toën. K-théorie et cohomologie des champs algébriques: théorèmes de Riemann-Roch, $\mathcal{D}$-modules et théorèmes "GAGA". Thesis, Toulouse (1999), see `arXiv:math/9908097`.
[138] B. Toën. Stacks and non-abelian cohomology. `http://nlab.mathforge.org/nlab/files/ToenStacksNAC.pdf` (2002).
[139] B. Toën. Vers une axiomatisation de la théorie des catégories supérieures. *K-theory* **34** (2005), 233-263.
[140] B. Toën, G. Vezzosi. Homotopical algebraic geometry I: Topos theory. *Advances in Math.* **193**, (2005), 257-372.
[141] B. Toën, G. Vezzosi. Homotopical algebraic geometry II: Geometric stacks and applications. *Mem. Amer. Math. Soc.* **193** (2008), no. 902.
[142] B. Toën. Higher and derived stacks: a global overview. Arxiv preprint math/0604504 (2006).
[143] B. Toën. Segal topoi and stacks over Segal categories. Arxiv preprint math/0212330 (2002).
[144] B. Toën. Homotopical and higher categorical structures in algebraic geometry. Arxiv preprint math.AG/0312262 (2003).
[145] B. Toën. Vers une interprétation galoisienne de la théorie de l'homotopie. *Cahiers Top. Géom. Diff. Catégoriques* **43** (2002), 257-312.
[146] B. Toën. Descente fidelement plate pour les n-champs d'Artin. *Compos. Math.* **147** (2011), 1382-1412.
[147] B. Toën, G. Vezzosi. Chern character, loop spaces and derived algebraic geometry. *Algebraic topology,* ABEL SYMP. **4**, Springer, Berlin (2009), 331-354.
[148] D. Toledo, Y. Tong. A parametrix for $\overline{\partial}$ and Riemann-Roch in Čech theory. *Topology* **15** (1976), 273-301.
[149] D. Treumann. Exit paths and constructible stacks. *Compositio Math.* **145** (2009), 1504-1532.
[150] N. Tsuzuki. Cohomological descent of rigid cohomology for proper coverings. *Inventiones* **151** (2003), 101-133.
[151] A. M. Uludag. Orbifolds and their uniformization. *Arithmetic and geometry around hypergeometric functions,* PROGR. IN MATH. **260**, Birkhäuser (2007), 373-406.
[152] G. Vezzosi. What is... a derived stack? *Notices A.M.S.* **58** (2011), 955-958.
[153] A. Vistoli. Intersection theory on algebraic stacks and on their moduli spaces. Invent. Math. **97** (1989), 613-670.
[154] A. Vistoli. Notes on Grothendieck topologies, fibered categories and descent theory. Preprint math/0412512 (2004).
[155] V. Voevodsky. The Milnor conjecture. Preprint, K-theory preprint arxives **0170** (1996).

[156] V. Voevodsky *et. al.* *The HoTT Book*, `http://homotopytypetheory.org/book/`, also preprint arXiv:1308.0729 (2013).
[157] J. Wallbridge. Tannaka duality over ring spectra. Preprint arXiv:1204.5787 (2012).
[158] S. Wiesenthal. *Segel der Hoffnung. Die geheime Mission des Christoph Columbus.* Walter, Olten-Freiburg (1972). French translation *La Voile de l'Espoir (la mission secrète de Christoph Colomb)* ed. Robert Laffont, Paris, (1992).
[159] J. Wirth. Fiber spaces and the higher homotopy cocycle relations. Doctoral dissertation (1965). See J. Wirth, J. Stasheff, Homotopy transition cocycles, *Journal of Homotopy and Related Structures* **1** (2006), 273-283.
[160] J. Wolfson. Descent for n-Bundles. Preprint arXiv:1308.1113 (2013).
[161] C. Zhu. n-Groupoids and stacky groupoids. *I.M.R.N.* **21** (2009), 4087-4141.

On Grothendieck's work on the fundamental group

Jacob P. Murre

Contents

1. Introduction

Grothendieck's work on the algebraic fundamental group is one of the first places where he demonstrated the power of his theory of algebraic *schemes*. Not only did he unify the two classical, clearly related but nevertheless very different theories of Galois theory of fields belonging to algebra on the one hand and the theory of the fundamental group belonging to topology on the other hand into one theory, but his most striking result was undoubtedly his work on the algebraic fundamental group of an algebraic curve in positive characteristic, a group which could be defined within the theory of algebraic varieties but which was in its deeper aspects entirely out of reach by the methods available in algebraic geometry in the epoch before the theory of schemes.

In topology the fundamental group can be approached along two different lines: one method is via "loops", the other method is via "unramified coverings". This second approach can be applied in algebraic geometry,

Mathematical Institute, Universiteit Leiden, P.O. Box 9512, 2300 RA Leiden, The Netherlands. murre@math.leidenuniv.nl.

namely by studying the "category of unramified coverings" of a given algebraic variety, and this method is clearly closely related to the usual Galois theory of fields in algebra.

In *characteristic zero* the algebraic fundamental group is then the completion of the topological fundamental group of the underlying analytic manifold, where "completion" means completion with respect to the topology induced in this group by the subgroups of finite index. This group can then be studied by using the deep Riemann Existence Theorem and topological methods. In this way, one obtains in particular a description of the fundamental group of an algebraic curve defined over the complex numbers of genus g as a group topologically generated by $2g$ generators with one well-known relation.

In *positive characteristic* one can still *define* the algebraic fundamental group of smooth algebraic varieties; this was done in particular by Zariski and Abhyankar, and these authors obtained many interesting results. However in positive characteristic, the *deeper* understanding of this group remains out of reach if one is restricted to the methods of algebraic *varieties*, i.e. if the tools of nilpotent elements are not available.

The study of this group, i.e. the study of the unramified coverings of a curve, is of great interest especially in the case of an algebraic curve defined over a finite field, because of the analogy with the theory of extensions of algebraic number fields. In his beautiful 1947 paper *"L'avenir des mathématiques"* (Coll. Papers, Vol. 1, p.366), André Weil wrote:

> *"...avant d'aborder la detérmination des extensions d'un corps de nombres algébriques par leurs propriétés locales, il conviendra peut-être de résoudre le problème analogue, déjà fort difficile, au sujet des fonctions algébriques d'une variable sur un corps de base fini, c'est-a-dire d'étendre à ces fonctions les théorèmes d'existence de Riemann..."*

and he went on (ibid):

> *"...Il n'est pas impossible que toutes les questions de ce genre puissent se traiter par une méthode uniforme, qui permettrait, d'un résultat une fois établi (par exemple par voie topologique) pour la caractéristique* 0*, de déduire le résultat correspondant pour la caractéristique p; la découverte d'un tel principe constituerait un progrès de la plus grande importance..."*

Weil's prophecy was fulfilled, and this progress was made by Grothendieck in 1959! By using his theory of schemes in an essential way, namely by allowing nilpotent elements in the structure sheaves, he did describe the algebraic fundamental group of an algebraic curve in positive characteristic p, or to be precise, at least the part dealing with coverings of degree prime to p. He reported on these results in his Bourbaki seminar #182 of 1959, and treated the subject in detail in his Séminaire SGA 1 of 1960-61.

The essential tool of his method is the lifting of the curve and its coverings from characteristic $p > 0$ to characteristic 0, first by infinitesimal liftings and then by using his deep existence and comparison theorem. Each of these tools is – by its very nature – completely out of reach in the theory of algebraic *varieties.*

On the first page #182-01 of this brilliant Bourbaki talk (full of new ideas!) Grothendieck starts by remarking that the "classical" dictionary between affine algebraic varieties on the one hand and algebras without nilpotent elements on the other hand can, without much difficulty, be generalized to the case of affine schemes and algebras which may have nilpotent elements. But then *he continues by saying*, and I quote (ibid):

"Jusqu'à présent, les géomètres s'étaient refusés à tenir compte de ces indications et se sont obstinés à se restreindre à la considération d'algèbres affines sans éléments nilpotents, i.e. d'espaces algébriques dans les faisceaux structuraux desquels il n'y a pas d'éléments nilpotents (et même le plus souvent des espaces "absolutement irréductibles"). Le conférencier pense que cet état d'esprit a été un obstacle sérieux au développement des méthodes vraiment naturelles en géométrie algébrique..."

In my opinion, the above quote, and especially the last sentence, gives precisely the key to understanding Grothendieck's way of doing mathematics! He does not strive – at least not in the first place – for *generality as such* but for *naturalness.* Grothendieck did not merely see that it was possible to *extend* the theory of algebraic *varieties* to a theory of *schemes*; for him it was clear that in order to *understand* the theory of algebraic *varieties* it was *necessary* to develop the theory of *schemes.* With his deep insight he does see what is essential and natural, and then he does not hesitate to develop with his formidable skills the tools necessary to move towards his goal, namely to "see how things truly are".[1]

---

[1]At this point I would like to insert a personal recollection.

Being a student of the school of Weil I was at first not at all happy with this new theory of schemes, as I had to learn the "foundations" (sic!) all over again! So at one of the first opportunities which arose for me to discuss privately with Grothendieck, I asked him why he developed this theory of schemes: the theory of algebraic varieties – i.e. the theory without nilpotent elements – was (and is!) a beautiful theory full of deep results and at the same time full of challenging open problems. Why this generalization?

Grothendieck answered me as follows. He said these nilpotent elements are there in algebraic geometry by nature! Neglecting them, i.e. destroying them, is an artificial "intervention" that obscures our vision and may lead to pathologies. On the other hand, if we take them into account then firstly this will resolve these apparent contradictions, and secondly these nilpotent elements provide us with powerful new technical tools to attack problems.

Gradually I learned to see how right he was (nowadays it is clear to everyone!). As an example of the first point, think of the algebraic group schemes which have only nilpotent elements in their structure sheaf but which provide us with kernels which we do not see otherwise, or think of the nilpotent elements which can appear in the Picard "scheme" (!) of an algebraic "variety" (!), or think of the (in)famous controversy in classical algebraic geometry between Enriques and Severi concerning the question of the "completeness of the characteristic series" (see [Mum], Introduction), a controversy which was only explained

The following is the written, slightly elaborated version of my lecture in the workshop held in Peyresq from August 24-30, 2008 on Grothendieck's life and mathematical work.

In this lecture I have tried to outline the ideas of Grothendieck's work on the fundamental group, and I hope it gives some impression of the richness, the depth and the originality of his work and of the many obstacles he had to overcome. Concerning the proofs of the main theorems, I have tried as far as possible to indicate the main points, but for the details one must of course go back to the original works, especially to SGA 1. For the precise references to SGA 1, I refer to the pages of the original edition, since one can trace these in both the original and the new editions.

Finally, I want to thank the organizers Leila Schneps, Pierre Lochak and Wilfried Scharlau for inviting me to this very interesting workshop. Also I want to thank Tamás Szamuely for some valuable remarks for improving the text, Bas Edixhoven for pointing out a number of inaccuracies in the original version and – finally again and last but not least – Leila Schneps for going carefully over the manuscript, improving not only the English but also the manuscript itself and for helping with the LaTeXing.

2. Period before Grothendieck

2.1. The beginning: Riemann, Enriques, Zariski and Abhyankar. As said above, in algebraic geometry the fundamental group is constructed via the finite unramified coverings of the given variety, or more generally for open varieties, via the finite coverings unramified outside the so-called branch locus.

Recall that a morphism of finite type $f : X \to Y$ is unramified at a point $x \in X$ if, setting $y = f(x)$, we have

(1) $m_y.\mathcal{O}_x = m_x$ and

(2) for the residue fields, $k(x)$ is a finite separable extension of $k(y)$.

2.1.1. *In characteristic zero.* The problem of studying finite coverings of the projective line unramified outside a finite number of given points in the complex case goes back to Riemann [Rie] and depends upon his famous "existence theorem" for analytic functions. Enriques [E] generalized this existence theorem to the case of two variables, i.e. to the case of finite coverings of the projective plane unramified outside a certain given branch curve. This existence theorem then gives the algebraic fundamental group of the corresponding open complement U as the completion of the topological fundamental group $\pi_1^{top}(U)$ completed with respect to the subgroups of finite index (of course there still remains the problem of descibing this topological fundamental group itself in terms of the given branch curve).

by Grothendieck's work. As for the second point concerning the power of the new tools, a striking example is Grothendieck's work on the algebraic fundamental group of an algebraic curve in positive characteristic.

Zariski worked intensively on this subject during the period 1927-37 (see [Za2], Collect.Papers, Vol.3). Zariski and others, later especially Abhyankar, obtained many interesting results. For these results we refer to [ibid] (see in particular p. 6 and following), to Zariski's book on algebraic surfaces ([Za1], chap.8 and its appendix 1 written by Abhyankar) and to Popp's lecture notes [Po]. These results all depend on topological and analytic tools and methods, notably on Riemann's famous existence theorem.

2.1.2. *In positive characteristic.* The foundation for a purely algebraic definition of the algebraic fundamental group for a *normal algebraic variety* defined over an algebraically closed field k of arbitrary characteristic was provided by Zariski's important theory of the "normalization" of a given normal variety in a field ([Za3] around 1937-1939). To be precise, given an irreducible, normal algebraic variety X defined over an algebraically closed field k, let L be a finite separable extension of the function field $k(X)$; then Zariski constructed a normal variety X' via the "integral closure" of X in L. This gives a finite covering $f : X' \to X$. Now consider such L for which $f : X' \to X$ is *unramified* over X, and let $k(X)_{nr}$ denote the union of these L inside the algebraic closure of $k(X)$. Then Abhyankar gave the following definition (see [Ab1],[Ab2] and see also the appendix by Abhyankar in Zariski's book on Algebraic Surfaces [Za1],chap.8).

Definition. Set

$$\pi_1^{\mathrm{alg}}(X) = Gal\big(k(X)_{nr}/k(X)\big).$$

This is clearly a profinite group, namely the projective limit of the projective system of finite groups $\mathrm{Gal}\big(L/k(X)\big)$, the limit running over the above mentioned fields L partially ordered by their inclusion in $k(X)_{nr}$.

More generally, let $U = X - D$ where X is a projective, irreducible, normal variety as above and D a subvariety of codimension 1 in X. Then Abhyankar similarly defined the *tame fundamental group* of U via fields L for which the normalization is unramified over U and at most "tamely ramified" over D ([ibid] appendix 1).

2.2. The case of an algebraic curve.

2.2.1. *In characteristic zero.* Now first assume again that we are in characteristic zero and that our field k is $\mathbb{C}$ (or more generally an algebraically closed subfield of $\mathbb{C}$, see [LS]). Then it follows from the Riemann Existence Theorem and the GAGA theorems that with the above definition we have

$$\pi_1^{\mathrm{alg}}(X) = \big(\pi_1^{top}(X^{an})\big)^\wedge,$$

where X^{an} is the analytic manifold underlying the algebraic variety X and where for a group G we denote by $G^\wedge$ the corresponding profinite group obtained from G by completion with respect to the finite index subgroups.

Example. Let X be an algebraic curve C minus a finite set of points, i.e. $X = C - \{p_1, ..., p_r\}$, where C is a smooth projective curve of genus g defined over the complex numbers (or more generally where k is an algebraically closed field contained in $\mathbb{C}$). Then the $\pi_1^{top}(X)$ of the corresponding analytic manifold (i.e. Riemann surface) X^{an} is the group G with $2g + r$ generators a_i, b_i $(i = 1, ..., g)$ and c_j $(j = 1, ..., r)$ satisfying the relation

$$a_1 b_1 a_1^{-1} b_1^{-1} .... a_g b_g a_g^{-1} b_g^{-1} c_1 .... c_r = 1,$$

and the $\pi_1^{\mathrm{alg}}(X)$ is the corresponding profinite group obtained from this $\pi_1^{top}(X^{an})$ by completing with respect to the subgroups of finite index.

2.2.2. *In characteristic p.* Next assume that the field k is still algebraically closed but of characteristic $p > 0$ and that we have a curve $X = C - S$ as before, i.e. C a curve of genus g and S a finite set of closed points on C. So now we are (almost) in the situation which was mentioned in Weil's 1947 paper (almost, because Weil mentioned the case where the base field k is in fact a finite field)!

Now consider the group $\pi_1^{(p)}(C - S)$ constructed via finite, connected, Galois coverings of C unramified as before over X but which also have *degree over C prime to p*. Then Abhyankar *conjectured* in 1957 that this group is the topological completion of a group G as above, i.e. with $2g+r$ generators and the same relation as above, but now completed with the subgroups of finite index prime to p. In other words he conjectured that *provided we restrict to coverings prime to p* (in order to avoid questions of inseparability), the situation is similar to the characteristic zero case.

However in spite of this restriction, a *proof* of this assertion remained (and – to my knowledge – *remains up to now*) entirely out of reach of the techniques and methods of the theory of algebraic *varieties* themselves, i.e. out of reach if we do not allow nilpotent elements to enter into the theory, and the above *conjecture* was actually only *proved* in 1959 by Grothendieck, using his theory of *schemes.*

3. Grothendieck's approach to the algebraic fundamental group

3.1. Finite, étale coverings. In the following, S is a scheme which (for simplicity) we always tacitly assume to be locally noetherian, connected and separated.

Consider the category $(\mathcal{FE}^{alg})_{S^{alg}}$ (or $\mathcal{FE}_S$ for short) of *finite étale coverings* of S (i.e. *"revêtements étales de S"*). The objects of $\mathcal{FE}_S$ are S-schemes $f : S' \to S$ with f finite and étale such that the morphisms $g : S_1' \to S_2''$ in this category are S-morphisms.

Recall that a morphism of finite type $f : X \to Y$ is *"étale"* at a point $x \in X$ if it is *unramified* there and is moreover *flat*, i.e. $\mathcal{O}_x$ is a flat $\mathcal{O}_y$-module, where $y = f(x)$. The morphism f is *étale as a morphism* if it is everywhere étale; i.e. a morphism is étale if it is *smooth* and of relative dimension zero.

Note that here, because of the more general situation, we need to add the extra technical condition "flat", in contrast to the situation in section 2 above. However if S is normal and irreducible and S' is finite over S, then S' is étale over S if and only if it is unramified, and moreover then the irreducible components are the normalization in the ring of rational functions $R(S')$ (see SGA 1, I, section 10); this explains the relation with the situation in section 2 above.

In SGA 1, I and EGA IV, pars. 17, 18 there is a detailed study of the properties of étale morphisms; in particular there is a local description of such morphisms (see SGA 1, I, section 7). From this it follows in particular that all morphisms $g : S_1' \to S_2''$ in $\mathcal{FE}_S$ are also themselves finite étale.

3.2. The category $\mathcal{FE}_S$. The category $\mathcal{C} = \mathcal{FE}_S$ has the following properties (see SGA1, V, sections 4 and 7).

- G1. $\mathcal{FE}_S$ has a *final* object e, and the fiber product of two objects over a third object exists (recall that an object e is final in $\mathcal{C}$ if we have that $\mathrm{Hom}_{\mathcal{C}}(X, e)$ consists of one element for all X).
- G2. Finite sums of objects exist, and the quotient by a finite group exists.
- G3. Every morphism $u : X \to Y$ in $\mathcal{C}$ factors into $u = u' \circ u''$ with u'' a strict epimorphism and u' a monomorphism, and any monomorphism $u : X \to Y$ is an isomorphism of X with a direct summand of Y.

Moreover, there is a covariant functor $F : \mathcal{FE}_S \to \{FiniteSets\}$ defined as follows. Let Ω be an algebraically closed field and let $a : Spec(\Omega) \to S$ be a *geometric* point of S. Now let $f : X \to S$ be in $\mathcal{FE}_S$, and define

$$F(X) = \{x : Spec(\Omega) \to X \text{ such that } f \circ x = a\}.$$

This functor has the following properties:

- G4. F transforms the final object into the final object (i.e. the set consisting of one element), and it commutes with fiber products.
- G5. F commutes with direct sums, transforms strict epimorphisms into epimorphisms and commutes with passage to quotients by action of finite groups.
- G6. If $u : X \to Y$ is a morphism in $\mathcal{C}$ such that $F(u)$ is an isomorphism, then u is itself an isomorphism.

3.3. Definition. Let $(\mathcal{C}, F)$ be a pair consisting of a category $\mathcal{C}$ and a covariant functor F from $\mathcal{C}$ to the category $\{FiniteSets\}$ having the properties G1,....,G6 above. Then $\mathcal{C}$ is called a *Galois category* and F is called a *fundamental functor*, or sometimes also a *fiber functor*.

3.4. Standard example. Let π be a *profinite group*, i.e. $\pi = lim_{\leftarrow}(\pi_i)$, where the limit is the projective limit over a partially ordered set I of *finite groups* π_i, and let $\mathcal{C}(\pi)$ be the category of π-sets, i.e. *finite sets* on which π operates continuously (i.e. through a finite quotient). Let $F : \mathcal{C}(\pi) \to \{FiniteSets\}$ be the *"forgetful functor"*, i.e. the functor forgetting about the π-action.

3.5. The Galois-Grothendieck correspondence. The following is the *main theorem* concerning Galois categories (SGA 1, V, thm. 4.1, p. 126).
Theorem. *Given a pair $(\mathcal{C}, F)$ with $\mathcal{C}$ a Galois category and F a fiber functor $F : \mathcal{C} \to \{FiniteSets\}$, there exists a profinite group π, unique up to isomorphism, and an equivalence of categories $\Phi : \mathcal{C} \to \mathcal{C}(\pi)$ such that* $F = (\text{forgetful}) \circ \Phi$.

In what follows we shall sometimes, by abuse of language, denote the functors F and Φ by the same letter.

3.6. Some definitions. Below we shall make this theorem more precise, but before doing so we need to recall a number of concepts.

3.6.1. *Galois objects in $\mathcal{C}$.* Let X be an object in $\mathcal{C}$ and assume it is connected, i.e. not the sum of disjoint objects. Consider the group of automorphisms $Aut_{\mathcal{C}}(X)$ operating on the right X, and hence also operating (on the right) on $F(X)$; then the number of elements $\#Aut_{\mathcal{C}}(X)$ is always at most $\#F(X)$ (see SGA 1, V, p. 121). The object X is called a *Galois object* in $\mathcal{C}$ if we have the equality $\#Aut_{\mathcal{C}}(X) = \#F(X)$, and this happens if and only if $Aut_{\mathcal{C}}(X)$ operates transitively on $F(X)$ (similarly to ordinary topology).

3.6.2. *Prorepresentability of a (covariant) functor.* This is a generalization of the more familiar notion of representability.

Recall that a covariant functor $F : \mathcal{C} \to Sets$ is *representable* by an object $P \in \mathcal{C}$ if there exists an element $p \in F(P)$ such that for all $X \in \mathcal{C}$ we have $Hom_{\mathcal{C}}(P, X) \simeq F(X)$, where the map is induced by the element $p \in F(P)$.

Now instead of a single object of $\mathcal{C}$ consider a "pro-object" $P = (P_i)$ of C. This is a projective system P_i of objects in C indexed by a partially ordered, filtered index set I together with a coherent system $\phi_{ji} \in Hom_{C}(P_j, P_i)$ defined for $i \leq j$ and with $\phi_{ii} = id_{P_i}$ (by abuse of notation we delete these morphisms in the description $P = (P_i)$ of P). The covariant functor $F : C \to (Sets)$ is *prorepresentable* by a pro-object P if there exist elements $p_i \in F(P_i)$ with $p_i = \phi_{ji}(p_j)$ for $i \leq j$ such that for all objects $X \in C$ we have

$$lim_{\to i} Hom_{C}(P_i, X) \simeq F(X),$$

where the limit is the inductive limit over I and the map is induced by the system of elements $p_i \in F(P_i)$.

Moreover, F is said to be *strictly* prorepresentable if one can take the ϕ_{ji} to be epimorphisms in C.

3.6.3. *Main Theorem: precise form.* After these preparations we can now state the *Main Theorem* in a more precise form (see SGA 1, V, thm. 4.1, p. 126).

Theorem. *Let $(\mathcal{C}, F)$ be a Galois category with fiber functor $F : \mathcal{C} \to \{FiniteSets\}$. Then F is strictly prorepresentable by a pro-object $P = (P_i)$ in $\mathcal{C}$, where the P_i are Galois objects in $\mathcal{C}$. Let $\pi_i = (Aut_{\mathcal{C}}(P_i))^{opp}$; these π_i operate on the left on the $F(P_i)$, and they form a projective system of finite groups. Let $\pi = lim_{\leftarrow}(\pi_i)$ be the projective limit, so π is a profinite group. Then we have*

$$lim_{\to i} Hom_{\mathcal{C}}(P_i, X) \simeq F(X)$$

where the limit is the inductive limit, and

$$(\mathcal{C}, F) \simeq (\mathcal{C}(\pi), \text{forgetful})$$

is an equivalence of categories.

3.7. Remark. Clearly one can introduce the category Pro-$\mathcal{C}$ of pro-objects of $\mathcal{C}$; the above P is then called a *fundamental pro-object* associated to F, and we have

$$\pi = (Aut_{Pro-\mathcal{C}})^{opp}.$$

3.8. Some indications about the proof. We do not give the proof here, but simply indicate some of the main steps. For the details we refer to [SGA 1, V.4] (see also [Mu] chap.IV, and one can look also into the very precise notes by Lenstra [Le], see chap. 3).

The first step is to show that every object of $\mathcal{C}$ is artinian (use G6). Next, the important consequence is that F is strictly prorepresentable by a pro-object of $\mathcal{C}$; this follows from another theorem of Grothendieck saying that on a category where the objects are artinian and where finite projective limits exists, a functor is strictly prorepresentable if and only if it is left exact, and these conditions are fulfilled in our case thanks to G3 and G4 (for this theorem see Grothendieck's Bourbaki seminar #195, p. 195-05). The following step is then to show that in the projective system $P = (P_i)$ there is a cofinal subsystem of objects (P_i) where the P_i are Galois objects in $\mathcal{C}$. Finally one needs to construct a functor $G : \mathcal{C}(\pi) \to \mathcal{C}$ that is "inverse" to the functor F; this G is given by $G(E) = P \times_\pi E$ where $E \in \mathcal{C}(\pi)$. Indeed, note that this product factors through P_i for some sufficiently large i, i.e. equals $P_i \times_{\pi_i} E$, and thus does give an element of $\mathcal{C}$.

3.9. The algebraic fundamental group of a scheme. Let us return to the category $\mathcal{FE}_S$. Let S be a locally noetherian connected scheme as before, and $a : Spec(\Omega) \to S$ a geometric point. Let F be the fiber functor as defined in 3.2 (as described between G3 and G4). For fixed S, we write $\mathcal{FE}$ for short instead of $\mathcal{FE}_S$. Then the pair $(\mathcal{FE}, F)$ is a Galois category with fiber functor F, and hence we can apply the Main Theorem (i.e. the theorem in 3.5, and in fact in its precise form as stated in 3.6.3). Hence this pair determines a pro-object $P = (P_i)$ with the $P_i \in \mathcal{FE}$ and a profinite group π which is the projective limit of the finite groups $\pi_i = (Aut_{\mathcal{FE}}(P_i))^{opp}$.

Definition. Set

$$\pi_1(S, a) := \pi := lim_{\leftarrow}(Aut_{\mathcal{FE}}(P_i))^{opp},$$

where the limit is the projective limit; this profinite group is called *the algebraic fundamental group of S with base point a.*

3.10. Change of base point. If we choose another base point $b : Spec(\Omega') \to S$, then we get another fundamental functor F', another pro-object P' and another equivalence of categories $(\mathcal{FE}, F') \simeq (\mathcal{C}(\pi'), \text{forgetful})$. As in the topological case, we see that we also get an isomorphism between the profinite groups π and π', which is determined up to an inner automorphism.

3.11. Special cases.

3.11.1. *Usual Galois theory.* If $S = Spec(k)$, then we get the usual Galois theory with $\pi_1(S, a) = \text{Gal}(k_s/k)$, where k_s denotes the separable algebraic closure of k; it is the usual Galois group of k_s over k from algebra.

3.11.2. *The case of a complex algebraic variety.* Let $S = X$ be a smooth algebraic variety defined over the complex numbers $\mathbb{C}$, and denote by X^{an} the corresponding analytic manifold. Then it is well-known (see [GAGA]) that by the Riemann Existence Theorem (generalized by Grauert-Remmert and Grothendieck), we have an equivalence of categories (see SGA 1, XII, thm. 5.1, p. 332):

$$\left(\mathcal{FE}^{alg}\right)_{X^{alg}} \simeq \left(\mathcal{FE}^{an}\right)_{X^{an}},$$

where $\left(\mathcal{FE}^{an}\right)_{X^{an}}$ is the category of analytic morphisms $f : X' \to X$ where the map f is finite and locally an isomorphism for the analytic topology on X' and X (i.e. a finite covering in the sense of classical topology). Hence, taking the same base point a, we get an isomorphism

$$((\pi_1)^{top}(X^{an}))^{\wedge} \simeq \pi_1(X),$$

where $((\pi_1)^{top})^{\wedge}$ is the profinite completion of the topological fundamental group of the complex analytic manifold X^{an} (and the profinite completion is the completion with respect to the topology defined by the subgroups of finite index). This result was "classical" (see for instance [Po], page 15).

3.11.3. *The case of a normal projective variety.* Let $S = X$ be a normal, quasi-projective irreducible variety defined over an algebraically closed field k. Then for $f : X' \to X$ finite and unramified, we have automatically that f is also étale, and hence Grothendieck's definition coincides with Abhyankar's definition from section 2.

3.12. Remark. Given a Galois category $\mathcal{C}$, i.e. a category satisfying G1, G2 and G3 and having a fundamental functor F satisfying G4, G5 and G6, there may be different fundamental functors, determined up to (non-unique) isomorphism, as we saw in the example of $\mathcal{FE}$ by taking different base points. Each F determines, and is determined by, a pro-object P as described above; such a pro-object is called a *fundamental pro-object.* It is easily seen that there is an anti-equivalence between the category of fundamental pro-objects and the category of fundamental functors (see SGA 1, p. 130). Therefore by 3.7 we have that

$$\pi \simeq Aut_{fiber\ functors}(F).$$

Hence in Grothendieck's approach the algebraic fundamental group is the automorphism group of the fiber functor F.

Grothendieck extended this idea to the study of fundamental functors for *tensor categories* satisfying certain conditions (so-called *Tannakian categories*), where now the fundamental functor takes its values in the category of finite-dimensional vector spaces over a field of characteristic zero; this theory plays an important role in the theory of *motives* (see section 6.2 below).

4. Main Properties

In this section we discuss some of the properties of the algebraic fundamental group. They were announced by Grothendieck in his 1959 Bourbaki seminar and treated and proved in detail in SGA 1, to which we refer for the proofs.

4.1. Behavior under morphisms.

4.1.1. *Base change.* Let $f : (T, b) \to (S, a)$ with $a = f(b)$ be a morphism of connected schemes S and T (always tacitly assumed to be separated and locally noetherian) with (corresponding) geometric points a and b. From this, by the pullback $f^* : \{S' \to S\} \to \{T' = T \times_S S' \to T\}$, we get a functor

$$f^* : \mathcal{FE}_S \to \mathcal{FE}_T$$

between the Galois categories, and from this a continuous homomorphism

$$\pi_1(f) : \pi_1(T, b) \to \pi_1(S, a).$$

An important case is when $T \to S$ is itself in $\mathcal{FE}_S$. Then we get $\pi_1(T, b) = H$ with H an open subgroup of $\pi_1(T, a)$, the stabilizer of the point $b \in F(T)$. In particular if T is a Galois object in $\mathcal{FE}_S$, then H is a normal subgroup, and we get an exact sequence

$$e \to H \to \pi_1(S, A) \to Aut_S(T) \to e.$$

4.1.2. *Change of the profinite group.* Let $u : \pi' \to \pi$ be a continuous homomorphism of profinite groups. Then we get an obvious functor

$$H_u : \mathcal{C}(\pi) \to \mathcal{C}(\pi').$$

This functor is left and right exact; conversely, one can show (SGA 1, V, cor. 6.2) that a functor that is left and right exact comes from such a continuous homomorphism.

4.1.3. *Correlation of the properties of the continuous homomorphism u with those of the functor H_u and vice versa.* We see for instance that clearly u is surjective if and only if for all E connected in $\mathcal{C}(\pi)$, $H_u(E) = E$ is connected in $\mathcal{C}(\pi')$.

More subtle is the translation of exactness of a sequence. Consider a sequence of continuous homomorphisms of profinite groups:

$$\pi'' \xrightarrow{u'} \pi' \xrightarrow{u} \pi.$$

This sequence is exact if and only if (i') and (i'') below hold:

(i') $u \circ u' = 0$ if and only if for all $X \in \mathcal{C}(\pi)$ we have that $(H_{u'}.(H_u)(X))$ is *completely decomposed* in $\mathcal{C}(\pi'')$;

(i'') $ker(u) \subset Im(u')$ if and only if for all open subgroups $U' \subset \pi'$ with $Im(u') \subset U'$, we also have that $ker(u) \subset U'$, and this holds if and only if for all pointed connected $X' \in \mathcal{C}(\pi')$ for which the $H_{u'}(X') \in \mathcal{C}(\pi'')$ is completely decomposed, there exists a pointed connected $(X, x_0) \in \mathcal{C}(\pi)$ such that $H_u(X) = X_1' \bigcup X_2'$ with $x_0 \in X_1'$ and X_1' connected in $\mathcal{C}(\pi')$ and such that we have a surjection $X_1' \to X'$ in $\mathcal{C}(\pi'')$.

For the proof see SGA 1, V, prop. 6.11 or [Mu], p. 94-98.

4.2. Infinitesimal lifting. A crucial role in Grothendieck's theory of the fundamental group is played by the *infinitesimal lifting* of finite étale morphisms.

Let S be a scheme (locally noetherian and connected as usual), let $j : S_0 \to S$ be a closed subscheme defined by a nilpotent ideal $\mathcal{N}$ in $\mathcal{O}_S$, and let a_0 be geometric point in S_0 and $a = j(a_0)$. Then we have the following result.

Theorem. *The functor*

$$j^* : \mathcal{FE}_S \to \mathcal{FE}_{S_0}$$

is an equivalence of categories, *and therefore*

$$\pi_1(j) : \pi_1(S_0, a_0) \to \pi_1(S, a)$$

is an isomorphism.

For the proof see SGA 1, I, section 8. The proof depends on the local description of étale morphisms. In fact, Grothendieck later used this infinitesimal lifting property as a starting point for the definition of étale morphisms, see EGA IV, pars. 17, 18; the above theorem is EGA IV, thm. 18.1.2.

Remark. In fact there is the slightly more general following theorem (SGA 1, IX, 4.10).

Theorem. *Let* $g : S' \to S$ *be a* finite, radicial, surjective *morphism. Then the functor*

$$g^* : \mathcal{FE}_S \to \mathcal{FE}_{S'}$$

is an equivalence of categories, and

$$\pi_1(g) : \pi_1(S', a') \to \pi_1(S, a)$$

is an isomorphism, where $a = g(a')$.

(Recall that a morphism is radicial if it is injective and if the residue field extensions are purely inseparable).

4.3. Application. Assume that A is a *artinian local* ring with residue field k; let $\bar{k}$ be the algebraic closure of k, and let k_s be the separable algebraic closure. Let $f : X \to S$ be a morphism of schemes with $S = Spec(A)$ and X connected. Consider the following diagram of cartesian squares:

$$\begin{array}{ccccc} X & \longleftarrow & X_0 & \longleftarrow & \bar{X}_0 \\ \downarrow{\scriptstyle f} & & \downarrow{\scriptstyle f_0} & & \downarrow{\scriptstyle \bar{f}_0} \\ S & \longleftarrow & \mathrm{Spec}(k) & \longleftarrow & \mathrm{Spec}(\bar{k}) \end{array}$$

where $X_0 = X \times \mathrm{Spec}(k)$ and $\bar{X}_0 = X \times \mathrm{Spec}(\bar{k})$. Furthermore let $\bar{a} : \mathrm{Spec}(\Omega) \to \bar{X}_0$ be a geometric point in $\bar{X}_0$, a the corresponding geometric point in X and $b = f(a)$.

Theorem. (SGA 1, IX, 6.1) *With the above notation and assumptions, we have an exact sequence*

$$e \to \pi_1(\bar{X}_0, \bar{a}) \to \pi_1(X, a) \to \pi_1(S, b) \to e,$$

and moreover $\pi_1(S, b) = \mathrm{Gal}(k_s/k)$.

Taking the special case where $A = k$ and writing $\bar{X}$ instead of $\bar{X}_0$, we get the following important corollary (which is the starting point of Grothendieck's so-called *anabelian geometry*, see 6.5 and [O], (5.3)).

Corollary. *The sequence*

$$e \to \pi_1(\bar{X}, \bar{a}) \to \pi_1(X, a) \to \mathrm{Gal}(k_s/k) \to e$$

is exact.

Using the facts from 4.2 concerning the infinitesimal lifting, we can reduce the proof to the case that $A = k$ with k a perfect field. Write k_s as the inductive limit of fields k', where k' runs through the finite Galois extensions of k. The theorem is obtained by taking the limit of the following special case, which is interesting in itself.

Corollary *The sequence*

$$e \to \pi_1(X_{k'}, a') \to \pi_1(X, a) \to \mathrm{Gal}(k'/k) \to e$$

is exact.

This sequence is a special case of the exact sequence mentioned at the end of 4.1.3.

4.4. Exact sequence of a fibration - enter the comparison theorem! In topology there exists a long exact sequence between the homotopy groups of a fibration. Of course in the framework of SGA 1, there was only the fundamental group, and so one could only hope for a part of this exact sequence. And indeed, in topology, the end of the long exact sequence does give a short exact sequence for the fundamental groups in the case of a fibration. To prove that this short exact sequence also holds in algebraic geometry, Grothendieck also used – apart from the previous results – his comparison theorem between the algebraic and the formal theory, for which we refer to EGA III (EGA III, thm. 4.1.5, p. 125; see also [Illu], section 8.2, p. 187).

Assume that $f : X \to Y$ is *proper* and *separable* (recall that separable implies that the fibers are reduced and remain so after every base extension). Let Y be connected and assume that $f_*(\mathcal{O}_X) = \mathcal{O}_Y$; our assumptions now imply that f has geometrically connected fibers. Let $y \in Y$ be a point, X_y (resp. $X_{\bar{y}} = \bar{X}_y$) be the corresponding fiber (resp. the geometric fiber) and let $a : Spec(\Omega) \to \bar{X}_y$ be a geometric point in $X_{\bar{y}}$ which we also consider as a geometric point in X_y and in X itself. Set $b = f(a)$.

We now have the following diagram of cartesian squares:

$$\begin{array}{ccccc} X & \longleftarrow & X_y & \longleftarrow & \overline{X}_y \\ \Big\downarrow f & & \Big\downarrow f_y & & \Big\downarrow \bar{f}_y \\ Y & \longleftarrow & y & \longleftarrow & \overline{y} \end{array}$$

Theorem. (SGA1, X, cor. 1.4 p.263) *Let* $\varphi : \pi_1(\bar{X}_y, a) \to \pi_1(X, a)$ *and let* $\psi : \pi_1(X, a) \to \pi_1(Y, b)$. *Under the above assumptions the sequence*

$$\pi_1(\bar{X}_y, a) \to \pi_1(X, a) \to \pi_1(Y, b) \to e$$

is exact. Moreover without the separability assumption, we still have $\psi \circ \varphi = 0$.

For the proof see SGA 1, X, par. 1.

The most difficult part is to prove that $Im(\varphi) \supset ker(\psi)$. For this, using the "correlation principle" explained in section 4.1.3, we see that we have to prove the following: if $g : X' \to X$ is a finite étale connected covering such that $\bar{X}'_y \to \bar{X}_y$ admits a section, then there exists a connected finite étale cover $h : Y' \to Y$ such that $X' = X \times_Y Y'$. It is in the proof of this part that the comparison theorem EGA III, thm. 4.1.5 is used.

We mention the following two important corollaries of the above theorem.

Corollary. *Let* X *and* Y *be connected schemes defined over an algebraically closed field* k *with* X *proper; take geometric points* a *(resp.* b*) in* X *(resp.* Y*). Then*

$$\pi_1(X \times Y, a \times b) = \pi_1(X, a) \times \pi_1(Y, b).$$

Corollary. *Let* X *be proper and connected over an algebraically closed field* k*, and let* $K \supset k$ *also be algebraically closed. Let* a *be a geometric point in* X_K *and for simplicity denote its image in* X *by the same letter. Then* $\pi_1(X_K, a) \simeq \pi_1(X, a)$ *is an isomorphism.*

Remark. These two corollaries were obtained earlier in 1957 by Lang and Serre [LS] for the case of normal algebraic varieties. Grothendieck mentions (SGA 1, X, remarque 1.10) that this work of Lang-Serre was a motivation for his work on the algebraic fundamental group.

4.5. Enter the existence theorem! Let A be a *complete local ring* with residue field $k = k(s_0)$, where s_0 is the closed point in $S = Spec(A)$. Let $f : X \to S$ be a proper morphism of schemes, and set $X_0 = X \times_A k$. Recall the (special case of) the *existence theorem* (see EGA III, thm. 5.1.4, p. 150; see also [Illu], section 8.4, p. 204) applied in our case (see SGA 1, IX, thm. 1.10, p. 231):

Theorem. *With the above notation and assumptions, let* $X' \to X$ *be a finite étale covering of* X *and set* $X'_0 = X' \times_X X_0$. *Then the functor*

$$\Phi : \mathcal{FE}_X \to \mathcal{FE}_{X_0}$$

given by $\Phi(X') = X'_0$ *is an equivalence of categories.*

Now let $a_0 : Spec(\Omega) \to X_0$ be a geometric point and let a be its image in X. Then (SGA 1, thm 2.1, p. 268):

Corollary. *The canonical morphism*

$$\pi_1(X_0, a_0) \simeq \pi_1(X, a)$$

is an isomorphism.

Now combining the above with the results from 4.3 (in particular the first corollary), we get the following theorem.

Theorem. (SGA 1, X, cor. 2.2) *Let the assumptions be as in the preceding theorem. Set* $\bar{X}_0 = X \times_A \bar{k}$ *with* $\bar{k}$ *the algebraic closure of* $k = k(s_0)$. *Assume moreover that* $\bar{X}_0$ *is connected, and let* $\bar{a}_0 : Spec(\Omega) \to \bar{X}_0$ *be a geometric point in the geometric fiber,* a *its image in* X *and* $b = f(a)$. *Then the sequence*

$$e \to \pi_1(\bar{X}_0, \bar{a}_0) \to \pi_1(X, a) \to \pi_1(S, b) \to e,$$

is exact, and $\pi_1(S, b) \simeq \mathrm{Gal}(k_s/k)$.

4.6. The specialization theorem. Semi-continuity of the fundamental group by a fibration. The *specialization theorem*, to be stated below, plays a crucial role in Grothendieck's theory of the fundamental group; it makes it possible to obtain – under certain conditions – information about an algebraic fundamental group in characteristic $p > 0$ from a topological fundamental group in characteristic zero. For this specialization theorem Grothendieck uses the full power of the theory of schemes: infinitesimal lifting, comparison theorem and existence theorem!

Theorem. (SGA 1, X, cor. 2.4, p. 270) *Let* $f : X \to Y$ *be a* proper *morphism of schemes, let* Y *be connected, and assume that* f *has geometrically connected fibers. Let* y_0 *and* y_1 *be points in* Y *such that* y_0 *is in the Zariski closure of* y_1, *i.e.* y_0 *is a* specialization *of* y_1. *Consider the geometric fibers* $\bar{X}_0$ *and* $\bar{X}_1$ *over* $\bar{y}_0$ *and* $\bar{y}_1$ *respectively, and let* $\bar{a}_o : Spec(\Omega_0) \to \bar{X}_0$ *and* $\bar{a}_1 : Spec(\Omega_1) \to \bar{X}_1$ *be geometric (base) points, with* Ω_0 *an algebraically closed field containing* $k(y_0)$ *and* Ω_1 *an algebraically closed field containing* $k(y_1)$. *Then there exists a continuous homomorphism (the* "specialization homomorphism"*)*

$$sp : \pi_1(\bar{X}_1, \bar{a}_1) \to \pi_1(\bar{X}_0, \bar{a}_0)$$

determined up to an inner automorphism. Moreover, this homomorphism is surjective if f *is separable (which in particular is the case if* f *is smooth).*

Remarks.

1. Note that the fields $k(y_0)$ and $k(y_1)$, and hence the fields Ω_o and Ω_1, may have different characteristics.

2. The inner automorphism is coming from an inner automorphism of $\pi_1(X, a_0)$, see below in the proof.

Proof. We give only some indications; for the details see SGA 1, X, p. 269.

Step 1. Using the fact (second corollary of 4.4) that the fundamental groups of the geometric fibers are independent of the algebraic closures of the fields $k(y_0)$ and $k(y_1)$, we can replace Y by a scheme of type $Spec(A)$ where A is a complete local ring and y_0 the closed point. So we can assume that we are in the situation of 4.5 with $Y = Spec(A)$.

Step 2. From the geometric points in the fibers we respectively get geometric points a_0 and a_1 in X itself, and then geometric points b_0 and b_1 in Y.

Now we have the following diagram (with the vertical arrow at the left still to be defined!):

$$\begin{array}{ccccccccc} & & \pi_1(\bar{X}_1, \bar{a}_1) & \xrightarrow{\varphi} & \pi_1(X, a_1) & \xrightarrow{\psi} & \pi(Y, b_1) & \longrightarrow & e \\ & & \downarrow{\scriptstyle \text{sp}} & & \downarrow & & \downarrow & & \\ e & \longrightarrow & \pi_1(\bar{X}_0, \bar{a}_0) & \longrightarrow & \pi_1(X, a_0) & \longrightarrow & \pi_1(Y, b_0) & \longrightarrow & e, \end{array}$$

where the lower horizontal row is exact, in the upper horizontal row the ψ is surjective, and $\psi \circ \varphi = 0$. The vertical arrows in the middle and at the right are isomorphisms, namely obtained by change of base points, but they are only determined up to an inner automorphism; however the inner automorphism of the middle one determines the inner automorphism of the right one, and therefore the diagram at the right is commutative. Now a diagram chase defines a (continuous) homomorphism (i.e. the *to-be-defined* vertical arrow at the left)

$$sp : \pi_1(\bar{X}_1, \bar{a}_1) \to \pi_0(\bar{X}_0, \bar{a}_0),$$

determined up to an inner automorphism of $\pi_1(X, a_0)$.

Moreover, if $f : X \to Y$ is separable the upper horizontal row is itself exact, and then the homomorphism *sp* is surjective. The *sp* is called the *specialization homomorphism*, and if it is surjective we have the *semi-continuity* of the algebraic fundamental groups of the fibers.

Remark. In the next subsection 4.7, we will see that if f is *smooth* and if the characteristic of $k(y_0)$ (and hence of $k(y_1)$) is zero, then this homomorphism *sp* is an isomorphism. However this is in general not the case if char $k(y_0) = p$ is positive (see SGA 1, X, p. 271). Nevertheless there is a result in this case, explained in section 4.7.

4.7. The specialization morphism in the case of coverings of degree prime to p. Let G be a *profinite group*, i.e. $G = lim_{\leftarrow}(G_i)$, where the G_i are *finite groups* and the limit is the projective limit taken over a partially ordered, filtered index set I.

Let p be a prime number, and let $G_i^{(p)}$ be the quotient group of G_i that is maximal with respect to having order prime to p, i.e. obtained by dividing out by the normal subgroup generated by the p-Sylow subgroups of G_i. Set

$$G^{(p)} = lim_{\leftarrow}(G_i^{(p)}),$$

where the projective limit is taken over the same index set I, so $G^{(p)}$ is a quotient group of G, with kernel the normal subgroup generated topologically by its p-Sylow subgroups.

Theorem. (SGA 1, X, cor 3.9, p. 283) *Let* $f: X \to Y$ *be a* proper, smooth *morphism of schemes with* geometrically connected fibers. *Let* y_0 *and* y_1 *be points in* Y *such that* y_0 *is in the Zariski closure of* y_1*, i.e.* y_0 *is a specialization of* y_1*. Let* $p = \operatorname{char} k(y_0)$*. Then for the geometric fibers* $\bar{X}_0$ *and* $\bar{X}_1$ *over the geometric points* $\bar{y}_0$ *and* $\bar{y}_1$ *respectively, with geometric base points* $\bar{a}_0$ *and* $\bar{a}_1$ *respectively, the above defined specialization homomorphism induces an* isomorphism

$$sp : \pi_1^{(p)}(\bar{X}_1, \bar{a}_1) \simeq \pi_1^{(p)}(\bar{X}_0, \bar{a}_0).$$

Remark. Therefore, as remarked at the end of the previous section, in characteristic 0 we have an isomorphism for the algebraic fundamental groups themselves.

Proof. (Indication: see SGA 1, X, section 3 for details).

Two new ingredients enter into the proof: firstly the theorem of Zariski-Nagata on the *purity of the branch locus*, saying that if we have a ramified covering of regular schemes then the irreducible components of the branch locus are all of codimension one, and secondly *"Abhyankar's lemma"*, which is a result on ramified coverings of discrete valuation rings.

Using the dictionary of 4.1.3, the proof of the theorem boils down to proving the following: suppose we are given a finite group G of order prime to p, and a finite étale Galois covering Z_1 of $\bar{X}_1$ with group G, i.e. a surjective homomorphism $f_1 : \pi_1(\bar{X}_1) \to G$: then we want to factor f_1 through the specialization homomorphism as $f_1 = f_0 \circ sp$ with a homomorphism $f_0 : \pi_1(\bar{X}_0) \to G$. One first reduces to the case $Y = Spec(A)$ with A a discrete valuation ring with fraction field K and residue field $k = \bar{k}$. Now $\bar{X}_0$ is the special fiber and $\bar{X}_1$ is the geometric generic fiber. We can then reduce to the case that the covering Z_1 is already defined over K itself. Let $F = R(X)$ and $F_1 = R(Z_1)$ be the function fields of X and Z_1 respectively. Now we extend the finite étale covering Z_1 of the generic fiber X_1 to a finite covering Z of all of X by taking the normalization of X in F_1. If this Z is unramified over X, then we are done by taking its restriction over the special fiber. On the other hand, if it is ramified, then by the purity theorem it must be ramified over all of X_0. So let us assume that it is ramified; then we have that the field extension F_1 of F is ramified over the discrete valuation ring V in F, where $V = \mathcal{O}_{Z_0,Z}$ with Z_0 the fiber in Z over X_0. Moreover F_1 over F is Galois, with group G of order prime to p. Now we are in a situation where we can apply Abhyankar's lemma saying, in our case, that we can extend the field K to a field K', hence obtaining an extension $F' = F \cdot K'$ of F, such that the field $F'_1 = F_1 \cdot F'$ is not ramified over the the corresponding valuation ring V' of F'. Then repeating the above procedure and constructing a finite covering Z' over the corresponding X', we get a covering which is now also no longer ramified over the closed fiber itself, and is therefore *everywhere* unramified by Zariski-Nagata.

5. The algebraic fundamental group of an algebraic curve

The most striking application of the above theory is Grothendieck's study of the algebraic fundamental group of an algebraic curve in characteristic $p > 0$. As we will see below, in order to do this he still did have to overcome formidable obstacles!

5.1. Lifting the curve from characteristic $p > 0$ to characteristic zero. Let C_0 be a *smooth, projective*, irreducible algebraic curve defined over an algebraically closed field k of characteristic $p > 0$, and let $g = g(C_0)$. Let $W = W(k)$ be the *Witt ring* corresponding to k and let K be its quotient field. Finally, let $Y = Spec(W)$ and let y_0 be the closed point of Y.

Theorem. (SGA 1, III, cor. 7.4, p. 85) *There exists a smooth projective curve $f : X \to Y = Spec(W)$ such that $X_0 = X \times_W k \simeq C_0$.*

Remark. Let $X_1 = X \times_W K$ be the fiber over the generic point of Y; note that $g(X_1) = g(X_0) = g(C_0) = g$.

Proof. (See SGA 1, III, section 7.)

We shall indicate here (very roughly!) the main points of the proof. There are three steps.

Step 1. *Infinitesimal lifting* (SGA 1, III, thm. 6.3)

Assume that we have a locally noetherian scheme S and a subscheme S_0 of S defined by a nilpotent ideal $\mathcal{N}$ such that $\mathcal{N}^2 = 0$, and assume that we are given a scheme X_0 over S_0 that we want to lift to a scheme X over S. In the general case, i.e dim $X_0 = n$, there is an *obstruction* in $H^2(X_0, \mathcal{D}_{X_0/Y_0})$, where $\mathcal{D}_{X_0/Y_0}$ is the sheaf of relative derivations on X_0 and $Y_0 = Spec(k)$. Since this is a coherent sheaf and since in our case $X_0 = C_0$ is a curve over a field, the obstruction is zero in our case.

Step 2. *Formal scheme.*

Lifting infinitesimally "step by step" as in step 1, we get a smooth *formal scheme* and in fact a formal curve: $\mathcal{X} \to \mathcal{Y}$ with $\mathcal{Y} = Spf(W)$.

Step 3. *Algebraization: from formal curve to "true" curve.*

Although we have now the formal curve, we still cannot immediately apply the existence theorem for coherent sheaves (to $\mathcal{O}_{\mathcal{X}}$, say) in order to get a true curve over $Y = Spec(W)$, because in order to apply the existence theorem we need to have an underlying scheme. However X_0 is projective, hence we have a very ample invertible sheaf $\mathcal{L}_0$. In order to lift this infinitesimally, we have an obstruction in $H^2(X_0, \mathcal{O}_{X_0})$, but again since $X_0 = C_0$ is a curve this obstruction vanishes, and hence $\mathcal{L}_0$ can be lifted step by step to a formal invertible sheaf $\mathcal{L}$ on $\mathcal{X}$ which gives an embedding of $\mathcal{X}$ into the formal projective space $\mathcal{P}_\mathcal{Y}$, which is itself the completion along the closed fiber of a "usual" algebraic projective space P_Y. Finally, we can now use the existence theorem for the "algebrization" of the formal morphism $\mathcal{X} \to \mathcal{P}$ into a morphism $X \to P_Y$ with $X \to Y$ a smooth curve lifting $X_0 = C_0$ (see EGA III, thm. 5.4, p. 150 for details).

Remark. It was pointed out to me by T. Szamuely that there is now also another way to lift a curve from characteristic $p > 0$ to characteristic zero using a moduli argument. Namely, one can use Deligne-Mumford's work on stable curves [DMu] and consider the Hilbert scheme H_g of tricanonically embedded curves and the fact that this is smooth over $\mathbb{Z}$; see p.190-191 of [Sz] for details. Note, however, that this method again uses Grothendieck's pioneering work in an essential way, for instance the theory of the Hilbert schemes.

5.2. The algebraic fundamental group of an algebraic curve. After all these preparations, Grothendieck was finally in a position to describe the structure of the algebraic fundamental group of an algebraic curve defined over an algebraically closed field k of positive characteristic p, or to be precise, the prime to p part of that group.

So if C is such a curve, then one lifts this curve by the theorem in 5.1 to a curve C_1 over a field K of characteristic zero. Next, one applies the specialization isomorphism *sp* of the theorem in 4.7, and then by *topological* and *transcendental* tools (the Riemann Existence Theorem!) one gets the description of $\pi_1^{(p)}(C)$.

In fact, Grothendieck extended this result – essentially by the same methods – also to the case of an open curve, i.e. to an open part $U = C - \Sigma$, where Σ consists of a finite number of closed points of C. He announced that result already in the Bourbaki seminar #182 (cor. to thm. 14), and it was worked out in detail in the exposés XII and XIII of SGA 1 (see XIII, cor. 2.12, p. 392). For this one needs in particular a generalization of the specialization theorem of 4.7 to a certain non-proper case, namely to the case that on X (which is now a smooth proper curve over Y) we have a divisor D that is étale and finite over Y, and we consider the complement U over Y (see SGA 1, XIII, section 2 and also [Sz], p. 192). These last two exposés also use results from the étale cohomology theory developed later; they were written up by Mme Raynaud, and also contain many interesting others results.

Using the notation introduced above in 4.7, the final result is the following.

Theorem. (SGA 1, XIII, cor. 2.12, p. 392) *Let C be a smooth, irreducible algebraic curve of genus g defined over an algebraically closed field k of* *characteristic* $p \geq 0$*, and let* Σ *be a finite set of* r *distinct closed points on* C*. Let* $U = C - \Sigma$*. Then*

$$\pi_1^{(p)}(U) = \pi^{(p)},$$

where π *is the profinite group topologically generated by* $2g + r$ *generators* a_i, b_i *with* $i = 1, .., g$ *and* c_j *with* $j = 1, .., r$ *subject to the relation*

$$a_1 b_1 a_1^{-1} b_1^{-1} ... a_g b_g a_g^{-1} b_g^{-1} c_1 ... c_r = 1$$

.

Remark. If the characteristic is zero, then the above $\pi_1^{(p)}(U)$ is of course the "usual" algebraic fundamental group $\pi_1(U)$.

5.3. An application: the topological finite generation of the algebraic fundamental group in general. We can use the above results for curves to get some information on the algebraic fundamental group of a scheme in any dimension.

Theorem. (SGA 1, X, theorem 2.9, p. 273) *Let* X *be a smooth, projective, connected variety defined over an algebraically closed field. Then the algebraic fundamental group of* X *is topologically finitely generated.*

Proof (Indication). One proceeds by induction on the dimension $\dim X = n$. For $n = 1$, it follows from the previous theorem (and in fact we need only the semi-continuity of the specialization homomorphism). If $n \geq 2$, one takes a smooth hyperplane section $Y = X \cap H$ of X. It remains to prove that $\pi_1(Y) \to \pi_1(X)$ is surjective, but this follows from a Bertini (type) theorem and Zariski's connectedness theorem (see SGA 1, X, p. 274 for details).

6. Remarks on some further developments

6.1. Theorems of Lefschetz type. In his 1962 seminar, Grothendieck extended a number of classical theorems of Lefschetz to the context of his algebraic fundamental group. In particular, he proved that for a smooth projective scheme X of dimension ≥ 3 and for a smooth hyperplane section $Y = X \bigcap H$, we have $\pi_1(Y) \simeq \pi_1(X)$ (see SGA 2, X, thm. 3.10, p. 123).

In order to do this, he had, in particular, to extend the comparison theorem and the existence theorem to the non-proper case, of course under certain conditions (see SGA 2, IX). The proofs of these results depend very much on the study of local cohomology groups.

6.2. Tannakian categories and motivic Galois theory. The theory of the algebraic fundamental group encloses and unifies both the usual Galois theory from algebra and the fundamental group from topology. It establishes a relation between finite étale covers of a scheme on the one hand, and continuous permutation representations of a certain profinite group, namely the algebraic fundamental group of that scheme, on the other. Grothendieck had the idea of extending this kind of relation much further, namely to introduce a profinite group scheme, the so-called *Tannakian fundamental group*, which would play a role in the theory of *motives* analogous to the role of the usual Galois group in algebra.

For this he introduced the notion of *Tannakian category.* Certain categories of motives (over a field, say) would be examples of such Tannakian categories *provided Grothendieck's standard conjectures are true!*

Without going in detail, and simplifying it greatly, the "situation" is roughly as follows.

Let k be a field. A Tannakian category consists of a k-linear abelian tensor category $\mathcal{C}$ (satisfying certain extra conditions) together with a faithful k-linear exact functor F from $\mathcal{C}$ to the category of finite dimensional k-vector spaces; F is the so-called fiber functor.

Example. The standard example of such a situation is the following: let G be a profinite affine group scheme, i.e. G is the projective limit of a system of group schemes that are affine and finite over k, and let $\mathcal{C} = Rep_{k(G)}$ be the category consisting of finite dimensional k-vector spaces on which G operates continuously. Let F be the functor from this $\mathcal{C} = Rep_{k(G)}$ to the category of finite dimensional k-vector spaces forgetting the action of G.

Now, returning to the general case, let $(\mathcal{C}, F)$ be a pair consisting of a Tannakian category $\mathcal{C}$ and an exact k-linear functor F from $\mathcal{C} \to Vec_{kf}$ where Vec_{kf} is the category of finite dimensional k-vector spaces. Then the *extended Galois-Grothendieck correspondence* establishes an equivalence between such a pair $(\mathcal{C}, F)$ and a pair $(Rep_{k(G)}, \text{forgetful})$, i.e. an equivalence between such a Tannakian category $\mathcal{C}$ and a category $Rep_{k(G)}$ for a certain profinite affine group scheme G, and this G is then the corresponding *Tannakian fundamental group.*

The Tannakian theory as such was worked out not by Grothendieck himself but by a student of his and Deligne's, namely by Saavedra Rivano [SaRi]; later this Tannakian theory was pushed much further by Deligne and others (see for instance [DM], [De1], [De2], [Br], ...).

Although the standard conjectures are still wide open, this "Tannakian philosophy" has turned out to be very inspiring and enlightening and has been the source of many results (see for instance [De1], [Se2], [An], [N2], ...).

6.3. Nori's fundamental group scheme. Let S be a connected, separated, locally noetherian scheme. Then Grothendieck's algebraic fundamental group is a profinite group G whose finite quotients give finite "constant" group schemes $G_i \times S$ over S. In $\mathcal{FE}_S$ there is a cofinal system of Galois covers over S, and they can be seen as *torsors* under a constant group scheme of the type mentioned above, i.e. defined by a suitable finite quotient of G.

Recall the definition of a *torsor*. Let $G \to S$ be a finite, flat group scheme. Then a scheme $f : X \to S$ with f a finite, locally free surjective morphism is a G-*torsor* if there is a group action $G \times_S X \to X$ such that

$$(g, \mathrm{id}) : G \times_S X \to X \times_S X$$

is an isomorphism.

In 1982 Nori considered more generally (under certain conditions for S) schemes over S which are *torsors* over S under group schemes that are finite and flat over S; the category of such torsors form a Tannakian category. The corresponding Tannakian fundamental group is *Nori's fundamental group scheme* $\pi^N(S)$. This therefore classifies torsors over S of the above-mentioned type; it is an affine pro-group scheme over S, i.e. the projective limit of group schemes that are finite and affine over S.

If the field is of characteristic zero, then it is isomorphic to Grothendieck's fundamental group, due to a theorem of Cartier saying that in characteristic zero every finite group scheme is reduced. In positive characteristic, however, it is "larger", and in fact Grothendieck's fundamental group is its *maximal pro-étale quotient.*

For recent work on Nori's fundamental group scheme see for instance papers by Esnault and Hai such as [EH].

6.4. Abhyankar's conjecture and work of Raynaud and Harbater. Let C be a smooth, projective curve of genus g defined over an algebraically closed field of positive characteristic p, and let U be the open curve obtained by deleting r closed points from C. Let G be a finite group and let $s(G)$ be the normal subgroup generated by the p-Sylow subgroups of G. Then Abhyankar conjectured in 1957 [Ab2] that G would be a quotient of $\pi_1(U)$ if and only if $G/s(G)$ is a quotient of $\pi_1^{(p)}(U)$ (with the notation from 4.7). If the order of G is prime to p then this was proved to be true by Grothendieck (see the theorem in section 5.2); however this still left open the case where $s(G)$ is different from zero. In 1994 Raynaud [Ra] proved Abhyankar's conjecture for the affine line, and somewhat later (but also in 1994) Harbater [H] extended this to the case of an arbitrary curve U. Raynaud's work was ignited by Serre's important Note aux Comptes Rendus from 1990, where he treated the solvable case (see [Se1], section 3). (For details, see the remark on p. 292 of the new edition of SGA 1; however be careful because the notation there is slightly different from the one used here.)

6.5. "Esquisse d'un Programme". Much later, in 1984, Grothendieck wrote his famous "Esquisse d'un Programme" which again contains many beautiful and deep ideas. Several of the topics deal with – or at least are related to – fundamental groups, such as for instance his idea of anabelian geometry (the starting point of which lies in the exact sequence in the first corollary to the theorem of 4.3), or else for instance his so-called "section conjecture". Since I have never worked or lectured on these topics I do not feel competent to report or comment on these ideas, nor on the subsequent work done later on these subjects. I refer therefore for these topics to the books [SchLo], [Sch] edited respectively by L. Schneps and P. Lochak and by L. Schneps, and to the recent book [Sz] by T. Szamuely.

References

[Gr] Grothendieck, A.:
SGA 1. *Revêtements étales et groupe fondamental*, Springer LNM **224** (1971); New Ed.: Doc. Math. SMF, Vol. 3 (2003).
SGA 2. *Cohomologie locale des Faisceaux Cohérents et Théorèmes de Lefschetz Locaux et Globaux*, North-Holland (1968), New Ed.: Doc. Math. SMF, Vol. 4 (2005).
FGA *Fondements de la Géométrie Algébrique*, Séminaire Bourbaki ,1957-62 , Séminaire 182 (1950) and Séminaire 195 (1960).
EGA IV (with Dieudonné, J.), *Eléments de géométrie algébrique*, Publ.Math. IHES, No. **32** (1967).
Esquisse d'un programme, see for instance in [SchLo].

[Ab1] Abhyankar, S.S., *On the ramification of algebraic functions*, Am.J. Math. 77 (1955).

[Ab2] Abhyankar, S.S., *Coverings of algebraic curves*, Am.J.Math.79 (1957).

[An] André, Y., *Une introduction aux motifs*, Panoramas et Synthèses, Vol. **17**, SMF 2004.

[Br] Breen, L., *Tannakian categories*, in "Motives", vol. **1**, Proc. conference Seattle, PSPM, vol. **55**, AMS 1994.

[De1] Deligne, P., *Catégories Tannakiennes*, in "The Grothendieck Festschrift", Vol.II, Progress in Math. vol. **87**, Birkhäuser, 1990.

[De2] Deligne, P., *Le groupe fondamental de la droite projective moins trois points*, in "Galois Groups", Y. Ihara, K. Ribet and J.-P. Serre, eds., MSRI publ. vol. **16**, 1989.

[DM] Deligne, P. and Milne, J.S., *"Tannakian Categories"*, in Springer LNM 900, (1982).

[DMu] Deligne, P. and Mumford, D., *The irreducibility of the space of curves of given genus*, Publ. Math. IHES, No. **36** (1969).

[E] Enriques, F., *Sulla costruzione delle funzioni algebriche de due variabili possendenti una data curva di diramazione*, Ann. di matematica pura et applicata, Vol. **1** (1923).

[EH] Esnault, H. and Hai, P.H., *The fundamental groupoid scheme and applications*, preprint arXiv:math/0611115.

[H] Harbarter, D., *Abhyankar's conjecture on Galois groups over curves*, Invent. Math. **117** (1994).

[Illu] Illusie, L., *Grothendieck's existence theorem in formal geometry*, part 4 in "Fundamental Algebraic Geometry, Grothendieck's FGA Explained", AMS Math. Surveys and Monographs, Vol. **123** (2005).

[LS] Lang, S. and Serre, J.-P., *Sur les rêvetements non ramifiés de variétés algébriques*, Am. J. Math. **79** (1957).

[Le] Lenstra, H.W. Jr., *Galois Theory of Schemes*, course notes available from the server of the Mathematics Department of the University of Leiden.

[Mum] Mumford, D., *Lectures on Curves on an Algebraic Surface*, Annals of Math. Studies, No. **59**, Princeton Univ. Press (1966).

[Mu] Murre, J.P., *Lectures on an introduction to Grothendieck's theory of the Fundamental Group*, Notes from the Tata Institute of Fundamental Research, No. **40**, Bombay 1969.

[N1] Nori, M., *On the representation of the fundamental group*, Comp. Math. **33** (1976).

[N2] Nori, M., *The fundamental group-scheme*, Proc. Indian Acad. Sci. Math. Sci. **91** (1982).

[O] Oort, F., *The algebraic fundamental group*, in [SchLo] (1997).

[Po] Popp, H., *Fundamentalgruppen algebraischer Mannigfaltigkeiten*, Springer LNM **176** (1970).

[Ra] Raynaud,M., *Rêvetements de la droite affine en caractéristique* $p > 0$ *et conjecture d'Abhyankar*, Invent.Math. **116** (1994).

[Rie] Riemann, B., *Theorie der Abelschen Funktionen*, Crelle Journal, Vol. **54** (1857).

[SaRi] Saavedra Rivano, N., *Catégories Tannakiennes*, Springer LNM 265 (1972).

[SchLo] Schneps, L. and Lochak, P. (eds.), *Geometric Galois Actions I*, London Math. Soc. LN **242** (1997) .

[Sch] Schneps,L. (ed.), *Galois groups and Fundamental groups,* MRSI Publ., Vol. **41**, 2003.

[GAGA] Serre, J.-P. ,*Géométrie algébrique et géométrie analytique*, Inst. Fourier, t. **6** (1955/56).

[Se1] Serre, J.-P., *Rêvetements des courbes algébriques* Séminaire Bourbaki #749 (1991/1992).

[Se2] Serre, J.-P., *Propriétés conjecturales de groupes de Galois motiviques et de représentations* ℓ*-adiques*, in "Motives", vol. **1**, Proc. Symp. Pure Math. Vol. **55**, AMS (1994).

[Sz] Szamuely, T., *Galois groups and Fundamental groups*, Cambridge studies in Adv. Math. Vol. **117** (2009).

[We] Weil, A., *L'avenir des mathématiques* (1947), in Collected papers Vol. **1**, Springer-Verlag.

[Za1] Zariski, O., *Algebraic Surfaces, 2nd ed.*, Erg. der Math. und ihrer Grenzgeb., Vol. **161**, Springer-Verlag (1935, 1971).

[Za2] Zariski, O.,*Collected Papers*, Vol. III, MIT, (1978).

[Za3] Zariski, O.,*Some results in the arithmetic theory of algebraic varieties*, Am.J. of Math. vol. **61** (1939).

An apprenticeship

Robin Hartshorne

I first saw Alexandre Grothendieck in Paris during the academic year 1959/60. I had just graduated from Harvard College, with a major in mathematics, and was spending one year studying as a "stagiaire étranger" at the Ecole Normale Supérieure (rue d'Ulm) in Paris. I attended courses by H. Cartan, C. Chevalley, and J.-P. Serre. Someone must have suggested that I go to hear the Séminaire Bourbaki talks. Grothendieck was lecturing about something that passed completely over my head. Looking at the dates of his talks, I presume it was his first talk on construction techniques: "Descente par morphismes fidèlement plats." Andy Gleason, who had been my undergraduate advisor at Harvard, was sitting next to me, and he couldn't tell what it was about either, so I didn't feel so bad.

The next year I was at Princeton. Then in the fall of 1961 I returned to Harvard and Grothendieck was there. He gave a course on local properties of morphisms, material that would later appear in EGA IV; a seminar on construction techniques, including the Hilbert scheme and Picard scheme; and a seminar on local cohomology. I had read Serre's FAC the year before, so this time I was ready to listen and appreciate what he was doing. Following the French tradition of students writing lecture notes for seminars, I wrote lecture notes for the local cohomology seminar. In the process, I expanded points that were not clear to me and rearranged the material. The notes were typed and printed as Harvard lecture notes, with a red paper cover. (They were reprinted later as #41 in the Springer Lecture Note Series.)

What do I remember of Grothendieck at Harvard that fall? He was everywhere, doing mathematics all the time. For me as a graduate student, he was an object of wonder, a bright star or meteor passing through the sky. While I absorbed as much as I could of his mathematical ideas, I had no sense of who he was as a person. All I can remember is a few anecdotes. Once I asked him if the reduced ring of a local Cohen-Macaulay ring was also Cohen-Macaulay. His answer was, "if it is not obvious, it is probably false." Well, of course it was not obvious to me, but I suppose what he meant was

---

Department of Mathematics, University of California, Berkeley, CA 94720-3840, USA. robin@math.hartshorne.net.

if it was not obvious to him, it was probably false. (He was right, because later Raynaud found a counterexample, and I also found one using a curve from a paper of Gallarati.)

One day in his course on local properties of morphisms, he stated a new theorem about excellent rings. To prove it, he first spent 15 or 20 minutes expanding the statement to its most general possible form. Then he spent half an hour reducing to one special case after another, until finally it boiled down to something apparently quite elementary. Suddenly, he stepped back from the blackboard, looked at what he had written, and said, "What am I telling you? This is false!" One had the impression that he had been making up the proof on the spot. When he realized it did not work, he had to start over from the beginning at the next lecture.

Then there was the time when he did not show up to teach his class. The rule at Harvard was if the professor was not there after ten minutes, you could leave. But in this case, those present, mostly graduate students and faculty, waited. He came about 45 minutes late, and lectured for an hour, as if nothing had happened. I suppose he had overslept.

The seminar on construction techniques got me thinking about the Hilbert scheme. Rather than listening to more abstractions and general results, Quot scheme, Picard scheme, etc, I wanted to see what this was like in an elementary situation. So I looked at the first interesting case, that of zero-dimensional schemes in the plane. Sets of distinct points were easy enough to visualize. But what were those thick schemes concentrated at a single point? By looking at their ideals and playing with the notion of flat families, I was able by explicit equations to show that any zero-scheme in the plane was a limit of a flat family whose general member was a set of distinct points. In other words, the Hilbert scheme of n points in the plane is irreducible. I did not write this up at the time. Later it came to be known as "Fogarty's theorem," because he published a more conceptual proof of a stronger result, that the Hilbert scheme of zero-schemes of length n in any nonsingular surface was smooth and irreducible.

I still remember clearly the feeling of elation when one day the next spring, after Grothendieck had returned to Paris, I realized that the same ideas that I had used studying zero-schemes in the plane could be extended to subschemes of any dimension in any projective space. The result was weaker, but it would show that for each Hilbert polynomial the corresponding Hilbert scheme was connected. There was still a lot of hard work to bring this vision into reality, but the certainty was there. The feeling of this kind of insight is that one day I am at a high point, like the top of a mountain, from which I can see clearly the landscape off into the distance. I see the distant peaks and the routes necessary to reach them. I try to remember everything as clearly as I can, because once the vision is gone, once I am again in the dark forests of the valley, I can no longer see where I am going, but have to work myself slowly, painstakingly, through brush and unknown terrain until I again arrive at a point from which I can see the goal.

When I had a written proof of this new result, I sent a copy to Grothendieck. He replied with four lines of appreciation of the result, and then four pages suggesting further questions to investigate, such as finding the irreducible components of the Hilbert scheme and their dimensions, the structure of the category of coherent sheaves over them, their Picard groups and so on. He was always thinking of what to do next. The result I found became my Ph.D. thesis. The other questions he proposed are still too hard to answer today.

Up to this point I had listened to Grothendieck's lectures; I had written lecture notes for one of his seminars; his newly defined Hilbert scheme had provided the subject material for my thesis; and he had been supportive of my work. But I was not his "student" and he was not my "advisor." My true apprenticeship came after my thesis with my work on his theory of duality.

Let me review the background of duality theorems in algebraic geometry. Over the complex numbers, topological methods and cohomology were already present, at least implicitly, in the work of Poincaré, Lefschetz and later Kodaira and Spencer. Sheaves were introduced by Leray. The notion of coherent sheaf was developed independently by Cartan and Oka in the 1940's, and the use of coherent analytic sheaves and their cohomology on complex manifolds was fully developed in the Cartan seminars of the early 1950's.

A first bit of duality appeared in the dual role played by $l(D)$ and $l(K-D)$ in the Riemann-Roch theorem of Kodaira and Spencer. Seizing upon this, Serre proved a duality theorem for a vector bundle on a compact complex manifold comparing its cohomology with that of the dual bundle tensored with the sheaf of top differential forms. His proof was analytic, using resolutions by sheaves of C^∞ differentiable forms and duality of Fréchet spaces. This is the model for later "Serre duality" theorems in algebraic geometry.

The introduction of sheaves and cohomology into abstract algebraic geometry (that is, over fields other than the complex numbers) is due to Serre in his seminal paper FAC. He defined cohomology using a Čech process, taking limits over finer and finer open coverings, but could prove the exact sequence of cohomology only for coherent sheaves on varieties.

Near the end of FAC, Serre proved a duality theorem comparing the cohomology of a coherent sheaf to certain Ext groups over the homogeneous coordinate ring of the ambient projective space. These Ext groups were defined in the book of Cartan and Eilenberg and were known at that time only for modules over a ring. The statement is awkward in comparison with the later version using Ext of shaves of modules, but the content is there.

Grothendieck's version in his Séminaire Bourbaki talk #149 was a significant improvement over Serre's result, though still for coherent sheaves on a nonsingular projective variety. By that time he had developed the theory of derived functors in an abelian category, so that both the cohomology of a sheaf and the Ext groups were interpreted as derived functors. Thus he was able to state the duality theorem intrinsically on a given variety. In the case of a singular variety with suitable conditions on its structure sheaf (what we now call a Cohen-Macaulay variety) he was able to prove the duality theorem replacing the sheaf of top differential forms with a suitable "dualizing sheaf."

Still, he was not satisfied. In a letter to Serre, 12 Jan 1956, which clearly shows his way of thinking, he says "Le rôle joué dans tout ça par l'espace projectif semble malheureusement encore excessif...J'ai aussi envie de regarder si on ne peut pas énoncer un théorème de dualité sur les variétés projectives pouvant avoir des singularités et si un énoncé plus général et plus technique ne pourrait se démontrer plus simplement encore que celui qu'on a en vue."

He gave some hints of this more general approach in his ICM talk of 1958. He recognized that a single "dualizing sheaf" would not suffice: one needed a "residual complex" to replace it. This would require an analogous local theory, and the residual complex would have to be constructed locally. Since it is not unique, this posed problems that the existing techniques of homological algebra, abelian categories and derived functors were simply not sufficient to handle. What was needed was a way of dealing with all the derived functors at once. Of course they come from a complex, but that complex is not unique. It is sometimes unique up to homotopy, but not always. It is however, unique up to "quasi-isomorphism," that is, maps of complexes that induce isomorphisms on the homology sheaves. Thus Grothendieck made the bold step of imagining a "derived category" whose objects were complexes and where all quasi-isomorphisms became isomorphisms. He suggested this to Verdier, who took it on as his thesis project. He soon developed the whole theory, introducing triangulated categories along the way. This provided a context in which to approach the more general theory.

Grothendieck himself was so busy with the writing of EGA with Dieudonné and the annual SGA seminars in Paris that he simply had not had the time to think any more about duality. Sometime in the spring of 1963 I offered to run a seminar and write notes if Grothendieck would explain his theory of duality to me. He agreed, and by August 1963 he had nearly completed a 250-page manuscript "Résidus et Dualité: prénotes pour un séminaire Hartshorne." I gave lectures on this material at Harvard in fall 63 and spring 64, aided by Mumford, Tate, Lichtenbaum, Fogarty, and others. At the time I wrote six exposés. Later I rewrote and completed them into the *Residues and Duality* [RD] published in 1966 as #20 in the new Springer Lecture Notes series.

The theory of duality as presented in the prenotes is a vast expansion and generalization of the earlier theory. The goal was no longer just duality for a projective variety over a field. It became duality for an arbitrary proper morphism of schemes, stated and proved in the language of derived categories. To break away from excessive dependence on projective space (though it does reappear near the end in the form of Chow's lemma), the theory includes a local theory of duality and local construction of the residual complex. To give sufficient flexibility to the theory, everything must be functorial and there are millions of functorial compatibilities to check along the way.

To begin the seminar, I had to swallow and digest the theory of derived categories. I had at my disposal Verdier's thesis (état 0), later published in SGA 4 $\frac{1}{2}$ (1977), but for many years my account in the first chapter of RD was the only published reference. Grothendieck's style in writing was to set each segment of the story in its own general framework, all the while suggesting further generalizations and problems to investigate. Besides the struggle of keeping up with so many new ideas, my task felt like trying to rein in a herd of wild horses who kept galloping off in different directions. To arrive at a useful result in a limited amount of time, I ruthlessly added finiteness restrictions (Noetherian schemes, finite Krull dimension, morphisms of finite type) to focus the narrative and keep it within bounds. In constructing the morphism called $f^!$ on derived categories, one takes an affine cover, but then the elements of the derived category on the open sets do not glue. Grothendieck's solution was to develop an entire theory of pseudo-complexes, defined by local elements of the derived category with gluing data. I did my best to avoid these, using restrictive hypotheses, and they do not appear in RD.

In the last year of rewriting the whole seminar, I would write one chapter at a time and send it to Grothendieck for comments. It would come back covered with red ink. I would make corrections and send it to him again. Again it would come back covered with red ink. This went on for some time, until one day I decided that was enough, and simply sent the typescript to Springer for publication. Grothendieck's response was that we agree the theory is not yet in a totally satisfactory state, but that it seemed I had done as good a job as possible at present.

A year or two after RD was published, Grothendieck observed to me that it was good rough account of the subject, but when was I going to write "the book" where everything was done in detail? I tried politely to explain that I had had enough of duality and wanted to move on to something else.

The last time I saw Grothendieck in person was I believe in Kingston, Ontario, in 1971. At that time, when asked to give a mathematical talk, he would request equal time to speak about his peace work and his organization Survivre. He was drifting away from mathematics and applying his brilliant mind to the problems of humanity. No matter how logically persuasive, I thought his efforts in that direction were politically naïve.

When my book *Algebraic Geometry* was published in 1977, I sent Grothendieck a copy, together with a note expressing my gratitude for all I had learned from him. A short note came in reply: "It looks like a nice book. Perhaps if one day I again teach a course in algebraic geometry, I will look at the inside." I was disappointed that he did not show more appreciation for my work. At the same time I became aware from his response how completely he had withdrawn from the mathematical community.

How do I feel now, looking back 50 years at my early encounter with Grothendieck? I was at the cusp of the schematic revolution: young enough to have avoided reading Weil's *Foundations*—that stultifying world of independent generic points and fields of definition—but old enough to have sufficient background in commutative algebra, topology and sheaf theory to be able to receive Grothendieck's message. His approach to algebraic geometry felt absolutely right to me, as it did to those around me. During the years of my apprenticeship I learned the craft and techniques of his methods. Using those skills I was then able to apply them to subjects that interested me, such as curves in projective space or vector bundles on projective space. While he continued in his seminars to develop ever more abstract structures and concepts, my interests became more and more classical.

I felt puzzled and confused by Grothendieck's leaving the world of mathematics. Even after reading sections of his long rambling reflections *Récoltes et Semailles* I still wonder how that sharp analytical mind became warped by feelings of rejection and lack of appreciation, when everyone knows of his unique and revolutionary contributions to algebraic geometry. For me, he remains unknown and unknowable by any "normal" standards of human psychology and behavior. I just feel a great sadness at the loss of the man who was the most important influence in my development as a mathematician.

Grothendieck et la cohomologie étale

Luc Illusie

De toute l'œuvre de Grothendieck, c'est sans doute la cohomologie étale qui aura exercé l'influence la plus profonde sur l'évolution de la géométrie arithmétique dans les cinquante dernières années. Cela tient autant à la multiplicité des applications et développements qu'elle a suscités qu'à la fécondité du concept de localisation qui en est à l'origine. C'est l'une des théories cohomologiques les plus achevées. Par sa puissance, sa souplesse, sa beauté, elle a souvent servi de modèle.

1. Les débuts

Il semble que ce soit à l'issue de l'exposé de Serre au séminaire Chevalley du 21 avril 1958 [S] que Grothendieck ait conçu l'idée de la topologie étale. Serre y introduisait une notion de fibrés algébriques plus large que celle des fibrés localement triviaux pour la topologie de Zariski, les fibrés *localement isotriviaux*, qui sont, localement pour la topologie de Zariski, trivialisés par un revêtement étale fini. Grothendieck, enthousiaste, confia à Serre qu'il voyait déjà comment la "localisation étale", en un sens qui devait être précisé plus tard, donnerait non seulement "le bon H^1" [1], mais aussi "les bons H^i supérieurs" d'une "cohomologie de Weil". Il l'annoncera la même année au congrès international d'Edimbourg [Gr2] : "Such an approach was suggested to me by the *connections between sheaf-theoretic cohomology and cohomology of Galois groups on the one hand, and the classification of unramified coverings of a variety on the other* $\cdots$, and by Serre's idea that a "reasonable" algebraic principal fiber space $\cdots$ should become locally trivial on some covering *unramified* over a given point." Cependant, quelques années s'écoulèrent avant que cette idée ne prenne réellement forme : Grothendieck ne voyait pas comment démarrer. Il avait aussi d'autres occupations :

Professeur émérite à l'Université Paris-Sud, Département de Mathématiques, Bât. 425, 91405 Orsay Cedex, France. Luc.Illusie@math.u-psud.fr.

[1]Un point technique : Raynaud a donné des exemples de torseurs sous des schémas abéliens qui sont localement triviaux pour la topologie étale, mais non localement isotriviaux [R, p. 199-200].

• ses exposés au séminaire Bourbaki sur les fondements de la géométrie algébrique [FGA] (géométrie formelle et géométrie algébrique, descente fidèlement plate, modules formels, construction de schémas de Hilbert et de Picard), qui se poursuivront jusqu'en mai 1962,

• sa série d'exposés au séminaire Cartan 1960-61 sur la géométrie analytique complexe,

• ses séminaires à l'IHÉS, SGA 1, 2, et 3, qui s'échelonnent de 1960 à 1963, et son séminaire à Harvard de l'automne 1961,

• la rédaction des ÉGA (ÉGA I sort en 1960, ÉGA II en 1961, ÉGA III en 1962-63, ÉGA IV, Première partie, en 1964).

D'autre part, des travaux de fondements étaient indispensables. D'abord, la notion même de *morphisme étale*, dans le cadre des schémas, n'avait pas encore été définie. Grothendieck le fera au début de SGA 1. Le choix de l'adjectif "étale", inspiré de la notion classique de "domaine étalé" en géométrie analytique complexe, sera, d'ailleurs, l'une de ses plus grandes réussites terminologiques. Les diverses définitions équivalentes des morphismes étales, lisses, et non ramifiés, leurs propriétés différentielles, les critères infinitésimaux serviront de base à tout l'édifice de la cohomologie étale. Les morphismes étales, analogues des isomorphismes locaux de la géométrie analytique complexe, permettent en quelque sorte de faire comme si l'on disposait, en géométrie algébrique, d'un théorème des fonctions implicites. Ce point de vue sera exploité quelques années plus tard par Artin dans sa théorie d'algébrisation (voir notamment [Ar2], [Ar3]). La localisation étale amenait à étudier l'opération inverse, le recollement, ou descente. Pour ce faire, Grothendieck introduira le langage des catégories fibrées ([FGA, Exp. 190], [SGA 1 VI]), que Giraud développera dans toute la généralité désirable ([Gi1], [Gi2]). Pour en revenir à la cohomologie étale, le premier objectif était l'étude du H^1, ce qui amenait à examiner le *groupe fondamental*. La construction, catégorique, que Grothendieck en donnera dans [SGA 1 V], ne sera pas dépassée. Elle sera imitée dans d'autres contextes (catégories tannakiennes, géométrie logarithmique, notamment). L'un des buts principaux de [SGA 1] était d'obtenir la structure du groupe fondamental *premier à p* d'une courbe lisse sur un corps algébriquement clos de caractéristique p, c'est-à-dire une description par générateurs et relations analogue à celle du cas transcendant [SGA 1 XIII 2.12]. En fait, les théorèmes de spécialisation pour le groupe fondamental qui sont à la base de cette description, en même temps que les théorèmes de type Lefschetz développés dans [SGA 2] donneront la clé des théorèmes fondamentaux de la cohomologie étale pour les coefficients de torsion.

Serre était sceptique sur la possibilité de comprendre les H^i supérieurs. D'ailleurs, comme le souligne Grothendieck dans [SGA 4 VII 2.1] : "On peut dire qu'en passant de la topologie de Zariski à topologie étale, "on a fait ce qu'il fallait" pour obtenir "le bon" H^1 [...] pour un groupe de coefficients constant fini G. *C'est un fait remarquable, qui sera démontré dans la suite de ce séminaire, que cela suffit également pour trouver les "bons"* $H^i(X, G)$ *pour tout groupe de coefficients de torsion (du moins si* G *est premier aux caractéristiques résiduelles de* X*)*". Je vois cependant quelques raisons à l'optimisme initial de Grothendieck. Grothendieck pensait toujours en termes relatifs : un espace au-dessus d'un autre. Une fois la cohomologie des courbes (sur un corps algébriquement clos) tirée au clair, on pouvait espérer des résultats similaires pour les images directes pour une courbe *relative* (les théorèmes de spécialisation du π_1 devaient le lui suggérer), et, "par dévissage" (fibrations en courbes, suites spectrales de Leray), atteindre les H^i supérieurs. Ces dévissages allaient faire intervenir la notion fondamentale de *faisceau constructible* [SGA 4 IX]. Elle était inspirée par celle, classique, de partie constructible [EGA O_{III} 9], dont le théorème de Chevalley (cf. [EGA IV 1.8.4]) avait montré l'intérêt. Les images directes font en effet sortir de la catégorie des faisceaux localement constants, tandis que la constructibilité devait être préservée.

Un autre ingrédient crucial allait être l'usage systématique de la *hensélisation*, introduite par Nagata ([Na1], [Na2], [Na3]) et développée par Grothendieck dans [EGA IV 18]. Cette notion avait été peu utilisée jusque là. C'est Grothendieck qui a mis en évidence son importance en géométrie algébrique, particulièrement pour la cohomologie étale. Les schémas *strictement locaux*, i. e. les spectres d'anneaux locaux strictement henséliens, sont en effet les objets locaux de la topologie étale, de même que les schémas locaux, i. e. les spectres d'anneaux locaux, sont les objets locaux de la topologie de Zariski. Ils se comportent topologiquement comme des "boules de Milnor". La fibre d'un faisceau étale en un point géométrique est l'ensemble de ses sections globales sur le localisé strict correspondant. Ce fait, combiné avec des théorèmes généraux de passage à la limite dans la cohomologie ([SGA 4 VI]), jouera un rôle capital dans tous les principaux théorèmes de la cohomologie étale. La hensélisation, qui fait intervenir une limite inductive filtrante, est à bien des égards plus maniable que la complétion. La descente du complété au hensélisé, dans le cas excellent, donnera lieu aux théorèmes d'approximation d'Artin [Ar4] et, plus tard, à ceux de Popescu ([P1], [P2]).

2. Les théorèmes fondamentaux pour les coefficients de torsion

Au printemps 1962, Artin donne à Harvard un séminaire sur les "topologies de Grothendieck". Il y présente la notion - toute nouvelle à l'époque - de topologie sur une catégorie, et le formalisme de faisceaux correspondant. Les mots "site" et "topos" n'y figurent pas encore. Ce sera le point de départ de la théorie développée plus sytématiquement, avec le concours de Giraud et Verdier, dans le premier volume de [SGA 4]. En même temps, Artin définit précisément la topologie étale : la considération des familles surjectives de morphismes étales arbitraires sera un point essentiel pour toute la suite. Il donne les premiers résultats sur la dimension cohomologique étale des schémas et le calcul de la cohomologie des courbes à coefficients dans un faisceau constructible (le cas des coefficients constants avait été abordé, dans un autre langage, par Kawada-Tate [KT]). Il examine aussi le cas d'une surface fibrée sur une courbe. Ce séminaire débloque complètement la situation. Grothendieck va désormais consacrer toute son énergie à la cohomologie étale. En moins d'un an, de septembre 62 à mars 63, Artin et Grothendieck auront établi les théorèmes fondamentaux pour les coefficients de torsion. Dressons-en brièvement la liste :

(i) structure de la cohomologie des courbes [SGA 4 IX 4.6]

(ii) changement de base propre [SGA 4 XII et XIII]

(iii) acyclicité locale des morphismes lisses [SGA 4 XV]

(iv) pureté relative et changement de base lisse [SGA 4 XVI]

(v) finitude pour un morphisme propre [SGA 4 XIV]

(vi) bornes pour la dimension cohomologique, Lefschetz affine [SGA 4 X, XIV]

(vii) comparaison avec la cohomologie de Betti [SGA 4 XI, XVI]

(viii) cohomologie à supports propres et dualité globale [SGA 4 XVII, XVIII]

(ix) dualité locale [SGA 5 I].

Les résultats (i), (ii) et (iii) sont en fait les piliers sur lesquels les autres reposent. Le théorème de Tsen et des énoncés de cohomologie galoisienne sont les outils qui permettent de calculer la cohomologie, à coefficients dans $\mathbb{G}_m$, d'une courbe propre et lisse sur un corps algébriquement clos, et donc aussi sa cohomologie à coefficients dans un faisceau constant $\mathbb{Z}/n\mathbb{Z}$. Plus tard, Grothendieck reviendra sur l'étude du *groupe de Brauer cohomologique* $H^2(X, \mathbb{G}_m)$ pour des schémas X plus généraux [Gr4]. Fibrations en courbes et suites spectrales de Leray ramènent (ii) à un théorème de spécialisation pour le π_1, sur une base S strictement locale noethérienne. Seul le cas où S est le spectre d'un anneau local noethérien complet était traité dans [SGA 1 X], Des arguments supplémentaires (délicats) sont donc requis [SGA 4 XIII]. Les théorèmes d'algébrisation d'Artin évoqués plus haut apporteront une simplification notable (cf. [SGA 4 1/2 Cohomologie étale : Les points de départ]). Des dévissages astucieux réduisent la vérification de l'acyclicité locale d'un morphisme lisse $X \to S$ au cas où X est la droite affine $S[t]$ sur S. La théorie de Lefschetz locale de [SGA 2 X] pour la section hyperplane $t = 0$, combinée au lemme d'Abyankhar et au théorème de pureté de Zariski-Nagata, permet alors de conclure.

Pour s'assurer que la cohomologie étale donnerait les "bons" H^i supérieurs, il y avait deux tests cruciaux. Le premier était la comparaison avec la cohomologie de Betti pour les schémas X séparés de type fini sur $\mathbb{C}$. Artin le fera d'abord, pour des coefficients constants du type $\mathbb{Z}/n\mathbb{Z}$, dans le cas où X est *lisse* sur $\mathbb{C}$, à l'aide d'une notion nouvelle qu'il dégage à cette occasion, celle de *bon voisinage* (et de *fibration élémentaire*) [SGA 4 XI]. Artin procède par réduction au cas d'une courbe (relative). Serre, de son côté, observe qu'un bon voisinage est un espace $K(\pi, 1)$, et donc que sa cohomologie étale se calcule de façon galoisienne (ce qui implique aisément la comparaison désirée). Cette remarque sera exploitée plus tard par Faltings dans sa théorie de Hodge p-adique [Fa 2]. Dans le cas général, pour des coefficients constructibles, et des $R^i f_*$ plutôt que des H^i, Artin, dans [SGA 4 XVI], fera appel à la résolution des singularités de Hironaka et au théorème de pureté pour les couples lisses (iv). Le second test était le calcul des $H^i(X, \mathbb{Z}/n\mathbb{Z})$ pour X l'espace affine épointé de dimension d sur un corps algébriquement clos (et n premier à la caractéristique), ou le localisé strict épointé correspondant : on devait trouver les mêmes valeurs que pour une sphère de dimension $2d-1$. Là encore, ce sera une conséquence facile de (iv) (le cas $d = 2$ avait déjà été traité, plus élémentairement, par Artin dans [Ar1]). D'ailleurs (iv) et (v), ainsi que les bornes pour la dimension cohomologique de schémas de type fini sur un corps [SGA 4 X], découlent sans grande difficulté de (i), (ii), et (iii). Il n'en va pas de même pour le théorème de Lefschetz affine (vi), dont la démonstration met en œuvre un subtil va-et-vient entre énoncés locaux et globaux. En fait, c'est dans l'énoncé dégagé là [SGA 4 XIV 3.1] qu'apparaît pour la première fois une condition de *perversité* (ou plutôt de *semi-perversité*). Il jouera un rôle important dans la théorie des faisceaux pervers [BBD]. Par ailleurs, la méthode de démonstration sera reprise par Gabber pour un morphisme affine sur un trait (au lieu d'un corps) (cf. [I2]) ; ce sera un ingrédient essentiel de sa démonstration de la conjecture de pureté absolue de Grothendieck (cf. [Fu2]).

Lorsqu'il s'attaque à la dualité en cohomologie étale, Grothendieck a déjà en tête un canevas, fourni par la dualité dans le cas des coefficients continus (faisceaux cohérents), traitée dans le cadre des catégories dérivées, même si ce n'est qu'à l'été 1963 qu'il rédigera les "prénotes" pour le séminaire de Hartshorne à Harvard de 1963-64 [H]. Le théorème de dualité globale de (*loc.cit*) avait inspiré à Verdier un analogue pour les coefficients discrets dans le cadre topologique ([V1], [V2]). Le passage au cadre étale posait cependant des problèmes sérieux. Il fallait d'abord définir la cohomologie à supports propres, et plus généralement, les images directes à supports propres. Cette définition n'allait pas de soi. La définition naïve, calquée sur celle du cas topologique, ne donnait pas les "bons" groupes de cohomologie. C'est un miracle que la définition hybride proposée par Grothendieck, à savoir $Rf_! := Rg_* \circ j_!$, où $f = gj$ est une compactification de f (j une immersion ouverte, g propre), ait finalement si bien fonctionné (ce foncteur était d'ailleurs noté initialement $R_! f$, pour indiquer qu'il ne s'agissait pas du foncteur dérivé du foncteur composé $g_* j_!$). Dans le séminaire oral, Grothendieck avait défini $Rf^! : D^+(Y, \mathbb{Z}/n\mathbb{Z}) \to D^+(X, \mathbb{Z}/n\mathbb{Z})$ pour $f : X \to Y$ *lissifiable*, i. e. de la forme gi, où g est lisse et i une immersion fermée, et n inversible sur Y, comme le composé $Ri^! \circ g^*[2d](d)$, où d est la dimension relative, et $Ri^!$ le dérivé du foncteurs "sections à supports dans Y", restreint à Y. L'indépendance du choix de la factorisation résultait facilement du théorème de pureté relatif (iv) (là aussi, la notation primitive était $R^! f$, ce foncteur n'étant pas non plus le dérivé d'un foncteur au niveau des faisceaux). La dualité globale exprime que le foncteur $Rf^!$ ainsi défini est adjoint à droite de $Rf_!$, ce qui fournit pour les schémas quasi-projectifs lisses sur un corps algébriquement clos, la dualité de Poincaré sous une forme analogue à celle du cas transcendant. La difficulté réside principalement dans la construction du morphisme d'adjonction (ou *morphisme trace*). La démonstration est esquissée dans l'article de Verdier [V3]. S'inspirant de la méthode utilisée par Verdier pour la dualité dans le cadre topologique, Deligne, dans [SGA 4 XVIII], définira *a priori* $Rf^!$ comme adjoint à droite de $Rf_!$. La difficulté est alors reportée dans le calcul de $Rf^!$ pour f lisse, où l'on doit retrouver la définition de Grothendieck. Deligne emploiera une méthode analogue dans l'appendice de [H]. Une présentation axiomatique de ces constructions, à partir d'une version triangulée d'un théorème de représentabilité de Brown, sera donnée par Neeman dans [Ne].

La dualité locale (ix), plus profonde, sera exposée par Grothendieck dans [SGA 5 I]. Il s'agit, comme dans le cas des coefficients continus [H], de construire, sur des schémas X "raisonnables", des *foncteurs dualisants* $D_X = R\mathcal{H}om(-, K)$, où K est un "complexe dualisant", de manière que $D_X D_X = \mathrm{Id}$ (et donc que D_X induise une anti-équivalence de la catégorie $D^b_c(X, \mathbb{Z}/n\mathbb{Z})$ avec elle-même) (n premier aux caractéristiques résiduelles, D^b_c désignant la sous-catégorie des complexes à cohomologie bornée et constructible). Cette construction sera possible sur les schémas excellents de caractéristique nulle, grâce à la résolution de Hironaka et aux résultats d'Artin dans [SGA 4 XIX], mais seulement conjecturale dans le cas excellent noethérien général. Une construction inconditionnelle dans ce cas ne sera donnée que très récemment, par Gabber [Ga]. Elle utilise, notamment, sa démonstration, mentionnée plus haut, de la conjecture de pureté absolue.

Si le théorème de finitude pour un morphisme propre (v), et plus généralement, la constructibilité des $R^q f_! F$ pour F constructible de torsion, s'avérait être une conséquence facile de (i) et (ii), il n'en allait pas de même pour la constructiblité des $R^q f_* F$. Tout d'abord, celle-ci requiert que la torsion de F soit première aux caractéristiques résiduelles. Pour un schéma X séparé de type fini sur un corps algébriquement clos k, la finitude des $H^q(X, F)$, pour F un faisceau constructible de $\mathbb{Z}/n\mathbb{Z}$-modules (n premier à la caractéristique de k) était connue, dans [SGA 4], dans le cas quasi-projectif lisse, comme conséquence de la dualité de Poincaré, mais n'était que conjecturale dans le cas général, subordonnée à la résolution des singularités. La constructibilité des $R^q f_* F$, pour $f : X \to Y$ un morphisme de schémas de type fini sur un corps (et F comme ci-dessus), ne sera établie qu'une dizaine d'années plus tard, par Deligne, dans [SGA 4 1/2, Th. de finitude]. Deligne traite plus généralement le cas d'un morphisme de schémas de type fini sur un schéma noethérien régulier de dimension ≤ 1, et établit la dualité locale dans ce cadre. Dans [SGA 4 XIX], Artin avait prouvé la constructibilité des $R^q f_* F$ pour f de type fini entre schémas excellents de caractéristique nulle (et F constructible de torsion), à l'aide de Hironaka. Gabber, exploitant les théorèmes d'altération de de Jong ([dJ 1], [dJ2]) réussira à lever cette restriction de caractéristique [Ga], i. e. prouver la constructibilité des $R^q f_* F$ pour f de type fini entre schémas quasi-excellents noethériens et F constructible de torsion première aux caractéristiques résiduelles. Il généralisera aussi, dans ce cadre, le théorème de Lefschetz affine (vi) et résoudra certaines conjectures sur la dimension cohomologique des corps posées dans l'exposé d'Artin [SGA 4 X].[2]

Indépendamment du problème de finitude pour Rf_*, le formalisme présenté dans [SGA 4] et [SGA 5], consistant en la définition des foncteurs $\otimes^L$, $R\mathcal{H}om$, f^*, Rf_*, $Rf_!$, $Rf^!$ et les relations entre eux, appelé, plus tard, "formalisme des six opérations", ainsi que le formalisme analogue développé dans [H], changeaient radicalement la vision que l'on avait jusque-là de la cohomologie. Ils allaient avoir un impact considérable, non seulement en géométrie algébrique, mais aussi, plus tard, dans des domaines apparemment éloignés, comme la théorie analytique complexe des $\mathcal{D}$-modules (Kashiwara-Schapira, Mebkhout, etc.) (cf. [KSc]) et son analogue algébrique (Beilinson, Bernstein) (cf. [Bo]) - qui, à leur tour, inspireront celle des $\mathcal{D}$-modules arithmétiques de Berthelot [Be].

3. La cohomologie ℓ-adique

Comme Grothendieck le rappelle dans l'avant-propos de [SGA 4], le but était de construire une "cohomologie de Weil" sur les schémas. Une telle théorie devait fournir des espaces vectoriels de dimension finie sur un corps de caractéristique nulle. Or la cohomologie étale n'est raisonnable que pour des coefficients *de torsion* (laquelle doit être de surcroît, le plus souvent, supposée première aux caractéristiques résiduelles) : pour X intègre et géométriquement unibranche, un calcul élémentaire montre que les $H^q(X, \mathbb{Z})$ sont de torsion pour $q \geq 1$. C'est ce qui a amené Grothendieck à définir la cohomologie ℓ-adique par passage à la limite sur les cohomologies à coefficients $\mathbb{Z}/\ell^n\mathbb{Z}$, par analogie avec la définition du module de Tate d'une variété abélienne. En particulier, pour X séparé de type fini sur un corps algébriquement clos k de caractéristique p, et l premier différent de p, les groupes de cohomologie à supports compacts de X à coefficients dans $\mathbb{Z}_\ell$ sont définis par

$$H^q_c(X, \mathbb{Z}_\ell) := \varprojlim_n H^q_c(X, \mathbb{Z}/\ell^n\mathbb{Z}).$$

Ce sont des $\mathbb{Z}_\ell$-modules de type fini, et les groupes de cohomologie à supports compacts à coefficients dans $\mathbb{Q}_\ell$, définis par $H^q_c(X, \mathbb{Q}_\ell) = \mathbb{Q}_\ell \otimes_{\mathbb{Z}_\ell} H^q_c(X, \mathbb{Z}_\ell)$, des $\mathbb{Q}_\ell$-espaces vectoriels de dimension finie. Plus généralement, Grothendieck propose une définition de $\mathbb{Z}_\ell$-faisceau (constructible) comme système projectif de faisceaux constructibles de $\mathbb{Z}/\ell^n\mathbb{Z}$-modules s'induisant les uns les autres par réduction modulo ℓ^n. Ceux d'entre eux pour lesquels les faisceaux de $\mathbb{Z}/\ell^n\mathbb{Z}$-modules sont localement constants - qu'il appelle

[2] Pour tous ces résultats, voir l'ouvrage *Travaux de Gabber sur l'uniformisation locale et la cohomologie étale des schémas quasi-excellents*, Séminaire à l'École polytechnique 2006-2008, dirigé par L. Illusie, Y. Laszlo et F. Orgogozo, à paraître dans Astérisque (SMF).

constants tordus (on dira *lisses* plus tard) - correspondent (dans le cas noethérien connexe) aux représentations du groupe fondamental dans des $\mathbb{Z}_\ell$-modules de type fini. Il prouve la stabilité (essentielle) des $\mathbb{Z}_\ell$-faisceaux par les $R^q f_!$. Ceci est expliqué dans les exposés de Jouanolou [SGA 5 V et VI]. Faute de disposer de théorèmes de finitude pour les $R^q f_*$, il n'était guère possible, à l'époque, d'aller beaucoup plus avant, du moins, autrement que conjecturalement. La théorie exposée dans (*loc. cit.*) suffira néammoins pour le formalisme des fonctions L.

Elle fournira aussi une *cohomologie de Weil* pour les schémas propres et lisses sur un corps algébriquement clos, en fait, une pour chaque nombre premier ℓ différent de la caractéristique (cf. [Kl]). La construction, par Grothendieck, de la classe de cohomologie associée à un cycle y joue un rôle crucial. Dans le cas des schémas singuliers, Grothendieck définira, à l'aide de la dualité locale, une théorie d'*homologie*, adaptant et généralisant la théorie de Borel-Moore du cas topologique (cf. [BM], [V5]). Ces constructions, qui avaient été exposées en détail dans le séminaire oral, ne seront pas reprises dans [SGA 5]. Elles font l'objet des exposés de Deligne [SGA 4 1/2, Cycle] et Laumon [La1]. La distinction entre homologie et cohomologie et les relations entre les deux étaient chères à Grothendieck. Elles se manifesteront de nouveau, dans [SGA 6], dans le jeu entre complexes "pseudo-cohérents" et complexes "parfaits", et entre les groupes de Grothendieck correspondants, désignés dans (*loc. cit.*) par $K_{\cdot}(X)$ et $K^{\cdot}(X)$. Bien entendu, Grothendieck avait aussi transposé, en cohomologie étale, la théorie des classes de Chern de [Gr1]. Elle est exposée dans [SGA 5 VII]. Il en donnera même une version plus générale, valable sur des topos de base, et en particulier applicable aux représentations linéaires des groupes discrets, dans [Gr5].

Restait cependant à étendre au cadre ℓ-adique le formalisme des six opérations. C'était la tâche assignée à Jouanolou pour sa thèse. Celle-ci ne sera pas publiée. A partir de 1973, disposant des théorèmes de Deligne de [SGA 4 1/2 Th. finitude], on pouvait espérer une solution inconditionnelle du problème pour les schémas séparés de type fini sur un schéma régulier de dimension ≤ 1. Celle-ci ne sera trouvée que plus tard, par Gabber. Elle ne sera pas rédigée. Une version équivalente sera obtenue indépendamment par Ekedahl, et, cette fois, publiée [E]. Une solution partielle, valable en tout cas pour les schémas séparés de type fini sur un corps fini ou un corps algébriquement clos, sera présentée par Deligne dans [D2]. Elle suffira pour les applications qu'il avait en vue, ainsi que la théorie ultérieure des faisceaux pervers [BBD]. Diverses extensions de la théorie peuvent être désormais envisagées : schémas - voire champs algébriques - excellents (grâce aux résultats récents de Gabber [Ga]), cf. [LO]. Les fondations sont encore inachevées !

4. Formule des traces et rationalité des fonctions L

L'idée de Weil, à la base de ses conjectures, était que le nombre de points rationnels d'une variété projective lisse X sur un corps fini $k = \mathbb{F}_q$ devait pouvoir s'exprimer comme somme alternée des traces (*nombre de Lefschetz*) de l'opérateur de Frobenius sur des groupes de cohomologie convenables de X : l'ensemble $X(k)$ des points rationnels de X sur k est en effet l'ensemble des points fixes de cet opérateur sur $X(\overline{k})$, où $\overline{k}$ est une clôture algébrique de k. Le fait que, pour ℓ premier distinct de la caractéristique de k, $X \mapsto H^*(X_{\overline{k}}, \mathbb{Q}_\ell))$ forme une "cohomologie de Weil" impliquait formellement une telle formule de Lefschetz (cf. [Kl]), à savoir :

$$(1) \qquad |X(k)| = \sum_i (-1)^i \operatorname{Tr}(F, H^i(X_{\overline{k}}, \mathbb{Q}_\ell)),$$

où F désigne le Frobenius *géométrique* ($a \rightarrow a^{1/q}$ sur $\overline{k}$). Grothendieck ira beaucoup plus loin. Il généralisera (1) au cas où le schéma X est seulement supposé séparé de type fini sur k et $\mathbb{Q}_\ell$ est remplacé par un $\mathbb{Q}_\ell$-faisceau arbitraire E, la cohomologie étant prise à supports compacts. La formule s'écrit alors :

$$(2) \qquad \sum_{x \in X(k)} \operatorname{Tr}(F, E_x) = \sum_i (-1)^i \operatorname{Tr}(F, H^i_c(X_{\overline{k}}, E)),$$

(au premier membre, E_x désigne la fibre de E au point géométrique $x \in X(\overline{k})$ fixe par F). Cette formule, qu'on appelle *formule des traces de Grothendieck*, entraîne non seulement la rationalité de la fonction zêta de X (cas $E = \mathbb{Q}_\ell$), mais aussi de la fonction L associée à E :

$$(3) \qquad L(X, E, t) = \prod_i \det(1 - Ft, H^i_c(X_{\overline{k}}, E))^{(-1)^{i+1}}.$$

À la différence de (1), elle ne résulte pas formellement des propriétés générales de la cohomologie ℓ-adique. Sa démonstration a donné lieu à malentendus et polémiques (cf. [RS]). L'histoire est la suivante. Des dévissages faciles montrent qu'il suffit de prouver (2) quand X est une *courbe propre et lisse* sur k. Dans ce cas, Grothendieck avait donné une démonstration (complète) par une méthode de *traces non commutatives* inspirée de travaux de Nielsen-Wecken. Cette démonstration est exposée par Bucur dans [SGA 5 XII], et sous une forme simplifiée, par Deligne dans [SGA 4 1/2, Rapport]. Grothendieck avait aussi esquissé, dans son exposé Bourbaki [Gr3], une autre approche, à partir de la *formule de Lefschetz-Verdier*. Cette formule très générale, due à Verdier, exprime, pour un schéma X *propre* sur le spectre S d'un corps algébriquement clos, le nombre de Lefschetz d'une *correspondance cohomologique* (c, u) sur X ($c : Z \rightarrow X \times X$, $u \in Hom(c_1^* L, c_2^! L)$, L un complexe de tor-dimension finie et à cohomologie constructible de $\mathbb{Z}/l^n\mathbb{Z}$-modules, ℓ inversible sur S) comme une somme de *termes locaux* attachés aux composantes de points fixes de c. Verdier avait montré [V4] que cette formule implique que si X est une courbe propre et lisse sur S, f un endomorphisme de X, L un $\mathbb{Q}_\ell$-faisceau sur X, $u : f^*L \rightarrow L$ un endomorphisme de L "au-dessus de f", alors, si les points fixes de f sont isolés et transversaux, on a

$$\sum_{x \in X^f} \operatorname{Tr}(u, L_x) = \sum_i (-1)^i \operatorname{Tr}((f, u), H^i(X, L)). \tag{4}$$

La formule désirée pour Frobenius en est un cas particulier. Le problème de cette méthode était le statut de la formule de Lefschetz-Verdier. Tout d'abord, la définition des termes locaux utilisait des propriétés non démontrées à l'époque (finitude, Künneth, dualité locale). Dans le cas d'une courbe propre et lisse X, celles-ci étaient cependant disponibles. Ensuite, la formule devait découler de compatibilités formelles, qui n'avaient pas été vérifiées. Ces difficultés seront levées dans la version publiée ([SGA 5 III, III B]) : les propriétés requises avaient été établies par Deligne dans [SGA 4 1/2 Th. finitude], et les compatibilités nécessaires sont prouvées ; diverses généralisations de (4) et [SGA 5 XII] y sont également données.

La formule (2) jouera un rôle fondamental dans les travaux de Deligne sur la conjecture de Weil (voir plus bas). Elle suggérera aussi à Deligne une conjecture sur les termes locaux de la formule de Lefschetz-Verdier, pour des correspondances cohomologiques composées avec une grande puissance de Frobenius, conjecture qui sera démontrée par Fujiwara [Fu1] et, indépendamment, Varshavsky [Va].

Par la méthode de Nielsen-Wecken, Grothendieck avait non seulement démontré le cas particulier de (2) où X est une courbe propre et lisse, mais aussi une généralisation de (4), ou plutôt de sa variante pour des coefficients de torsion, sans hypothèse de transversalité sur les points fixes, cf. [SGA 5 XII]. Les termes locaux qui apparaissent sont analogues à ceux que Grothendieck avait découverts pour la formule donnant la caractéristique d'Euler-Poincaré $\chi(X, L)$, qui font intervenir des conducteurs de Swan. Cette dernière formule, qu'on appelle maintenant *formule de Grothendieck-Ogg-Shafarevitch*, est le fruit d'un extraordinaire échange épistolaire entre Grothendieck et Serre, cf. [GS, 139-144]. Elle inspirera un grand nombre de travaux ultérieurs. Une généralisation de celle-ci, de type Riemann-Roch-Grothendieck, reste un problème ouvert, activement étudié aujourd'hui (voir notamment les travaux récents d'Abbes, Kato, Saito [AS], [KS]).

Pour X propre et lisse sur $k = \mathbb{F}_q$ et $E = \mathbb{Q}_\ell$, (3) s'écrit

$$Z(X, t) = \prod_i \det(1 - Ft, H^i(X_{\overline{k}}, \mathbb{Q}_\ell))^{(-1)^{i+1}}.$$

L'équation fonctionnelle de $Z(X, t)$, conjecturée par Weil, découle aisément de la dualité de Poincaré (cf. [D1, 2.6]). Quelques années après, Deligne démontrera que le polynôme $\det(1 - Ft, H^i(X_{\overline{k}}, \mathbb{Q}_\ell))$ est à coefficients entiers, indépendant de ℓ, et que les inverses de ses racines (valeurs propres de F) sont des entiers algébriques dont tous les conjugués complexes sont de valeur absolue $q^{i/2}$, achevant ainsi la preuve des conjectures de Weil (voir [D1] pour le cas projectif, [D2] pour le cas général).

Les conjectures de Weil, et le formalisme des poids (que Grothendieck appelait “yoga des poids”, conjectural à l'époque), développé par Deligne dans [D2] puis dans la théorie des faisceaux pervers [BBD], auront d'immenses applications, dont la description sort du cadre de cet exposé. Bornons-nous à mentionner :

• la théorie de Lusztig des faisceaux caractères [Lu]

• la transformation ℓ-adique de Fourier-Deligne et la méthode de la phase stationnaire ℓ-adique de Laumon (inspirée d'une démonstration de Witten des inégalités de Morse) ([La3], [KL], [I1])

• la construction, par Lafforgue, de la correspondance de Langlands pour GL_n sur les corps de fonctions [Laf]

• la preuve, par Laumon-Ngo et Ngo, du lemme fondamental de Langlands ([LaN], [Ng]).

5. Cycles évanescents et monodromie

Dans l'avant-propos de [SGA 4], Grothendieck mentionne l'influence sur la cohomologie étale des travaux d'Igusa sur les cycles évanescents. Il annonce qu'il en sera question dans un séminaire ultérieur, à savoir [SGA 7]. Celui-ci, continuation logique de [SGA 5], ne commencera toutefois qu'un an plus tard, après [SGA 6]. Dans sa lettre à Serre du 30 octobre 1964 [GS, p. 214], Grothendieck introduit déjà les foncteurs Φ^n et les calcule dans un cas crucial. Ces foncteurs, qui “mesurent” la différence entre la cohomologie de la fibre générale et de la fibre spéciale, seront l'objet principal de la théorie développée dans [SGA 7]. Le livre de Milnor [M] ne paraît qu'en 1968. Grothendieck devait cependant en avoir eu connaissance quelque temps auparavant. La conjecture de (*loc. cit.*) sur la quasi-unipotence de la monodromie d'un point critique isolé a en effet été pour lui une motivation, en même temps qu'un test, pour sa théorie générale des foncteurs $R\Psi$ (*cycles proches*) et $R\Phi$ (*cycles évanescents*) (la terminologie “cycles proches” ne lui est cependant pas due, ce n'est que vers la fin des années 70 qu'elle deviendra d'un usage courant).

Les deux résultats majeurs que Grothendieck allait présenter dans [SGA 7] sont :

(1) le théorème de réduction semi-stable pour les variétés abéliennes [SGA 7 IX 3.6],

(2) le théorème de monodromie locale [SGA 7 I 1.3].

La démonstration de (1) donnée par Grothendieck dans (*loc. cit.*) s'appuie d'une part sur un théorème d'orthogonalité pour les modules de Tate [SGA 7 IX 2.4], d'autre part sur un résultat sur l'échelon de quasi-unipotence du H^1 des courbes, qui était établi dans sa lettre à Serre citée plus haut (il figure, avec sa démonstration originelle, dans le résumé rédigé par Deligne [SGA 7 I 3.5]). Mumford donnera, à peu près au même moment une démonstration indépendante, à l'aide de sa théorie des fonctions thêta, valable en caractéristique résiduelle $\neq 2$). Deligne et Mumford montreront que (1) implique le théorème de réduction semi-stable pour les courbes [DM,

2.4], dont, un peu plus tard, Artin et Winters donneront une démonstration indépendante de (1) [AW].

Grothendieck donna deux démonstrations de (2), l'une géométrique et conditionnelle (dépendant de la pureté et de la résolution des singularités), l'autre, arithmétique et inconditionnelle. Cette dernière repose sur un résultat élémentaire de quasi-unipotence pour des représentations ℓ-adiques d'un corps local à corps résiduel pas trop gros [ST, Appendix], obtenu bien avant (cf. lettre de Grothendieck à Serre du 24.9.64 [GS, p. 182]). Une variante, inconditionnelle, de la première démonstration, sera donnée par de Jong à la fin des années 90 [dJ1].

Les théorèmes (1) et (2) et le formalisme des foncteurs $R\Psi$ et $R\Phi$ seront extrêmement féconds. La conjecture de Milnor citée plus haut est un corollaire facile de (2). La théorie cohomologique des pinceaux de Lefschetz, développée dans la seconde partie de [SGA 7], sera un ingrédient essentiel de la première démonstration, par Deligne, de la conjecture de Weil sur les valeurs propres de Frobenius [D1]. A la suite de (1), la notion de réduction semi-stable prendra une importance considérable dans les questions de compactifications de problèmes de modules, par le biais de la théorie des variétés toriques, et plus tard, de la géométrie logarithmique. Une généralisation en dimension relative supérieure du théorème de réduction semi-stable pour les courbes, en caractéristique nulle, sera obtenue par Mumford et al. [KKMS]. On peut considérer le théorème "d'altération semi-stable" de de Jong [dJ1] comme un substitut efficace à ce dernier résultat en caractéristique positive ou mixte. Par ailleurs, une généralisation de (1) au cas d'une base normale, due à Gabber [D3], sera un ingrédient important de la démonstration, par Faltings [Fa1], de la conjecture de Mordell.

Le calcul des cycles évanescents dans le cas de réduction semi-stable, en caractéristique positive ou mixte, et pour le faisceau constant, conditionnel dans [SGA 7 I], sera effectué quinze ans après, par Rapoport et Zink [RZ]. Réduction semi-stable et cycles évanescents joueront un rôle majeur en théorie de Hodge classique et, à partir de la fin des années 80, dans la théorie de Hodge p-adique. La conjecture dite de *monodromie-poids* ("pureté de la filtration de monodromie") reste à ce jour l'un des problèmes ouverts majeurs du sujet. Rappelons que celle-ci n'est établie pour l'instant qu'en égale caractéristique positive ([D2, 1.8.4], [It1]) ou en égale caractéristique nulle [St], [Sa, 4.2]) (mis à part le cas de dimension relative ≤ 2 [RZ] ou de variétés admettant une uniformisation p-adique ([dS], [It2])).[3]

Le foncteur $R\Psi$ globalise, le long de la fibre spéciale, les cohomologies des "fibres de Milnor". Il n'est défini toutefois que sur une base de dimension 1. Au début des années 80, Deligne proposera une construction sur des bases de dimension quelconque [La2] et formulera des conjectures de finitude qui n'ont été établies que tout récemment [Or]. L'étude de ce foncteur $R\Psi$ généralisé

[3]Voir [P. Scholze, *Perfectoid spaces*, Publ. math. de l'IHÉS 116 (2012), no. 1, 245–313] pour une démonstration de cette conjecture dans le cas des intersections complètes.

est encore embryonnaire. Signalons cependant que la construction de Deligne intervient dans les travaux de Gabber cités à la fin du numéro 3.

6. Cohomologies ℓ-adiques et motifs

C'est sans doute le fait que les cohomologies ℓ-adiques pour ℓ premier aux caractéristiques résiduelles se comportent toutes de la même manière qui a amené Grothendieck à concevoir sa théorie des motifs. Cependant, en dépit de tous les développements qu'elle a suscités depuis quarante ans (cf. [An], [Mot]), les questions d'indépendance de ℓ qu'elle était censée expliquer restent encore ouvertes pour la plupart. On ignore par exemple si les nombres de Betti ℓ-adiques (à supports propres ou sans supports) d'un schéma séparé de type fini sur un corps algébriquement clos (ℓ différent de la caractéristique) sont indépendants de ℓ. Le cas où X est propre et lisse est cependant connu, comme conséquence du théorème principal de Deligne [D2], ainsi que l'analogue pour les nombres de Betti de la cohomologie d'intersection dans le cas propre, d'après Gabber [Fu]. Voir [Ka] et [I3] pour un aperçu sur ces problèmes, et [Z] pour quelques progrès récents.

Je tiens à remercier M. Artin, P. Deligne, S. Kleiman, B. Messing, F. Orgogozo, M. Raynaud et J-P. Serre pour leurs remarques et suggestions sur une première version de ce texte.

Bibliography

[An] Y. André, *Une introduction aux motifs*, Panoramas et synthèses 17 (2004), 1-258, SMF.

[Ar1] M. Artin, *Grothendieck Topologies*, Harvard mimeographed notes, 1962.

[Ar2] M. Artin, *Algebraization of formal moduli I*, in Global Analysis (papers in honor of K. Kodaira), Univ. Tokyo P. (1969), 21-71.

[Ar3] M. Artin, *The implicit function theorem in algebraic geometry*, Algebraic Geometry (Int. Coll. Tata Inst. Fund. Res., Bombay, 1968), Oxford Univ. Press, London, 1969, 13-34.

[Ar4] M. Artin, *Algebraic approximation of structures over complete local rings*, Pub. Math. IHÉS 36 (1969), 23-58.

[AS] A. Abbes, T. Saito, *The characteristic class and ramification of an ℓ-adic sheaf*, Invent. math. 168 (2007), 567-612.

[AW] M. Artin, G. Winters, *Degenerate fibres and stable reduction of curves*, Topology 10 (1971), 373-383.

[Be] P. Berthelot, *Introduction à la théorie arithmétique des $\mathcal{D}$-modules*, Cohomologies p-adiques et applications arithmétiques II, P. Berthelot, J.-M. Fontaine, L. Illusie, K. Kato, M. Rapoport, Éds., Astérisque 279 (2002), SMF, 1-80.

[Bo] A. Borel et al., Algebraic $\mathcal{D}$-Modules, Perspectives in Math. 2, J. Coates, S. Helgason, eds., 1987, Academic Press Inc.

[BBD] A. Beilinson, J. Bernstein and P. Deligne, *Faisceaux pervers*, Astérisque 100, SMF, 1982.

[BM] A. Borel, J. C. Moore, *Homology theory for locally compact spaces*, Mich. Math. J. 7 (1960, 137-159.

[D1] P. Deligne, *La conjecture de Weil. I*, Pub. Math. IHÉS 43 (1974), 273-307.

[D2] P. Deligne, *La conjecture de Weil. II*, Pub. Math. IHÉS 52 (1980), 137-252.

[D3] P. Deligne, *Le lemme de Gabber*, Séminaire sur les pinceaux arithmétiques : la conjecture de Mordell, par L. Szpiro, Astérisque 127, 1985, 131-150.

[DM] P. Deligne and D. Mumford, *The irreducibility of the space of curves of given genus*, Pub. math. IHÉS 36 (1969), 75-110.

[dJ1] A. J. de Jong, *Smoothness, semi-stability and alterations*, Pub. Math. IHÉS 83 (1996), 51-93.

[dJ2] A. J. de Jong, *Families of curves and alterations*, Ann. Inst. Fourier 47 (1997), 599-621.

[dS] E. de Shalit, The p-adic monodromy-weight conjecture fpr p-adically uniformized varieties, Comp. Math. 141 (2005), 101-120.

[E] T. Ekedahl, *On the Adic Formalism*, The Grothendieck Festschrift, vol. II, Progress in Math. 87, Birkhäuser, 1990, 197-218.

[Fa1] G. Faltings, *Endlichkeitsätze für abelsche Varietäten über Zahlkörpern*, Invent. math. 73 (1983), 349-366.

[Fa2] G. Faltings, *p-adic Hodge theory*, J. of the AMS 1 (1988), 255-299.

[Fu1] K. Fujiwara, *Rigid geometry, Lefschetz-Verdier trace formula and Deligne's conjecture*, Invent. math. 127 (1997), 489-533.

[Fu2] K. Fujiwara, *A Proof of the Absolute Purity Conjecture (after Gabber)*, Algebraic Geometry 2000, Azumino (Ed. S. Usui, M. Green, L. Illusie, K. Kato, E. Looijenga, S. Mukai, S. Saito), Advanced Studies in Pure math. 36 (2002), 153-183.

[Fu3] K. Fujiwara, *Independence of ℓ for Intersection Cohomology (after Gabber)*, Algebraic Geometry 2000, Azumino (Ed. S. Usui, M. Green, L. Illusie, K. Kato, E. Looijenga, S. Mukai, S. Saito), Advanced Studies in Pure math. 36 (2002), 145-151.

[Ga] O. Gabber, *Finiteness theorems for étale cohomology of excellent schemes*, Conference in honor of P. Deligne on the occasion of his 61st birthday, IAS, Princeton, October 2005.

[Gi1] J. Giraud, *Méthode de la descente*, Bull. SMF, Mémoire 2, 1964.

[Gi2] J. Giraud, *Cohomologie non abélienne*, Grundl. der math. 179, Springer-Verlag, 1971.

[Gr1] A. Grothendieck, *La théorie des classes de Chern*, Bull. SMF 86 (1958), 137-154.

[Gr2] A. Grothendieck, *The cohomology theory of abstract algebraic varieties*, Proc. Internat. Congress Math. (Edinburgh, 1958), 1960, 103-118, Cambridge Univ. Press, New York.

[Gr3] A. Grothendieck, *Formule de Lefschetz et rationalité des fonctions L*, Sém. Bourbaki 1964/65, 290, dans *Dix exposés sur la cohomologie des schémas*, A.

Grothendieck and N. Kuiper, eds., Noth Holland Pub. Co., Masson et Co., 1968, 31-45.

[Gr4] A. Grothendieck, *Le groupe de Brauer I, II, III*, dans *Dix exposés sur la cohomologie des schémas*, A. Grothendieck and N. Kuiper, eds., Noth Holland Pub. Co., Masson et Co., 1968, 46-188.

[Gr5] A. Grothendieck, *Classes de Chern et représentations linéaires des groupes discrets*, dans *Dix exposés sur la cohomologie des schémas*, A. Grothendieck and N. Kuiper, eds., Noth Holland Pub. Co., Masson et Co., 1968, 215-305.

[GS] *Correspondance Grothendieck-Serre*, P. Colmez, J-P. Serre Eds., Documents Mathématiques 2, SMF, 2001.

[H] R. Hartshorne, *Residues and Duality*, SLN 20, Springer-Verlag, 1966.

[I1] L. Illusie, *Deligne's ℓ-adic Fourier transform*, Algebraic Geometry Bowdoin 1985, Proc. Symp. in Pure Math. 46, Part 2, S. Bloch, ed., AMS 1987, 151-163.

[I2] L. Illusie, *Perversité et variation*, Manuscripta. math. 112 (2003), 271-295.

[I3] L. Illusie, *Miscellany on traces in ℓ-adic cohomology : a survey*, Japanese J. of Math. 1, 107-136 (2006).

[It1] T. Ito, *Weight-monodromy conjecture over equal characteristic local fields*, Amer. J. Math. 127 (2005), 647-658.

[It2] T. Ito, *Weight-modromy conjecture for p-adically uniformized varieties*, Invent. math. 159 (2005), 607-656.

[Ka] N. M. Katz, *Review of ℓ-adic Cohomology*, Motives, Proc. of Symp. in Pure Math. 55 , Part I, U. Jannsen, S. Kleiman, J-P. Serre, Eds., AMS (1994), 21-30.

[Kl] S. Kleiman, *Algebraic cycles and the Weil conjectures*, dans *Dix exposés sur la cohomologie des schémas*, A. Grothendieck and N. Kuiper, eds., Noth Holland Pub. Co., Masson et Co., 1968, 359-386.

[KKMS] K. Kempf, F. Knudsen, D. Mumford, B. Saint-Donat, *Toroidal embeddings I*, SLN 339, Springer-Verlag, 1973.

[KL] N. Katz, G. Laumon, *Transformation de Fourier et majoration de sommes exponentielles*, Pub. math. IHÉS 62 (1986), 361-418.

[KS] K. Kato, T. Saito, *Ramification theory for varieties over a perfect field*, Ann. of Math. 168 (2008), 33-96.

[KSc] M. Kashiwara, P. Schapira, *Sheaves on Manifolds*, Grundlheren der math. Wiss. 292, Springer-Verlag, 1990.

[KT], Y. Kawada, J. Tate, *On the Galois Cohomology of Unramified Extensions of Function Fields in One Variable*, Amer. J. Math. 77 (1955), 197-217.

[La1] G. Laumon, *Homologie étale*, Séminaire de géométrie analytique A. Douady - J.-L. Verdier, Astérisque 36-37, SMF, 1976, 163-188.

[La2] G. Laumon, *Vanishing cycles over a base of dimension > 1*, Algebraic Geometry (Tokyo-Kyoto 1982), SLN 1016, Springer-Verlag, 1983, 143-150.

[La3] G. Laumon, *Transformation de Fourier, constantes d'équations fonctionnelles et conjecture de Weil*, Pub. Math. IHÉS 65 (1987), 131-210.

[Laf] L. Lafforgue, *Chtoucas de Drinfeld et correspondance de Langlands*, Invent. math. 147 (2002), 1-241.

[LO] Y. Laszlo, M. Olsson, *The six operations for sheaves on Artin stacks II : adic coefficients*, Publ. Math. IHÉS 107 (2008), 169-210.

[LaN] G. Laumon, B. C. Ngo, *Le lemme fondamental pour les groupes unitaires*, Annals of Math., 168 (2008), 477–573.

[Lu] G. Lusztig, *Character sheaves I, II, III*, Adv. in Math. 56 (1985), 193-237, 57 (1985), 226-265, 57 (1985), 266-315.

[M] J. Milnor, *Singular points of complex hypersurfaces*, Ann. of Math. Studies 61, Princeton Univ. Press, 1968.

[Na1] M. Nagata, *On the theory of Henselian rings*, Nagoya Math. J. 5 (1953), 45-57.

[Na2] M. Nagata, *On the theory of Henselian rings II*, Nagoya Math. J. 7 (1954), 1-19.

[Na3] M. Nagata, On the theory of Henselian rings III, Mem. Coll. Sci. Univ. Kyoto, Ser. A, Math., vol. 32 (1959), 93-101.

[Ne] A. Neeman, *The Grothendieck duality theorem via Bousfield's techniques and Brown representability*, J. Amer. Math. Soc. 9 (1996), 205-236.

[Ng] B. C. Ngo, *Le lemme fondamental pour les algèbres de Lie*, Pub. Math. IHÉS 111 (2010), 1–169.

[Or] F. Orgogozo, *Modifications et cycles proches sur une base générale*, IMRN, vol. 2006, ID 25315, 38 p., 2006.

[P1] D. Popescu, *General Néron desingularization and approximation*, Nagoya Math. J. 104 (1986), 85-115.

[P2] D. Popescu, *Letter to the editor, General Néron desingularization and approximation*, Nagoya Math. J. 118 (1990), 45-53.

[R] M. Raynaud, *Faisceaux amples sur les schémas en groupes et les espaces homogènes*, SLN 119, Springer-Verlag, 1970

[RS] A. Grothendieck, *Récoltes et semailles, Réflexion et témoignage sur un passé de mathématicien*, Univ. Sc. et Techniques du Languedoc, Montpellier, 1985.

[RZ] M. Rapoport, T. Zink, *Ueber die lokale Zetafunktion von Shimuravarietäten, Monodromiefiltration und verschwindende Zyklen in ungleicher Charakteristik*, Inv. Math. 68 (1982), 21-201.

[Sa] M. Saito, *Modules de Hodge polarisables*, Publ. RIMS, Kyoto Univ. 24 (1988), 849-995.

[S] J-P. Serre, *Espaces fibrés algébriques*, Séminaire Claude Chevalley, t. 3 (1958), Exp. 1, 1-37, Secr. math. Paris, 1958.

[St] J. Steenbrink, *Limits of Hodge structures*, Inv. math. 31 (1976), 229-257.

[ST] J-P. Serre J.-P. and J. Tate, *Good reduction of abelian varieties*, Ann. of Math. 88 (1968), 492-517.

[Va] Y. Varshavsky, *Lefschetz-Verdier trace formula and a generalization of a theorem of Fujiwara*, Geom. Funct. Anal. 17 (2007), 271-319.

[V1] J.-L. Verdier, Le théorème de dualité de Poincaré, C. R. Acad. Sci. Paris 256 (1963), 2084-2086.

[V2] J.-L. Verdier, Dualité dans la cohomologie des espaces localement compacts, Séminaire Bourbaki, Vol. 9, Exp. 300, 337-349, SMF Paris; 1995.

[V3] J.-L. Verdier, *A Duality Theorem in the Etale Cohomology of Schemes*, Proc. of a Conf. on Local Fields, Driebergen 1966, Ed. T. A. Springer, Springer-Verlag, 1967, 184-198.

[V4] J.-L. Verdier, *The Lefschetz Fixed Point Formula in Etale Cohomology*, Proc. of a Conf. on Local Fields, Driebergen 1966, Ed. T. A. Springer, Springer-Verlag, 1967, 199-214.

[V5] J.-L. Verdier, *Classe d'homologie associée à un cycle*, Séminaire de géométrie analytique A. Douady - J.-L. Verdier, Astérisque 36-37, SMF, 1976, 101-151.

[Z] W. Zheng, *Sur l'indépendance de ℓ en cohomologie ℓ-adique sur les corps locaux*, Ann. Sci. Éc. Norm. Supér. (4) 42 (2009), 291–334.

[EGA] *Éléments de Géométrie Algébrique*, par A. Grothendieck, rédigés avec la collaboration de J. Dieudonné, Pub. Math. IHÉS 4, 8, 11, 17, 20, 24, 28, 32, and Grundlehren 166, Springer-Verlag, 1971.

[FGA] A. Grothendieck, *Fondements de la géométrie algébrique. Extraits du Séminaire Bourbaki 1957-1962*, Secr. math. Paris, 1962.

[Mot] *Motives,* Proc. of Symp. in Pure Math. 55 , Parts I, II, U. Jannsen, S. Kleiman, J-P. Serre, Eds., AMS (1994).

[SGA 1] *Revêtements étales et groupe fondamental*, Séminaire de Géométrie Algébrique du Bois-Marie 1960/1961, dirigé par A. Grothendieck, SLN 224, Springer-Verlag, 1971, et Documents mathématiques 3, SMF, 2003.

[SGA 2] *Cohomologie locale des faisceaux cohérents et théorèmes de Lefschetz locaux et globaux*, Séminaire de Géométrie Algébrique du Bois-Marie 1962, dirigé par A. Grothendieck, North-Holland Pub. Comp., Masson et Cie, 1968, et Documents mathématiques 4, SMF, 2005.

[SGA 3] *Schémas en groupes I, II, III*, Séminaire de géométrie algébrique du Bois-Marie 1962-64, dirigé par M. Demazure et A. Grothendieck, SLN 151, 152, 153, Springer-Verlag, 1970.

[SGA 4] *Théorie des topos et cohomologie étale des schémas*, Séminaire de géométrie algébrique du Bois-Marie 1963-64, dirigé par M. Artin, A. Grothendieck, J.-L. Verdier, SLN 269, 270, 305, Springer-Verlag, 1972, 1973.

[SGA 4 1/2] *Cohomologie étale*, par P. Deligne, SLN 569, Springer- Verlag, 1977.

[SGA 5] *Cohomologie ℓ-adique et fonctions L*, Séminaire de Géométrie Algébrique du Bois-Marie 1965/66, dirigé par A. Grothendieck, SLN 589, Springer-Verlag, 1977.

[SGA 6] *Théorie des intersections et théorème de Riemann-Roch*, Séminaire de Géométrie Algébrique du Bois-Marie 1966/67, dirigé par P. Berthelot, A. Grothendieck, L. Illusie, SLN 225, Springer-Verlag, 1971.

[SGA 7] *Groupes de monodromie en géométrie algébrique*, Séminaire de Géométrie Algébrique du Bois-Marie 1967-1969, I dirigé par A. Grothendieck, II par P. Deligne et N. Katz, SLN 288, 340, Springer-Verlag 1972, 1973.

The Grothendieck-Serre correspondence

Leila Schneps

The Grothendieck-Serre correspondence[1][2] is a very unusual book: one might call it a living math book. To retrace the contents and history of the rich plethora of mathematical events discussed in these letters over many years in any complete manner would require many more pages than permitted by the notion of a book review, and far more expertise than the present author possesses. More modestly, what we hope to accomplish here is to render the flavor of the most important results and notions via short and informal explanations, while placing the letters in the context of the personalities and the lives of the two unforgettable epistolarians.

The exchange of letters started at the beginning of the year 1955 and continued through to 1969 (with a sudden short burst in the 1980's), mostly written on the occasion of the travels of one or the other of the writers. Every mathematician is familiar with the names of these two mathematicians, and has most probably studied at least some of their foundational papers—Grothendieck's "Tohoku" article on homological algebra, Serre's FAC and GAGA, or the volumes of EGA and SGA. It is well-known that the work of these two mathematicians profoundly renewed the entire domain of algebraic geometry in its language, in its concepts, in its methods and of course in its results. The 1950's, 1960's and early 1970's saw a kind of heyday of algebraic geometry, in which the successive articles, seminars, books and of course the important results proven by other mathematicians as consequences of their foundational work—perhaps above all Deligne's finishing the proof of the Weil conjectures—fell like so many bombshells into what had previously been a well-established classical domain, shattering its

Analyse Algébrique, Institut de Mathématiques de Jussieu, Université Pierre et Marie Curie, Case 247/4, place Jussieu, F-75252 Paris 05, France. leila.schneps@imj-prg.fr.

[1]Published by the Société Mathématique de France in 2001, then translated and edited as a bilingual version by the American Mathematical Society in 2003.

[2]A much shorter version of this paper appeared in the *Mathematical Intelligencer* **29** No. 4, in 2007. For the present version, I am grateful to S. Kleiman for sharing with me many useful comments based on his deep knowledge of the ideas—with all the richness of their historical context—that peopled the mathematical world of Serre and Grothendieck. Some of his remarks appear verbatim as footnotes.

concepts to reintroduce them in new and deeper forms. But the articles themselves do not reveal anything of the actual creative process that went into them. That, miraculously, is exactly what the correspondence does do: it sheds light on the *development* of this renewal in the minds of its creators. Here, unlike in any mathematics article, the reader will see how Grothendieck proceeds and what he does when he is stuck on a point of his proof (first step: ask Serre), share his difficulties with writing up his results, participate with Serre as he answers questions, provides counterexamples, shakes his finger, complains about his own writing tasks and describes some of his theorems. The letters of the two are very different in character and Grothendieck's are the more revealing of the actual creative process of mathematics, and the most surprising for the questions he asks and for their difference with the style of his articles, whereas Serre's letters for the most part are finished products which closely resemble his other mathematical writings, a fact which in itself is almost as surprising, for it seems that Serre reflects directly in final terms. Even when Grothendieck surprises him with a new result, Serre responds with an accurate explanation of what he had previously known about the question and what Grothendieck's observation adds to it.

They tell each other their results as they prove them, and the responses are of two types. If the result fits directly into their current thoughts, they absorb it instantly and, for the most part, add to it. Otherwise, there is a polite acknowledgement ("That sounds good"), sometimes joined to a confession that they have had no time to look more closely. The whole of the correspondence yields an extraordinary impression of speed, depth and incredible fertility. Most of the letters, especially at first, are signed off with the accepted Bourbaki expression "Salut et fraternité".

At the time the correspondence began, in early 1955, Jean-Pierre Serre was twenty-eight years old. A young man from the countryside, the son of two pharmacists, he had come up to the Ecole Normale Supérieure in 1945 at the age of 19, then defended an extraordinary thesis under the direction of Henri Cartan in 1951, in which he applied Leray's spectral sequences, created as a tool to express the homology groups of a fibration in terms of those of its fibre space and base space, to study the relations between homology groups and homotopy groups, in particular the homotopy groups of the sphere. After his thesis, Serre held a position in the Centre National des Recherches Scientifiques (CNRS) in France before being appointed to the University of Nancy in 1954, the same year in which he won the Fields Medal. He wrote many papers during this time, of which the most important one, largely inspired by Cartan's work and the extraordinary atmosphere of his famous seminar, was the influential "FAC" (Faisceaux Algébriques Cohérents, published in 1955), developing the sheaf theoretic viewpoint (sheaves had been introduced some years earlier by Leray in a very different context) in abstract algebraic geometry. Married in 1948 to a brilliant chemist who had been a student at the Ecole Normale Supérieure for girls, Serre was the father of a small daughter, Claudine, born in 1949.

In January 1955, Alexandre Grothendieck had just arrived in Kansas to spend a year on an NSF grant. Aged twenty-six, his personal situation was chaotic and lawless, the opposite of Serre's in almost every possible way. His earliest childhood was spent in inconceivable poverty with his anarchist parents in Berlin; he then spent five or six years with a foster family in Germany, but in 1939 the situation became too hot to hold a half-Jewish child, and he was sent to join his parents in France. The war broke out almost immediately and he spent the war years interned with his mother in a camp for "undesirables" in the south of France; his father, interned in a different camp, was deported to Auschwitz in 1942 and never returned. After the war, Grothendieck lived in a small village near Montpellier with his mother, who was already seriously ill with tuberculosis contracted in the camp; they lived on his modest university scholarship, complemented by his occasional participation in the local grape harvest. He, too, was the father of a child: an illegitimate son from an older woman who had been his landlady. His family relations—with his mother, the child, the child's mother, and his half-sister who had come to France to join them after a twelve-year separation, was wracked with passion and conflict. He was stateless, with no permanent job and the legal impossibility to hold a university position in France, so that he was compelled to accept temporary positions in foreign countries while hoping that some suitable research position in France might eventually be created. After Montpellier, he spent a year at the Ecole Normale in Paris, where he met Cartan, Serre, and the group that surrounded them, and then, on their advice, he went to do a doctoral thesis under Laurent Schwartz in Nancy. His friends from his time in Nancy and after, such as Paulo Ribenboim, remember a young man deeply concentrated on mathematics, spending his (very small amount of) spare time taking long walks or playing the piano, working and studying all night long. During his whole life, Grothendieck would keep his mathematical activities sharply separate from his private affairs, of which next to nothing appears in the letters. And during his whole life, he would spend his nights working and writing.

At the time of his visit to Kansas, Grothendieck already had his dissertation and nearly twenty publications to his credit, on the subject of topological vector spaces, their tensor products, and nuclear spaces. To summarize Grothendieck's pre-Kansas research in brief, Laurent Schwartz had given him the following subject as a thesis topic: put a good topology on the tensor product of two locally convex spaces. Schwartz was at that time studying $\mathcal{D}$, the space of smooth real-valued functions with compact support, and the dual space of distributions $\mathcal{D}'$. Wanting to extend the space to functions with values in any locally convex space F, Schwartz was trying to put a topology on $\mathcal{D}' \otimes F$ whose completion would naturally be the topological space $\mathcal{L}(\mathcal{D}, F)$ of linear maps from $\mathcal{D}$ to F. Grothendieck began by discovering two natural topologies on a tensor product of locally convex spaces, which discouraged him greatly at the start[3] because of its depressing lack of canonicity. He overcame this difficulty by restricting attention to the important class of spaces for which the two topologies coincide, which he dubbed nuclear spaces. Grothendieck's discovery that $\mathcal{D}'$ was a nuclear space, and his development of the general theory of nuclear spaces, set the whole subject into its wider context. He completed his thesis in 1953 and then spent the years 1953-1955 in São Paulo where he continued to work on the subject. His move to Kansas marked the beginning of the first of several major shifts in his mathematical interests.

1955–1957: Two mathematicians in their twenties

From the very first letter of the correspondence with Serre, dated January 1955, the words homology, cohomology and sheaf make their appearance, as well as a plethora of inductive and projective limits. These limits and their duality to each other, now a more-than-familiar concept even for students, were extremely new at the time. Although projective limits of topological groups had been studied and the notions of Krull topology and profinite groups were known, projective limits were not yet considered as a general "procedure" which could be applied to projective systems of any type. As for inductive limits, they were studied only in the early 1950s. Their introduction into homological algebra, together with the notion that the two types of limit are dual to each other, dates to very shortly before the exchange of the earliest letters of the correspondence.

From the naturalness with which Grothendieck and Serre juggle inductive and projective limits of homology and cohomology groups in the earliest letters, it is not easy to realize that they are largely creating, not just learning, the elements of the theory. In the letter of February 26, 1955, for example, Grothendieck states a theorem which, as he observes, has already been proved "umpteen times in all kinds of special cases" and which is now considered a basic result:

Theorem: *Let $(A_i)_{i\in I}$ and $(B_i)_{i\in I}$ be two projective systems of groups, let (ϕ_i) be a homomorphism from the first to the second, ϕ the homomorphism from $\varprojlim A_i$ to $\varprojlim B_i$ defined by the ϕ_i, and (N_i) the "kernel" of (ϕ_i). Assume that I contains a cofinal sequence and that (N_i) satisfies the following property:*

(A) The N_i have Hausdorff topologies, compatible with the group structure, for which the maps $N_i \to N_j$ $(i > j)$ are continuous, such that for any i there exists a $j \geq i$ such that for any $k \geq j$, the image of N_k in N_i is dense in the image of N_j.

[3]According to Laurent Schwartz's autobiography *A Mathematician grappling with his century*, Birkhäuser 2000 (English translation), p. 283.

Under these conditions, an element $b \in B$ is contained in the image of ϕ if and only if for all i, its component b_i in B_i is an element of the image of ϕ_i.

Here is another way of saying the same thing: consider the following property h_1 of a projective system (N_i): for any exact sequence $0 \to (N_i) \to (A_i) \to (B_i) \to 0$ of projective systems, the corresponding sequence of projective limits is exact, i.e. $\varprojlim A_i \to \varprojlim B_i$ is surjective[...] The theorem then says that condition (A) ("approximation") implies h_1.

At the time of these early letters, Grothendieck considers that he is merely learning (as opposed to creating) homological algebra: "For my own sake, I have made a systematic (as yet unfinished) review of my ideas of homological algebra. I find it very agreeable to stick all sorts of things, which are not much fun when taken individually, together under the heading of derived functors." This remark is the first reference to a text that would eventually grow into his famous Tohoku article. He wants to study and teach a course on Cartan and Eilenberg's new book but cannot get hold of a copy, so that he is compelled to work everything out for himself, following what he "presumes" to be their outline.

The Tohoku paper gives an introduction to abelian categories, extracting the main defining features of some much-studied categories such as abelian groups or modules. Grothendieck introduces the essential notion of having "enough injectives", generalizing Baer's 1940 construction of injectives so that it works as well for the cateogry of sheaves on any space, and the criterion for them to exist. This allows him to extend Cartan-Eilenberg's notion of derived functors of functors on the category of modules to a completely general notion of derived functors. If $F : \mathcal{A} \to \mathcal{B}$ is a left-exact functor between two abelian categories and $0 \to A \to B \to C \to 0$ is a short exact sequence of objects of $\mathcal{A}$, then $0 \to F(A) \to F(B) \to F(C)$ is exact in $\mathcal{B}$. What Grothendieck showed is that if $\mathcal{A}$ has enough injectives, then there is a canonical, repeatable derivation operation $F \to R^1F$ giving a new functor such that we can continue this exact sequence to a long exact sequence

$$\begin{aligned}0 &\to F(A) \to F(B) \to F(C) \to R^1F(A) \to R^1F(B) \to R^1F(C)\\ &\to R^2F(A) \to \cdots\end{aligned}$$

as was already known to Cartan and Eilenberg for modules.

This long exact sequence was a grand generalization of the familiar long exact sequence of cohomology groups associated to an exact sequence of modules, so that cohomology now comes under the "heading of derived functors". Grothendieck's use of abelian categories, injective resolutions and derived functors allowed him to extend Serre duality—a relation between the sheaf cohomology groups H^i and H^{n-i} associated to a non-singular

n-dimensional projective variety—to more general situations, including possibly singular algebraic varieties (Grothendieck duality).

In his own words (letter to Serre of Feb. 26, 1955): "I have noticed that, formulating the theory of derived functors for more general categories than modules, one obtains at the same time the cohomology of a space with coefficients in a sheaf with little extra effort; take the category of sheaves on a given space X, consider the functor $\Gamma(F)$ which takes values in the category of abelian groups, and consider its derived functors. Their existence follows from a general criterion, in which fine sheaves will play the role of "injective" modules. One also obtains the fundamental spectral sequence as a special case of the delectable and useful general spectral sequences. But I am not yet sure if everything works out so well for non-separated spaces[...] Moreover, all this is probably contained more or less explicitly in the Cartan-Eilenberg book, which I have not yet had the pleasure of seeing." (At the end of this letter, Grothendieck hopefully mentions that he has heard that some associate professorships open to foreigners are going to be created in France: does Serre know anything about it?) A couple of weeks later, the answer from Serre arrives: "Your stuff on topologies on projective limits looks very nice, and it is a great pleasure to see the limit process in the theory of Stein varieties finally swallowed up by a general argument[...] The fact that sheaf cohomology is a special case of derived functors (at least in the paracompact case) is not in Cartan-Sammy. Cartan was aware of it, and had told Buchsbaum to do it, but it appears that he never did." (And in the same letter: "We have no details of the associate professorships. How many will there be in the whole of France? A mystery[...] In any case, you may be sure that if there is an opening for you we will jump at it...")

Grothendieck goes on working by himself throughout the ensuing months. In June, he sends his write-up to Serre: "You will find enclosed a neat draft of the outcome of my initial reflections on the foundations of homological algebra." Serre takes the draft to Bourbaki with him, and answers Grothendieck in July: "Your paper on homological algebra was read carefully and converted everyone (even Dieudonné, who seems to be completely functorized) to your point of view..." but it "raises a totally disjoint question, namely that of publishing it in a journal," because Buchsbaum had betweentimes added an appendix to Cartan-Eilenberg containing some overlap with Grothendieck's work. Serre and the others have obviously thought about the best solution to the problem, and tactfully suggest that Grothendieck cut out some parts: "As you could use Buchsbaum for all the trivial results on classes, you would basically only need to write up the interesting part, and that would be very good..."

It is really striking to see how some of the most typical features of Grothendieck's style over the coming decade and a half are already totally visible in the early work discussed in these exchanges: his view of the most general situations, explaining the many "special cases" others have worked on, his independence from (and sometimes ignorance of) other people's written work, and above all, his visionary aptitude for rephrasing classical problems on varieties or other objects in terms of morphisms between them, thus obtaining incredible generalizations and simplifications of various theories.

Six months later, in December, Grothendieck is back in Paris with a temporary job at the CNRS, while Serre is on leave from the University of Nancy, spending some time in Princeton and working on his "analytic=algebraic diplodocus", which would become the famous GAGA, in which he proved the equivalence of the categories of algebraic and analytic coherent sheaves, obtaining as applications several general comparison theorems englobing earlier partial results such as Chow's theorem (a closed analytic subspace of projective space is algebraic). The comparison between algebraic and analytic structures in any or every context is at this point one of the richest topics of reflection for both Grothendieck and Serre. Grothendieck finds a more general form for the major duality theorem from Serre's FAC, which he expresses as a canonical bijection between the dual of $H^p(X,F)$, where X is an n-dimensional projective algebraic variety and F is a coherent algebraic sheaf on X, and $\mathrm{Ext}^{n-p}_{\mathcal{O}}(X,F,\Omega^n)$, where Ω^n is the sheaf of germs of differential n-forms on X. Serre's delighted response: "I find your formula very exciting, as I am quite convinced that it is *the* right way to state the duality theorem in both the analytic case and the algebraic case..." And in mid-January 1956, after explaining that recent results of Cartan have allowed him to prove that $H^i(X,k^*)=0$ for $i\geq 2$ if X is a algebraic variety without singularities, Grothendieck observes: "This proves in particular that a projective algebraic bundle over a base without singularities comes from a vector bundle (as I thought I had already shown last summer); in the case where X is a projective complex variety without singularities, it is known that algebraic classification=analytic classification, and thus one gets an answer to a question of Kodaira's..." only to rectify humbly two weeks later, no doubt in response to an objection by Serre, "As for the algebraic classification=analytic classification question for projective bundles, I confess that I was taking it on trust that the usual embedding of the projective group into the linear group has a rational section, since everybody seemed convinced that that should always happen for a fibration by a linear group (you see what an irresponsible individual I am!). It is true that Chevalley does not know of any theorem in this direction[...]and it would be very sad if you already had a counterexample for the projective group." And Serre, the expert, soon responds: "On the subject of bundles whose group is the projective group: I am now practically certain that if this group is embedded into the linear group, there is no rational section" (and he turned out to be right). The letter goes on: "I have no comments on the rest of your letter...because I have hardly had the time to study it in detail. I am busy with my blasted analytic paper (I write horribly slowly)..." GAGA would be published during the course of that same year.

The following month finds Grothendieck eagerly telling Serre about some recent ideas of Cartier's on coalgebras, which Cartan adapted to compute the homological structure of the loop space and the loop space of the loop space of the sphere—but remarks that this formalism is not yet refined enough to yield $\pi_6(S^3)$, an allusion to one of the most difficult results obtained in Serre's doctoral thesis.

A year earlier, Grothendieck was working on topological vector spaces; in Kansas in 1955-56, he began his synthesis of homological algebra, and in March 1956, we find that he has "gone back to the classification of analytic bundles over the Riemann sphere with semi-simple structural group, and I have more or less proved my conjecture...Do your constructions also show that over a complex algebraic projective variety, the analytic and algebraic classification of bundles, for example with structural group $\mathrm{SO}(n)$, are not the same?" Grothendieck's work on classification of analytic bundles over the Riemann sphere can be described as his first foray into algebraic geometry: he showed that every such bundle is the direct sum of a certain number of tensor powers of the tautological line bundle. Serre, always concerned with analytic=algebraic theories, replies: "Congratulations on your classification of analytic bundles over the Riemann sphere with semi-simple structural group. How do you do that? I suppose that at the same time you show that they are actually algebraic?" Both young men, more involved in creating new than in studying older mathematics, were apparently unaware of the fact that others (Segre, Birkhoff, Hilbert, Dedekind, Weber) had already considered various versions of the question of classifying bundles over the Riemann sphere.

Serre has finished writing up GAGA by this time ("For information on the relationship between the algebraic and analytic classifications, I refer you to my paper...") and by early April, Grothendieck has finished writing up his classification of analytic bundles on the Riemann sphere. In July, Serre receives a small letter in which Grothendieck proposes to write a supplement to the analytic=algebraic diplodocus (GAGA) showing that "algebraic=analytic" for algebraic coherent sheaves on a complete compact algebraic variety not assumed to be projective—providing that someone does not in the meantime discover how to embed a complete algebraic variety into projective space, in which case the supplement would be useless! As it happens, no one did; Nagata[4] showed later that on the contrary, one can find complete algebraic varieties which cannot be embedded into projective space.

September 1956 finds Grothendieck trying to finish his enormous article on homological algebra and puzzled about where to publish something so long — not in France, where he recently published another long article, not in the American Journal which just accepted the bundles over the Riemann

[4]On the embedding problem of abstract varieties in projective varieties, *Mem. Coll. Sci. Kyoto (A)* **30** (1956), 71-82

sphere, not in the Transactions, because Sammy (Eilenberg) demands further work on the presentation which Grothendieck is reluctant to comply with. Serre scolds him for this ("I find your objections to publishing in the Transactions idiotic...armed with a little patience and a little glue, it would surely take you no more than a day...") but instead, Grothendieck tells him that he has proposed it to Tannaka, who accepted it for the Tohoku Journal.

In November, Serre is working on stability of cohomology groups of algebraic varieties with values in a coherent algebraic sheaf under blowups at a simple point, and Grothendieck writes to ask him a "stupid question on which I am stuck: let X be an algebraic curve, take a representation of its fundamental group by unimodular matrices with integral coefficients, which gives a holomorphic bundle on X whose fiber is the group $\mathbb{C}^{*n}$. Is it true that this bundle will almost never be an algebraic variety? Obviously, I am looking for an algebraic definition of the fundamental group, and I want to be sure that my idea cannot give anything..." This striking sentence reveals Grothendieck's very earliest thoughts on a subject which was to become one of the essential topics of SGA. Serre's answer: "I have no idea about your holomorphic bundles with fiber $\mathbb{C}^{*n}$. I do not see how to prove that they are 'almost never' algebraic." Grothendieck later answered his own question (SGA 3).

Questions and answers are one of the most lively aspects of the correspondence; the one that makes reading it such a different experience from reading a mathematical article. The most common situation is that Grothendieck asks a question, and Serre either answers or provides a counterexample, though of course there are also questions whose answer he doesn't know. Grothendieck's questions are sometimes quite easy and the answers apparently well-known: "Is a finitely generated projective module over the ring in question (a local Noetherian ring) free? Is this easy to see in special cases?" (Serre answers that it is always free, giving a ten-line proof); "This would mean that if G is a group of automorphisms of a semi-local ring $\mathcal{O}$ which acts simply transitively on its maximal ideals, then $H^p(G, \mathcal{O}) = 0$ for all $p > 0$[...] I realize that I did not give a proof in my papers, and I can't seem to improvise one..." (Serre answers "Set your mind at rest, it is indeed true...", sketching a five-line proof); "What is the Pontryagin square?" (Serre admits that it is something of a mystery to him and refers him to Cartan). Some of Grothendieck's remarks about his own ignorance of mathematics are most refreshing: "I have been reviewing class field theory, of which I finally have the impression that I understand the main results (but not the proofs, of course!) But to my shame I have been unable to find the 'corollary' stating that all ideals of K become principal in the maximal abelian extension unramified at finite places..." or "As messy as it is, Lang's report was very helpful for my understanding what unramified means; I had previously more or less imagined that it meant that the action of the Galois group on the maximal ideals of $\mathcal{O}'$ is fixed-point free!"

During the period covered by these early letters, the notion of a scheme was just beginning to make its appearance (see below). It does not seem that Grothendieck paid particular attention to it at the time, but a scattering of early remarks turns up here and there. Already at the beginning of 1955 Grothendieck wrote of FAC: "You wrote that the theory of coherent sheaves on affine varieties also works for spectra of commutative rings for which any prime ideal is an intersection of maximal ideals. Is the sheaf of local rings thus obtained automatically coherent? If this works well, I hope that for the pleasure of the reader, you will present the results of your paper which are special cases of this as such; it cannot but help in understanding the whole mess." Later, of course, he would be the one to explain that one can and should consider spectra of *all* commutative rings. A year later, in January 1956, Grothendieck mentions "Cartier-Serre type ring spectra," which are nothing other than affine schemes, and just one month after that he is cheerfully proving results for "arithmetic varieties obtained by gluing together spectra of commutative Noetherian rings"—schemes! A chatty letter from November 1956 gives a brief description of the goings-on on the Paris mathematical scene, containing the casual remark "Cartier has made the link between schemes and varieties," referring to Cartier's formulation of an idea then only just beginning to make the rounds: *The proper generalization of the notion of a classical algebraic variety is that of a ringed space* $(X, \mathcal{O}_X)$ *locally isomorphic to spectra of rings.* Over the coming years, Grothendieck would make this notion his own.

1957–1958: Riemann-Roch... and Hirzebruch... and Grothendieck

The classical Riemann-Roch theorem over the complex numbers, stated as the well-known formula

$$\ell(D) - \ell(K - D) = \deg(D) - g + 1$$

concerns a non-singular projective curve over the complex numbers equipped with a divisor D; the formula equates a difference of the dimensions of two vector spaces of meromorphic functions on the curve with prescribed behavior at the points of the divisor D (the left-hand side) with an integral expression in numbers associated topologically with the curve and the divisor (the right-hand side).

In the early 1950's, Serre reinterpreted the left-hand side of the Riemann-Roch formula as a difference of the dimensions of the zero-th and first cohomology groups associated to the curve, and generalized this expression to any n-dimensional non-singular projective variety X equipped with a vector bundle E as the alternating sum of dimensions of cohomology groups $\sum(-1)^i \dim H^i(X, E)$.

In 1953, Hirzebruch gave a generalization of the classical Riemann-Roch theorem to this situation, by proving that Serre's alternating sum was equal to an integer which could be expressed in terms of topological invariants of the variety called the Chern class (associated to E) and the Todd class (associated to X).

It seems that the idea of trying to prove a general algebraic version of Riemann-Roch was in Grothendieck's mind from the time he first heard about Hirzebruch's proof. In February 1955, he writes "I do not see why it would not be possible to introduce Chern et al. classes via universal spaces[5] (from the homological point of view, as explained in Borel's thesis) and classifying spaces, which would then play the same role in an algebraic Riemann-Roch as in the one (due to Hirzebruch) that you vaguely explained to me, which works in the complex case." Serre's answer: "No, one should not try to define 'Chern classes' as elements of certain $H^q(X)$ with coefficients in coherent sheaves, since these are vector spaces over the base field and the aim is to be able to define intersections with integral coefficients. Moreover, the 'last' Chern class is already known, namely the canonical class, and it is a divisor class, defined up to linear equivalence. It is absolutely certain that it is possible to define the other classes as equivalence classes of algebraic cycles up to 'numerical' or 'algebraic' equivalence. And this should not even be difficult."

In the end, what Grothendieck brought to the Riemann-Roch theorem is one of the basic features of all of his mathematics, and was already visible in his Tohoku article: the transformation of statements on *objects* (here, varieties) into more general statements on *morphisms* between those objects. He reinterpreted both sides of the formula that Hirzebruch proved in the framework of morphisms $f : X \to Y$ between varieties. The right-hand side (the integer defined in terms of Chern and Todd classes) can naturally be generalized to this situation. However, in order to reinterpret the left-hand side, the alternating sum of dimensions of cohomology groups, in the framework of a morphism between varieties, Grothendieck introduced what came to be called the Grothendieck group $K(Y)$, a quotient of the free abelian group generated by all bundles (locally free sheaves) on Y (up to isomorphism) which generalizes the abelian monoid of such sheaves under direct sum. The Grothendieck group later gave rise to the entire domain of K-theory. In this framework, Grothendieck was able to prove a statement of the Riemann-Roch theorem which freed it completely from the need to work specifically over the complexes[6] ("You will find enclosed a very simple proof of Riemann-Roch independent of the characteristic", November 1, 1957),

[5]Grothendieck later wrote up an algebraic theory of Chern classes, which was published as an appendix to the Borel-Serre paper.

[6]For curves, the Riemann-Roch theorem was "freed" from the complex numbers by Dedekind and Weber in 1882, and from characteristic 0 by F.K. Schmidt in 1929.

and could be expressed as the commutativity of a single square diagram

$$\begin{array}{ccc} K(X) & \longrightarrow & K(Y) \\ {\scriptstyle \mathrm{ch}_X(-)\mathrm{td}(X)}\downarrow & & \downarrow{\scriptstyle \mathrm{ch}_Y(-)\mathrm{td}(Y)} \\ H^*(X) & \longrightarrow & H^*(Y) \end{array}$$

where the two horizontal maps are derived directly from the given morphism $f : X \to Y$, the symbols ch and td refer to Chern and Todd classes, and Hirzebruch's formula is recovered by taking Y to be a point. Serre sent him several corrections, to which Grothendieck responded in detail in a letter from November 12, 1957—and then dropped the whole matter! The letter ends with the words "...at the moment I have just dropped research in order to finally start writing up the varieties, of which I hadn't written a single word until the day before yesterday!" The 'varieties' he refers to is one of the anonymous papers written by all the Bourbaki members (this one, as it happens, never attained publication...)

Grothendieck did this work between 1954 and 1957. He wrote up something (RRR, "rapport Riemann-Roch") that he considered a mere preliminary and sent it to Serre, then in Princeton; Serre organized a seminar around it, and then, as Grothendieck was clearly onto other things and not going to publish, Serre wrote the proof up together with Borel and published it in the Bulletin de la Société Mathématique de France in 1958. This article begins "What follows consists in the notes taken at a seminar held in Princeton in the autumn of 1957 on the work of GROTHENDIECK; the new results are due to him, our contribution is uniquely for the writing-up. The 'Riemann-Roch' theorem here is valid for (non-singular) algebraic varieties over a field of any characteristic; in the classical case, over the base field **C**, this theorem contains as a special case the theorem proved some years ago by Hirzebruch." Grothendieck finally included his original RRR at the beginning of SGA 6, held in 1966-67 and published only in 1971, at the very end of his established mathematical career.

What is not revealed in the letters is that Grothendieck's mother was dying at the very time of these exchanges. He does add as a postscriptum to the letter of November 1, "You are moving out of your apartment; do you think it might be possible for me to inherit it? As the rent is not very high, if I remember rightly, I would then be able to buy some furniture (on credit). I am interested in it for my mother, who isn't very happy in Bois-Colombes, and is terribly isolated." But Hanka Grothendieck was suffering from more than isolation. She had been nearly bedridden for several years, a victim of tuberculosis and severe depression. After their five-year separation during his childhood, she and Alexandre had grown inseparable in the war and post-war years, but during the last months of her life, she was so ill and so bitter that his life had become extremely difficult. She died in December 1957. Shortly before her death, Grothendieck encountered, through a mutual friend, a young woman named Mireille who helped him care for his mother during her last months, and fascinated and overwhelmed by his powerful personality, fell in love with him. At the same time, the Grothendieck-Riemann-Roch theorem propelled him to instant stardom in the world of mathematics.

1958–1960: Schemes and EGA

The idea of schemes, or more generally, the idea of generalizing the classical study of coordinate rings of algebraic varieties defined over a field to larger classes of rings, appeared in the work and in the conversation of various people—Nagata, Serre, Chevalley, Cartier—starting around 1954. It does not appear, either from his articles or his letters to Serre, that Grothendieck paid overmuch attention to this idea at first. However, by the time he gave his famous talk at the ICM in Edinburgh in August 1958, the theory of schemes, past, present and future, was already astonishingly complete in his head. In that talk, he presents his plan for the complete reformulation of classical algebraic geometry in these new terms:

"I would like, however, to emphasize one point[...], namely, that the natural range of the notions dealt with, and the methods used, are not really algebraic varieties. Thus, we know that an *affine* algebraic variety with ground field k is determined by its co-ordinate ring, which is an arbitrary finitely generated k-algebra without nilpotent elements; therefore, any statement concerning affine algebraic varieties can be viewed also as a statement concerning rings A of the previous type. Now it appears that most of such statements make sense, and are true, if we assume only A to be a commutative ring with unit, provide we sometimes submit it to some mild restriction, as being noetherian, for instance[...] Besides, frequently when it seemed at first sight that the statement only made sense when a ground field k was involved,[...] further consideration of the matter showed that this impression was erroneous, and that a better understanding is obtained by replacing k by a ring B such that A is a finitely generated B-algebra. Geometrically, this means that instead of a single affine algebraic variety V (as defined by A) we are considering a 'regular map' or 'morphism' of V into another affine variety W, and properties of the variety V then are generalized to properties of a morphism $V \to W$ (the 'absolute' notion for V being obtained from the more general 'relative' notion by taking W reduced to a point). On the other hand, one should not prevent the rings having nilpotent elements, and by no means exclude them without serious reasons." There follows the full definition of affine pre-schemes and schemes, then general pre-schemes and schemes, then several theorems concerning (quasi-)coherent sheaves on schemes, then several open problems together with a hint that they are open only because the techniques are so new that no one had time to solve them yet, and finally the so-typical conclusion: 'As a quite general fact, it is believed that a better insight in any part of even the most classical Algebraic Geometry will be obtained by trying to re-state all known facts and problems in the context of schemata. This work is now begun, and will be carried on in a treatise on Algebraic Geometry which, it is hoped, will be written in the coming years by J. Dieudonné and myself..."

By October 1958, the work is underway, with Grothendieck sending masses of rough—and not so rough—notes to Dieudonné for the final writing-up: "I have started writing up some short papers and commentaries for Dieudonné, who seems to have gotten off to a good start on writing up schemes. I hope that by spring, the first four chapters will be written up..." He goes on to give more or less the same plan for the coming chapters that appears in the introduction of EGA I, with a chronology, however, that turned out to be wildly off target. Indeed, as it turned out, EGA I was published in 1960, EGA II in 1961, EGA III in 1961 (first part) and 1963 (second part), EGA IV in 1964 (first part), 1965 (second part), 1966 (third part) and 1967 (fourth part). As for EGA V, a set of very partial prenotes for this volume were typed up by Grothendieck and are covered with his handwritten corrections. He distributed them informally – S. Kleiman had a copy already in 1966-67 – but they lay unused for several years. Grothendieck sent them to Piotr Blass in 1985, and Piotr Blass and Stan Klasa read, translated and gave a final cleaning up to a sizeable section of these notes which they published between 1992 and 1995 in the Ulam Quarterly Journal. It is however clear that this very partial EGA V is far from what Grothendieck had in mind when he called the projected volume "Procédés élémentaires de construction de schémas" in 1958. The rest of EGA was never written at all, although a large portion of the topics it was meant to cover appear in the SGA, and in the FGA (Fondements) which is nothing but the collection of Grothendieck's Bourbaki seminars.

In this period, the exchanges between Serre and Grothendieck become less intense as their interests diverge, yet they continue writing to each other frequently with accounts of their newest ideas, inspiring each other without actually collaborating on the same topic. In the fall of 1958, Zariski invited Grothendieck to visit Harvard. He was pleased to go, but made clear to Zariski that he refused to sign the pledge not to work to overthrow the American government which was necessary at that time to obtain a visa. Zariski warned him that he might find himself in prison; Grothendieck, perhaps mindful of the impressive amount of French mathematics done in prisons (think of Galois, Weil, Leray...) responded that that would be fine, as long as he could have books and students could visit.

The Harvard letters show that Grothendieck is mulling over fundamental groups: "I would like to prove the following result: Let X be a scheme over Y, proper over Y, whose 'tangent map is everywhere surjective'. Show that the 'geometric' fundamental group of the fiber $f^{-1}(y)$ is independent of y...Have you ever thought about questions of this flavor?" This eventually turned into Chap. X of SGA 1. However, Serre, back in Paris and now working hard on local fields as well as on write-ups for Bourbaki, replies simply "I am beginning to feel guilty for not having replied to your letters sooner: the sad truth is that I have nothing serious to say about them..." Undeterred, Grothendieck fires back "I have the impression I am making progress with the π_1. It seems to me that one of the fundamental things to prove is the following: Let X be proper over a *local* ring $\mathcal{O}$, and let F be the geometric fiber of the origin in Spec($\mathcal{O}$). Then the homomorphism $\pi_1(F) \to \pi_1(X)$ is *injective*." This result is equivalent to the statement that every covering of F is dominated by the restriction to F of a covering of X.

Serre's response: "Your letter makes me want to take stock of what I am doing with local fields," and he presents an analog of Grothendieck's injectivity statement for fundamental groups in the framework of class field theory. Namely, fixing a connected commutative algebraic group G defined over an algebraically closed field, he notes that the collection of isogenies $G' \to G$ where G' is also a connected algebraic group form a projective system; he sets $\overline{G}$ to be the projective limit, and defines $\pi_1(G)$ as the kernel of the surjective homomorphism $\overline{G} \to G$. "These constructions yield formal results of the usual kind. The only surprising result is the following: If $G \subset G'$, then $\pi_1(G) \to \pi_1(G')$ is *injective*..." This result proves the exact analog of Grothendieck's desired result stated above, in the situation of commutative algebraic groups, with coverings replaced by isogenies. Serre's ten-page, perfectly written letter goes on to classify isogenies of the groups of units of a discrete valuation field K with algebraically closed residue fields in terms of the abelian extensions of K and study the behavior of the conductor of a finite extension L/K. The results ended up published in Serre's articles on algebraic groups and local fields.

A break of several months in the letters, due no doubt to the presence of both the correspondents in Paris, brings us to the summer of 1959. During the gap, Grothendieck's job problem had been solved once and for all when he accepted the offer of a permanent research position at the IHES (Institut des Hautes Etudes Scientifiques), newly created in June 1958 by the Russian businessman Léon Motchane as the French answer to Princeton's Institute for Advanced Study. He and Mireille had also become the parents of a little girl, Johanna, born in February 1959. The letters from this period show that Grothendieck was already thinking about a general formulation of Weil cohomology (planned for chapter XIII of EGA, now familiarly referred to as the Multiplodocus), while still working on the fundamental group and on writing the early chapters. The progress of the write-up is still seriously overestimated: "Next year, I hope to get a satisfactory theory of the fundamental group, and finish up the writing of chapters IV, V, VI, VII (the last one being the fundamental group) at the same time as categories. In two years, residues, duality, intersections, Chern and Riemann-Roch. In three years, Weil cohomology and a little homotopy, God willing... Unless there are unexpected difficulties or I get bogged down, the multiplodocus should be ready in 3 years' time, or 4 at the outside. Then we can start doing algebraic geometry!"

In the fall, Serre writes from Princeton, where under Weil's influence no one can help thinking about the Weil conjectures (see below); he is, however, also participating in a seminar on the part of the multiplodocus that has already been written, and sends detailed letters containing remarks, criticisms, corrections and suggestions. Although he does not naturally talk about his own interests in terms of schemes, he is very willing to do so in order to get a point across to his one-track minded friend: "Another interesting question raised by Weil is that of the restriction of scalars. This problem can be perfectly well expressed in terms of schemes: given two "base preschemes" S and T, and a morphism $T \to S$, one would like to associate to any T-prescheme V an S-prescheme $R_{T/S}(V)$ and a morphism $p : R_{T/S}(V) \times_S T \to V$ such that for every S-prescheme U, the obvious map from $\mathrm{Morph}_S(U, R_{T/S}(V))$ to $\mathrm{Morph}_T(U \times_S T, V)$ is bijective. This probably doesn't exist except under very strong conditions..." One can imagine that Weil did not express the restriction of scalars studied in his famous 1956 paper (or descent theory, as it later came to be called) in these terms! In fact, Weil and Grothendieck appear to have been rather allergic to each other's style (if not to each other); to a later remark in the same letter: "Tate gave a very pretty lecture on adelic points from the point of view of schemes; they are the same as Weil's in the separated case," Serre added a footnote when the Correspondence was published, commenting "I have a very vivid memory of attempting to explain to Weil (resp. Grothendieck) that Grothendieck's (resp. Weil's) definition is equivalent to his. Neither of them wanted to listen: his definition was obviously 'the right one', why go looking for another one?"

To Serre's addition to the above question: "Lang says he finds all this insufficient, and that one should find a big bag containing all this stuff together with the Greenberg functor...Have you done this?" Grothendieck replies that: "I have indeed fully clarified the Weil and Greenberg operations...as Lang said, they go in the same bag, otherwise the proofs have to be repeated..." This is followed by a lengthy and complete resolution of the question Serre had posed—itself followed by an anxious query from Lang, worried that Grothendieck would publish something on the Greenberg functor before Greenberg himself did! Grothendieck reassures him; he plans to include his result in EGA V which will not be out for what he supposes to be a year or two. At this point, he is still devoted to the foundations of scheme theory. However, as the whole exchange of letters around this time clearly show, the Weil conjectures were the major inspiration for the development of the new language of algebraic geometry, and sketching possible proofs was an irresistible temptation, even if it did run ahead of the inexorable plan for EGA.

1959–1961: The Weil conjectures: first efforts

The Weil conjectures, first formulated by André Weil in 1949, were very present in the minds of both Serre and Grothendieck at least from the early

1950's. Weil himself proved his conjectures for curves and abelian varieties, and reformulated them in terms of an as yet non-existent cohomology theory which, if defined, would yield his conjectures as natural consequences of its properties. This was the approach that attracted Grothendieck; as he explained at the very beginning of his 1958 ICM talk, the precise goal that initially inspired the work on schemes was to define, for algebraic varieties defined over a field of characteristic $p > 0$, a 'Weil cohomology', i.e. a system of cohomology groups with coefficients in a field of characteristic 0 possessing all the properties listed by Weil that would be necessary to prove his conjectures.

As Weil first stated them, the conjectures do not appear overtly cohomological. For a non-singular projective variety X defined over the field $k = \mathbb{F}_q$ of q elements, let $\overline{X}$ be the same variety considered to be defined over the algebraic closure $\overline{k}$ of k, and for each $r \geq 1$, let N_r denote the number of points in $\overline{X}$ whose coordinates lie in the field $\mathbb{F}_{q^r}$. Define the zeta function of X by

$$Z(X,t) = \exp\Big(\sum_{r=1}^{\infty} N_r \frac{t^r}{r}\Big).$$

Then a rough statement of the Weil conjectures is as follows: (1) $Z(X,t)$ is a rational function, in fact $Z(X,t)$ has a decomposition as

$$Z(X,t) = \frac{P_1(t)P_3(t)\cdots P_{2n-1}(t)}{P_0(t)P_2(t)\cdots P_{2n}(t)}$$

where $P_0(t) = 1-t$, $P_{2n}(t) = 1-q^n t$ and each of the $P_i(t)$ is a product $P_i(t) = \prod_j (1-\alpha_{ij}t)$ where all of the α_{ij} are algebraic integers (so that the $P_i(t)$ are defined over $\mathbb{Z}$); (2) it satisfies a simple functional equation of the form $Z(X, 1/q^n t) = \pm q^{nE/2} t^E Z(X,t)$ where E is an integer associated to the geometry of X known as the Euler characteristic of X; (3) the roots α_{ij} satisfy $|\alpha_{ij}| = q^{i/2}$ for all i, j; (4) the degrees $B_i(X)$ of the $P_i(t)$ have two important properties: $\sum_i (-1)^i B_i(X) = E$ and if X is the reduction mod p of a variety Y defined in characteristic 0, then $B_i(X) = \dim H^i(Y, \mathbb{Z})$.

The idea of reformulating the conjectures in cohomological terms, as simple consequences of the properties of a suitable cohomology with coefficients in characteristic zero, was first initiated by Weil himself, as he tried to prove his conjecture—and succeeded in proving it for curves and abelian varieties—by making use of the Frobenius morphism F which maps a point of $\overline{X}$ given by $P = (a_1, \ldots, a_n) \in \overline{k}^n$ to the point $F(P) = (a_1^q, \ldots, a_n^q)$. Then P is a fixed point of F if and only if the $a_i \in k$, and more generally P is a fixed point of the iterate f^r if and only if the $a_i \in \mathbb{F}_{q^r}$. Thus the number N_r measures exactly the number of fixed points of f^r on $\overline{X}$.

Suppose we are in possession of a 'Weil cohomology' associated to a scheme of finite type X over an algebraically closed field k of characteristic $p > 0$. This is a set of vector spaces $H^i(X, K)$ over some characteristic 0 field K such that $H^i(X,K) = 0$ unless $0 \leq i \leq 2n$, which satisfy a number of good properties of which we mention only a few here. The

$H^i(X,K)$ should be contravariant functors in X, and should be equipped with a cup-product. Under the assumption that X is smooth and proper, the $H^i(X,k)$ should be finite-dimensional and satisfy Poincaré duality, and for a morphism $f: X \to X$ with isolated fixed points, they should satisfy a Lefschetz fixed-point formula:

$$\#\text{fixed points of } f = \sum_{i=0}^{2n}(-1)^i \mathrm{Tr}(f^*; H^i(X,K)),$$

where for each i, $f^*: H^i(X,K) \to H^i(X,K)$ is the linear map induced by f, and the fixed points are counted with their multiplicity.

Now, if we have such a cohomology, the Lefschetz fixed-point formula applied to the iterated Frobenius morphism f^r associated to $\overline{X}$ as above says that

$$N_r = \sum_{i=0}^{2n}(-1)^i \mathrm{Tr}\big((f^r)^*; H^i(\overline{X},K)\big).$$

Thus, Weil's zeta function can be written

$$Z(X,t) = \exp\Big(\sum_{r=1}^{\infty}\sum_{i=0}^{2n}(-1)^i \mathrm{Tr}\big((f^r)^*; H^i(\overline{X},K)\big)\frac{t^r}{r}\Big)$$

$$(*) \qquad = \prod_{i=0}^{2n}\Big[\exp\Big(\sum_{r=1}^{\infty}\mathrm{Tr}\big((f^r)^*; H^i(\overline{X},K)\big)\frac{t^r}{r}\Big)\Big]^{(-1)^i}.$$

An elementary lemma shows that in general, if $\phi: V \to V$ is a linear map on a finite-dimensional vector space V defined over a field k, then as power series in t with coefficients in k, we have

$$\exp\Big(\sum_{r=1}^{\infty}\mathrm{Tr}(\phi, V)\frac{t^r}{r}\Big) = \det(1-\phi t, V)^{-1}.$$

Since the right-hand side is a rational function in t, we immediately obtain from this and from $(*)$ that

$$Z(X,t) = \frac{P_1(t)P_3(t)\cdots P_{2n-1}(t)}{P_0(t)P_2(t)\cdots P_{2n}(t)}$$

where $P_i(t) = \det(1 - f^*t, H^i(\overline{X},K))$. This shows that $Z(X,t)$ is a rational function (the first Weil conjecture), although much more work is needed to prove the exact nature of the $P_i(t)$ predicted by Weil. The degrees of the $P_i(t)$ are equal to the dimensions of the $H^i(\overline{X},K)$ (Betti numbers), and Poincaré duality implies that $Z(X,t)$ satisfies the functional equation (second conjecture).

This simplified discussion shows that defining a Weil cohomology would prove at least a large part of the Weil conjectures, and this is the approach that was taken up by both Serre and Grothendieck, who set to work on trying to discover the right cohomology theory. Serre used Zariski topology and tried cohomology over the field of definition of the variety; even though this field was in characteristic p, he hoped at least to find the right Betti numbers, but didn't. Then he tried working with the ring of Witt vectors, so that he was at least in characteristic zero, but this too failed to yield results. At the beginning of his ICM talk, Grothendieck describes these efforts and concludes "Although interesting relations must certainly exist between these cohomology groups [those studied by Serre] and the 'true ones', it seems certain now that the Weil cohomology has to be defined by a completely different approach." In October 1959, Serre writes from Princeton: "I have spent my time thinking about the Artin representation and the Weil conjectures (particularly the L series formalism). I have nothing precise to tell you. I am wondering (and have not yet been able to decide) if the generalization of the Weil formula $\sigma(X.X') \geq 0$ might not be $\sigma_n(X.X') \geq 0$, where X is a algebraic correspondence on V (non-singular of dimension n), $X' = \text{transp}(X)$ and σ_n denotes the trace of the homology representation in dimension exactly n. A priori, this is not ridiculous, and would lead to a natural plan of a proof of the Weil conjectures on the absolute value of the eigenvalues of Frobenius: in dimension n, the equation $F.F' = q^n{\cdot}1$ ($F =$ Frobenius), together with the positivity of the trace, shows that this eigenvalue is $q^{n/2}$, as it should be: in dimension $< n$, a Lefschetz-type theorem will hopefully allow us to reduce to a hyperplane section; in dimension $> n$, Poincaré duality allows us to reduce to the previous case. It is very tempting, but one should at least check that $\sigma_n(X.X')$ is ≥ 0 in the classical case, by a Kählerian argument, and I have not managed either to do it or to find a counterexample!" Although this definition of X' does not quite work to generalize Weil's formula, Serre soon found a way that did, and published it in a three-page note—an extract from a letter to Weil – called *Analogues kähleriens des conjectures de Weil* (Annals of Math. 1960).

Grothendieck's blunt response to Serre's remarks: "I have no comments on your attempts to generalize the Weil-Castelnuovo inequality; I confess that these positivity questions have not really penetrated into my yoga yet[7]; besides, as you know, I have a sketch of a proof of the Weil conjectures based on the curves case, which means I am not that excited about your idea." His mind still running on several simultaneous tracks, he adds: "By the way, did you receive a letter from me two months ago in which I told you about the fundamental group and its infinitesimal part? You probably have nothing to say about that either!" The impression is that the two friends are thinking along different lines, with an intensity that precludes their looking actively at each other's ideas. Yet it is only a question of time. Just a few years later, Serre's *Analogues kählériens* would play a fundamental role in Grothendieck's reflections aiming at a vast generalization of the

[7]In fact, Grothendieck had already published a paper containing a proof of a generalization of the Weil-Castelnuovo inequality in 1958 (*Sur une note de Mattuck-Tate*, J. reine angew. Math. **200** (1958), 208-215), from a product of curves to an arbitrary surface.

Weil conjectures, making them even more "geometric", extending them to a statement on general endomorphisms rather than the all-pervasive but restrictive Frobenius. Serre's ideas stimulated him to formulate the famous 'standard conjectures' on algebraic cycles which were formulated in order to lead towards an unconditional theory of motives (see below), and whose proof would have fulfilled Grothendieck's plan for proving the third Weil conjecture.

On November 15, 1959 came the news which Michel Raynaud, a 21-year-old student at the time, describes as a thunderclap: "First of all, a surprising piece of news: Dwork phoned Tate the evening of the day before yesterday to say he had proved the rationality of zeta functions (in the most general case: arbitrary singularities). He did not say how he did it (Karin took the call, not Tate)...it is rather surprising that Dwork was able to do it. Let us wait for confirmation!" writes Serre. To quote Katz and Tate's memorial article on Dwork in the March 1999 Notices of the AMS: "In 1959 he electrified the mathematical community when he proved the first part of the Weil conjecture in a strong form, namely, that the zeta function of *any* algebraic variety over a finite field was a rational function. What's more, his proof did not at all conform to the then widespread idea that the Weil conjectures would, and should, be solved by the construction of a suitable cohomology theory for varieties over finite fields (a 'Weil cohomology' in later terminology[8]) with a plethora of marvelous properties." Dwork did, however, make use of the Frobenius morphism and detailed p-adic analysis in a large p-adic field.

It is hard to assess the effect this announcement had on Grothendieck, because he did not respond (or his response is missing). However, one thing is absolutely clear: Dwork's work had little or no effect on his own vast research plan to create an algebraic-geometric framework in which a Weil cohomology would appear naturally. In August 1960 he writes: "I have remarked that my duality theory for coherent sheaves will be a wonderful guide to constructing a general duality theory encompassing it together with duality theory for algebraic groups or group schemes and duality of Weil cohomology. This has led me to expand my planned program by so much, since such questions now seem much more amenable to attack than previously. Once this theory has been developed, I hope the Weil conjectures will come out all by themselves. I will work on this next year..." This was the year in which he began running his famous SGA (Séminaire de Géométrie Algébrique). The first year of the seminar, 1960-61, was devoted to the study of the fundamental group and eventually published as SGA 1.

---

[8]Notice that Grothendieck, ahead of his time as usual, had used this exact terminology in his 1958 paper cited above.

1961: Valuations—and War

October 1961 finds Grothendieck happily ensconced at Harvard—married, now, to Mireille, as this made it easier for the couple to travel to the US together, and the father of a tiny son born in July, named Alexandre and called Sasha after Grothendieck's father, who perished in Auschwitz in 1942. His letters show him to be full of ideas and surrounded by outstanding students and colleagues: John Tate, Mike Artin, Robin Hartshorne, David Mumford. "The mathematical atmosphere at Harvard is absolutely terrific, a real breath of fresh air compared with Paris which becomes gloomier every year. There is a good number of intelligent students here, who are beginning to be familiar with the language of schemes, and ask for nothing better than to work on interesting problems, of which there is obviously no lack. I am even selling (the little I know of) Weil sheaves and Weil cohomology with the greatest of ease, including to Tate [...] Meanwhile, Mike Artin is getting excited about global degeneracy phenomena for elliptic curves...which he wants to understand in terms of Weil cohomology..." By this time, Grothendieck's vision of the right way to do mathematics is strong and clear, and he is intolerant of other views. Valuations, for some reason, provoke intense annoyance; in his criticism of Bourbaki's draft for *Commutative Algebra*, he expostulates: "Chaps. VI and VII appear to me to be unworthy of Bourbaki. I have proposed several times in vain that Chap. VI (Valuations) should be purely and simply thrown out; even though I have come to understand why resolution of singularities is useful since then, I am still of the opinion that VI should be removed, or at the very least moved from its current position to the end of the book, among the things 'not to be read.' Its current position will mislead the reader as to the right ideas and methods...Chap. VII is also copied from Krull, and appears to me to be far removed from both geometric intuition (which is a good guide) and from actual practice...one can only understand *properly* if a geometric language is available...even if Bourbaki does not launch into such things, it would be nice if he at least had the right 'yoga'. At the moment, Chap. VII reeks of dusty academics." Serre is not impressed: "You are very harsh on Valuations! I persist nonetheless in keeping them, for several reasons, of which the first is practical: n people have sweated over them, there is nothing wrong with the result, and it should not be thrown out without very serious reasons (which you do not have). Of course, if it were proved to be of no use and misleading, this first argument would not hold water. But that is not the case. Even an unrepentant Noetherian needs discrete valuations and their extensions...Nor do I really agree with your objections to the Krull chapter...It is clear that these two chapters are basically an insertion into Bourbaki of 'papa's Commutative Algebra', as de Gaulle would say. But I am much less 'fundamentalist' than you on such questions (I have no pretension to know 'the essence' of things) and this does not shock me at all."

This is the first time that a pinch of annoyance can be felt in Serre's tone, underlying the real divergence between the two approaches to doing mathematics. Serre was the more open-minded of the two; any proof of a good theorem, whatever the style, was liable to delight him, whereas obtaining even good results "the wrong way"—using clever tricks to get around deep theoretical obstacles—could infuriate Grothendieck. These features became more pronounced in both mathematicians over the years; the author still recalls Serre's unexpected reaction of spontaneous delight upon being shown a very modest lemma on obstructions to the construction of the cyclic group of order 8 as a Galois group, simply because he had never spotted it himself, whereas Grothendieck could not prevent himself, later, from expressing bitter disapproval of Deligne's method for finishing the proof of the Weil conjecture, which did not follow his own grander and more difficult plan.

Grothendieck, ever the idealist, fires back a response also tinged with irritation and again making use of his favorite word 'right' as well as the picturesque style he uses when he really wants to get a point across: "Your argument in favor of valuations is pretty funny. The principle generally respected by Bourbaki is rather that there should be very good reasons for including a huge mess, especially in a central position; the fact that n people have sweated over it is certainly not a good reason, since these n people had no idea of the role of the mess in commutative algebra, but had simply received an order to figure out, Bourbakically, some stuff that they unfortunately did not bother to examine critically as part of a whole. Your comments on rank zero valuations constitute an argument for removing it from where it is now. Indeed, the right point of view for this is not commutative algebra at all, but absolute values of fields (archimedean or not). The p-adic analysts do not care any more than the algebraic geometers (or even Zariski himself, I have the impression, as he seems disenchanted with his former loves, who still cause Our Master to swoon) for endless scales and arpeggios on compositions of valuations, baroque ordered groups, full subgroups of the above and whatever. These scales deserve at most to adorn Bourbaki's exercise section, as long as no one uses them."

These very same letters, as well as a famous one dated October 22, 1961 and addressed to Cartan, contain a fascinating exchange of views on the situation in France connected with the Algerian war and the necessity of military service. By October 1961, the end of the Algerian war of independence was thought to be in sight: already in 1959, de Gaulle had pronounced the fatal words 'self-determination', admitting the possibility that France would eventually have to give up Algeria, and in May 1961 he had negotiated a cease-fire set to begin in March 1962. However, hostilities continued during the interim period, with violent terrorist acts on the part of Algerian independence factions, and even more violent repression from the French police and anti-independence groups such as the OAS (Secret Army Organization), in both France and Algeria. Between August and October 1961, eleven policemen were killed and many more wounded in Algerian bomb attacks in Paris. The French prefect of police responded with "For every blow received, we will respond with ten," and Algerians living in France were subjected to harassment, imprisonment, torture and 'disappearing'. On October 5, a curfew concerning all "French Muslims from Algeria" was announced. On October 17, thousands of Algerians poured into the streets of Paris to protest. The massacre that occurred on that day left dozens of bloody bodies piled in the streets or floating down the Seine, where they were still to be seen days later.

Grothendieck's letter to Cartan was written from Harvard just four days after this event. Surprisingly for a man whose extreme antimilitarist, ecological views were to become his dominating preoccupation ten years later, when he left the IHES with fracas because he discovered that a small percentage of its funding was of military origin, the tone he adopts to criticize the effect of the mandatory two years' military service on budding mathematicians is quite moderate. Rather than lambasting military service on principle, he emits more of a lament at its effect on mathematics students. "I am starting to realize that the long military service has a disastrous influence. Surely it is not necessary for me to explain to you that an enormous effort and a continual tension are necessary for the beginner to be able to absorb a mass of very diverse technical ideas in order to get to the point where he may be able to do something useful, maybe even original. For our part, we use up enough chalk and saliva until the moment finally arrives when the fellow can pull his own weight. Alas, that is precisely the moment when he is called upon to serve his country, as they say, and the beautiful enthusiasm and subtle cerebral reflexes acquired by years of studying and meditation will be put aside for two years, provided the General consents not to keep him at the shooting range for even longer. With such a prospect in view, I quite understand that a budding Mathematician is inhibited before he starts, and his natural enthusiasm is blunted. Whether he manages to hastily cobble a thesis together before his military service, or plays it smart and enrolls early, he will be useless for several years as an 'insider' or at least as a 'Parisian', i.e. someone who contributes to the fertility of the scientific atmosphere of Paris. Cartier has not been a Parisian for ages, and even his lectures at the Collège have not changed this, being nothing but the fugitive appearances of a fellow on leave carrying out a social duty of no importance. Just when Gabriel is beginning to be interesting, off he goes to the army, and when he comes back he will go to Strasbourg, which I feel as a serious loss for Paris. Apparently we cannot even invite him to the IHES immediately, as this would not go down well with the University, which does not have enough professors! The situation is absolutely grotesque. With some difficulty, I have managed to scrape together four or five ex-Normaliens for my algebraic geometry seminar at the IHES, who are just beginning to have some vague glimmers of understanding, and one or two of whom even appeared to be about to start on some useful and even urgent work, namely Verdier and Giraud. Nothing doing: unless I am mistaken, both of them, and certainly Verdier, are enrolling early, and in the end someone else (myself if necessary) may end up doing the work for them. If I do not actually have the impression of preaching in the wilderness in Paris, I am at least certain of building on sand.

"This situation does not exist in the USA, where at least the State is intelligent enough not to waste its 'brain-power' on military exercises. There is no difficulty for a talented student to get exempted from the draft on the grounds of being 'indispensable to the defense of the nation', a euphemism which has probably never fooled a single American civil servant. This is exactly the point I wanted to make in this letter. We cannot require the soldiers or the politicians or the princes that govern us to be aware of the psychological subtleties of scientific research, or to realize that it affects the scientific level of a country when the development of their young researchers is halted or put on hold for two critical years of their training. If they need to be informed of this fact, the only people who can do so with a certain degree of authority are yourself and our colleagues. (I personally am in any case completely out of it.) I am thinking particularly of you, because of your position at the Ecole Normale, which does after all potentially carry with it non-standard duties towards your present and former students. What is more, as you are not suspected of any political 'partiality', you are in a better position to do something about it than Schwartz would be, for example: something like writing a series of articles in 'Le Monde', or a personal letter to the President, or whatever. In any case, if you do not speak out, I really wonder who will..."

Cartan's response is not included in the Correspondence, but Cartan showed this letter to Serre, who responded to Grothendieck directly, in very typical, simple and pragmatic terms, which probably resonate with the majority: "What is certainly [...] serious is the rather low level of the current generation ('orphans', etc.) and I agree with you that the military service is largely responsible. But it is almost certain we will get nowhere with this as long as the war in Algeria continues: an exemption for scientists would be a truly shocking inequality when lives are at stake. The only reasonable action at the moment—we always come back to this—is campaigning against the war in Algeria itself (and secondarily, against a military government). It is impossible to 'stay out of politics'." It is not certain whether Serre himself took any kind of action against the war in Algeria, but other mathematicians, above all Laurent Schwartz—whose apartment building was plastic bombed by the OAS—certainly did.

Grothendieck replied to Serre in the same measured terms as before: "I do not agree with you that nothing should be done against the military service—for gifted scientists in particular—before the end of the Algerian war. To start with, as far as injustice is concerned 'when lives are at stake', if it is an injustice to exempt certain people from national service, then the difference between doing so during or after a time of guerrilla war is one of degree and not of essence. I do not think that the danger of losing one's life is such, at this point, that it has become more important than the loss of two years of training (for any young person, scientist or otherwise), leaving aside entirely the moral question (to which most people are apparently indifferent). The minimal probability of being killed does not seem to me to make a big difference. On the other hand, if certain Academics brought the effects of military service (and, by implication, of the Algerian war) on the scientific level of the country to the attention of the public and the authorities, and required some reforms, it would not exclude the possibility of classical scholars, technicians, firemen and lamp-lighters grouping together to require analogous reforms for themselves, on analogous and to my mind equally valid grounds. Any action in this direction, even if very limited, will contribute to making people realize the consequences of the militarization of the country, and might create a precedent for analogous and vaster actions. But in this case it is obvious that it is only by limiting the problem and the proposals to a restricted situation which from many points of view is 'ideal' that there is any chance for rapid success, especially if it is done by Academics without political affiliations, such as Cartan. Note that the arguments being put forward are just as valid in wartime, if not more so, I mean from the government's point of view, as it is quite obvious that the Americans, for example, are even more careful to keep from removing their scientists and their high-class technicians from their laboratories in wartime than in peacetime. —And finally, I have a very down to earth point of view on the military service, namely catch as catch can, and the more people there are who, by whatever means, be it conscientious objection, desertion, fraud or even knowing the right people, manage to extricate themselves from this idiocy, the better."

Few if any of Grothendieck's French colleagues shared his views, however, and even after the Algerian war wound down, military service remained mandatory in France until 2001.

1962–1964: Weil conjectures more than ever

The letters of 1962 are reduced to a couple of short exchanges in September; they are rather amusing to read, as the questions and answers go so quickly that letters cross containing the same ideas. These letters concern the questions that Serre was working on at the time; Lie algebras associated to groups of Galois type, Galois cohomology and 'good' groups. Many of the results he mentions were eventually published in a form almost identical to the letters (compare for instance the letter dated Friday evening 1962 with the exercises given in *Galois Cohomology*, Chap 1, §2.6). Serre's letter deals with "the relation between the cohomology of 'discrete' groups and of 'Galois' groups (i.e. 'ordinary' cohomology and 'Grothendian' cohomology)". He considers homomorphisms $G \to K$ where G is discrete and K totally disconnected compact, the basic case being where K is just the profinite completion $\widehat{G}$ of G, and distinguishes 'good' groups as being those such that $H^i(G, M) \simeq H^i(\widehat{G}, M)$ for all i and all topological $\widehat{G}$-modules M. The expression "Grothendian cohomology" is probably in view of the case where G is a topological fundamental group and its profinite completion the associated geometric fundamental group much studied by Grothendieck.

The next letters date from April 1963. By this time, Grothendieck had already developed many of the main properties of etale and ℓ-adic cohomology, which he would explain completely in his SGA lectures of 1963-64 (etale, SGA 4) and 1964-65 (ℓ-adic, SGA 5). The ℓ-adic cohomology was developed on purpose as a Weil cohomology, and indeed, in his Bourbaki seminar of December 1964, Grothendieck stated that using it, he was able to prove the first and the fourth Weil conjectures in April 1963, although he published nothing on the subject at that time.

Serre must have been aware of this result, so that it is never explicitly mentioned in the letters of April 1963 or in any others, leaving one with the same disappointed feeling an archeologist might have when there is a hole in a newly discovered ancient parchment document which must have contained essential words. But it was one of Grothendieck's distinguishing features as a mathematician that he was never in a hurry to publish, whether it be for reasons of priority, or credit, or simply to get the word out. Each one of his new results fitted, in his mind, into an exact and proper position in his vast vision, and would be written up only when the write-up of the vision had reached that point and not before (as for actual publishing, this often had to wait for several more years, as there was far too much for Grothendieck to write up himself and he was dependent on the help of a large number of more or less willing and able students and colleagues).

Let us take the time here to sketch out something of Grothendieck's further work on the Weil conjectures even though it does not appear explicitly in the letters, because it sheds light on everything he was thinking about at the time. For this, we need to briefly explain the development of ℓ-adic etale cohomology. It fits very naturally into Grothendieck's vision in general. To start with, Grothendieck's idea of a scheme S was far from what the commun run of mortals might associate with that notion (for instance, an ordinary algebraic variety defined by polynomial equations over a field). His notion of a scheme carried with it, automatically, the whole range of associated properties and objects: covers, morphisms, mod p versions, cohomology... He uses the words "set of yardsticks" for the cohomology groups and "fan" for the family of mod p versions, but "fan" is a very good image in general: when spread, the fan gives a clear picture of S, whereas the classical attitude of considering just one or two aspects (such as geometric points, for instance) shows only a very tiny slice of the full fan.

Although this quintessentially Grothendieckian point of view may be difficult to adopt for oneself, it is not conceptually out of reach. The next idea, however, is certainly one of Grothendieck's most astonishing contributions to mathematics. Grothendieck's work on generalizing the notion of fundamental group of an algebraic variety, in particular the geometric fundamental group which classifies unramified covers, to scheme theory led him naturally to consider at a single blow the whole collection of etale (flat, unramified—the schematic generalization of the topological notion of being a local homeomorphism, or finite covering map of an open set) morphisms to a given scheme S. The leap consists in considering this collection of maps—or, more generally, *any* collection of maps to a given object S satisfying a few good properties—as a new, more general kind of *topology* on S. In this view, open sets are now replaced by these maps $X \to S$, inclusions of open sets are replaced by compositions of morphisms $Y \to X \to S$, and the basic axioms of a topology are easily generalized. Such a topology is now known as a Grothendieck topology, and the initial example is the etale topology on a scheme S given by the collection of all etale morphisms to S. The scheme equipped with this topology is called the etale site of S, S_{et}. Grothendieck saw that this new type of topology had the power to yield results which the Zariski topology was not fine enough to give, even in Serre's able hands.

The next step was to define sheaves and sheaf cohomology for this new type of topological space, which presented no real problem since Grothendieck topology possesses all the axiomatic properties of ordinary topology. So he was able to simply apply the derived functor definition of sheaf cohomology which had already enchanted him back in 1955, obtaining a definition of etale cohomology groups $H^i(S_{et}, \mathcal{F})$ for a sheaf $\mathcal{F}$ on S_{et}.

The example of elliptic curves already yields some very important information about what kind of sheaves have to be used in order to build up a cohomology theory with the desired properties of a Weil cohomology. First of all, one needs to use *torsion* sheaves, for instance, the constant sheaf on a finite group. Secondly, the order of this group has to be prime to p, the characteristic of the field of definition of the algebraic variety in the Weil conjectures. So Grothendieck considered the simplest case, where the abelian group is $\mathbb{Z}/\ell\mathbb{Z}$ for a prime ℓ different from p, defining the etale cohomology groups $H^i(S_{et}, \mathbb{Z}/\ell\mathbb{Z})$. But the coefficients are in characteristic ℓ, whereas they should be in characteristic 0. Not to worry: Grothendieck defined the *ℓ-adic cohomology group* $H^i_{et}(S, \mathbb{Q}_\ell)$ as the $\mathbb{Q}_\ell$-vector space obtained by taking the inverse limit of the $H^i(S_{et}, \mathbb{Z}/\ell^n\mathbb{Z})$ as $n \to \infty$ to get a $\mathbb{Z}_\ell$-module and tensoring it with $\mathbb{Q}_\ell$.

Instead of writing explicitly about his results on the Weil conjectures, Grothendieck's letters from April 1963 are concerned with recasting the Ogg-Shafarevitch formula expressing the Euler characteristic of an algebraic curve in his own language, and generalizing it to the case of wild ramification. To do this, he looks for "local invariants" generalizing the terms of the Ogg-Shafarevitch formula, but although he sees what properties they should have, he doesn't know how to define them. His letter asking Serre this question bears fruit just days later, as Serre recognized that the desired local invariants can be obtained using the Swan representation, allowing Grothendieck to establish his Euler-Poincaré formula for torsion sheaves on an algebraic curve. Grothendieck did not get around to publishing this result either; it eventually appeared in his student Michel Raynaud's Bourbaki seminar of February 1965 (Raynaud recalls a slight feeling of panic the day before the seminar, when Grothendieck lightheartedly suggested that he talk about a grand generalization of what he had already carefully prepared.)

Grothendieck continued to work on the second and third Weil conjectures throughout 1963 and 1964; at one point he tried to plan a proof based on showing that every variety is birationally a product of curves... until Serre sent him a counterexample (letter of March 31, 1964). He then had another idea, to judge by Serre's letter of April 2, in which Serre writes "Good luck for your 'second attack' on the Weil conjectures. I may have been a bit too pessimistic on the telephone; it is not entirely out of the question that it works." But Serre's intuition (as usual) was right, because the reply arrives the very next day: "I have convinced myself that my second approach to the Weil conjectures cannot work..." followed, undeterred, by yet another possible attack. However, in Grothendieck's Bourbaki seminar of December 1964, he states that the second and third Weil conjectures are still open. The functional equation was surely proved by the end of 1966, when Grothendieck finished giving the SGA lectures that eventually became the SGA 5 volume on ℓ-adic cohomology. This seminar contained proofs that the ℓ-adic cohomology satisfies all the properties of a Weil cohomology—Poincaré duality in particular—and it is studied there in great depth.

Rather than attacking the remaining conjecture directly, Grothendieck sketched out a vast generalization of the Weil conjectures and stated the difficult 'standard conjectures' (see motives, below) which remain unproven to this day. However, in 1975, Deligne managed to get around the standard conjectures and prove the remaining Weil conjecture by using deep and subtle properties of the ℓ-adic cohomology and original, far-from-obvious techniques[9].

1964: Good and bad reduction of abelian varieties over local fields

The Weil conjectures on algebraic varieties over finite fields, and all of the mathematics that grew up around them, stimulated great interest in the study of algebraic varieties defined over *local* fields, and consequently also the study of the different types of reduction modulo the prime ideal of the local field. The sudden flush of letters exchanged over the fall of 1964 very largely concerns this theme, concentrating especially on elliptic curves and abelian varieties (to Serre's delight and Grothendieck's annoyance: "It might perhaps be possible to get at least the abelian variety case by this method [...]This would at least get us a bit further than the sempiternal

[9]In particular, Deligne made use of Rankin's method, which was far from Grothendieck's style.

elliptic curves via Tate's sempiternal construction... The irritating thing is that one never seems to be able to get past abelian varieties!")

It all began in August 1964 at the Woods Hole Summer Institute, which Serre attended but Grothendieck didn't. On his return, Serre went off on vacation to the south of France, and from there, sent Grothendieck a very long letter describing, in detail, the main new ideas from what must have been a very lively meeting. Reading over the interests, conjectures and recent results of the mathematicians he names: Shimura, Atiyah, Bott, Verdier, Mumford, Ogg, Bombieri, Tate makes it abundantly clear how the Weil conjectures motivated much of the work in number theory and algebraic geometry at that time. Local fields, however, now play a major role.

At Woods Hole, Shimura expressed a conjectural Lefschetz-type formula expressing the trace of f in terms of its fixed points rather than the contrary; so many people started working on it that the formula was proved by the end of the Institute and became known as the "Woods Hole fixed-point formula" (Grothendieck's unenthusiastic reaction: "the fixed point theorem seems to me to be nothing more than a variation on a well-worn theme." And indeed, using his formalism of six operations, he had simultaneously proved a more general fixed-point formula.) Ogg studied the module of ℓ-torsion points on elliptic curves defined over local fields and showed that a certain invariant was independent of ℓ. Serre and Tate generalized this work by conjecturing the independence from ℓ of ℓ-adic representations given by ℓ-adic cohomology groups. This independence (a constant preoccupation in Grothendieck's mind and one that foreshadows the idea of motives), would generalize Ogg's results to abelian varieties. Tate added his own conjectures on ℓ-adic cohomology, algebraic cycles and poles of zeta functions, and Serre and Tate worked on lifting from characteristic p to characteristic 0.

In the months following this report, the exchange of letters between Serre and Grothendieck is exceptionally rich, with almost twenty letters exchanged over the summer and fall of 1964. Even though both epistolarians were in France, the ideas they wanted to share were too complex to discuss only over the phone, and the twenty or so kilometers separating the Collège de France from the IHES in Bures prevented them from seeing each other on a daily basis.

In the first letters following the Woods Hole report, Serre writes to Ogg and Grothendieck about one of the major topics of the rest of the correspondence: the reduction of an abelian variety A defined over a local field K with algebraically closed residue field k of characteristic p. He rephrases results of Ogg to show that A has good reduction if and only if the action of the absolute Galois group of K on the associated Tate module $T_\ell(A)$ for $\ell \neq p$ is trivial. This result, known as the Ogg-Néron-Shafarevitch criterion, was published by Serre and Tate, following the conventions of a pre-publish-or-perish time, in which it was occasionally those who strove to understand a given result who published it— often contributing substantially to the original idea—rather than the originator (cf. the case of Borel and Serre's article on Grothendieck's Riemann-Roch theorem). Using this criterion, Serre recovers known results about the reduction of elliptic curves, and shows that in the presence of complex multiplication (CM) abelian varieties have good reduction everywhere after finite extension of the base field. In the following letter, he adds another consequence of the Ogg-Néron-Shafarevitch criterion: if A has good reduction and B is an abelian variety over K equipped with a K-homomorphism $B \to A$ with finite kernel, then B also has good reduction. This result had already been proved by Shimura and Koizumi, and then by Grothendieck, who responds to Serre's request "I believe you had a proof of this result in your language. Am I right, and what is it?" by posting back a ten-line proof indeed in his own language, i.e. full of words like "closed flat pre-scheme", "general fiber" and "representable".

Serre also asks a precise question about bad reduction: can one always find a finite extension of K such that the (Néron model of the) abelian variety now reduces to an extension of an abelian variety by a torus? ("This may be a stupid question. Do you see a counterexample?") This result, which can be re-stated as the existence of a semi-stable model after finite extension of the base field, was soon proved by Mumford when $p \neq 2$, but on September 24, Grothendieck sends Serre a letter proving the general case. In fact, his theorem is quite general, giving conditions on an ℓ-adic vector space E equipped with an action of the Galois group $\pi = \mathrm{Gal}(\overline{K}/K)$ (where K is the fraction field of a discrete valuation ring) under which there exists an open subgroup of the inertia group which acts unipotently on E. But his conditions apply to the case where E is the Tate module $T_\ell(A)$ associated to an abelian variety A defined over K. Since a unipotent Galois action on $T_\ell(A)$ corresponds to existence of a semi-stable model after finite extension of the base field, just as the trivial action corresponds to good reduction (see above), this answers Serre's question completely.

Serre's response to Grothendieck's letter is simultaneously excited, suspicious and as usual, precise: "Your theorem on the action of the inertia group is terrific—if you really have proved it. In fact, I cannot see how you generalize the argument which proves corollary 2; in the general case, you need Frobenius automorphisms which *normalize* the group I, and I really do not see how you go about it." He adds "Note also that the theorem (and even corollary 2) are false as stated"—in these two statements Grothendieck had spoken of the whole inertia group and not just an open subgroup—"all that you prove (but it is quite sufficient to enchant me) is that there is an *open* subgroup (as in corollary 1) where everything works; indeed, look at your proof of corollary 2, you will see that the so-called q^N-th roots of unity you obtain are in fact $(q^N - 1)$-th roots of unity, and $q^N - 1$ can very well be divisible by ℓ." Coming from Grothendieck, this kind of error is not in the least surprising—nor is the fact that in spite of it, his insight still yielded a powerful theorem. He answers reassuringly on October 5: "I just received your letter. All right, I had confused q^N and $q^N - 1$ and my conclusion thus has to be modified as you say. Here is how one deals with the general case..."

and goes on to give full details of the proof. On October 30—after an interlude devoted to L and Z-functions—he posts off another long letter, proving another result apparently conjectured by Serre ("I was disappointed that you did not find an expert on surfaces to solve your conjecture on abelian varieties...and therefore, finding myself in a healthy temper, I broke off my current reflections to find a proof myself, which I hand you fresh out of the oven.") Namely, he shows that in the case of curves, the unipotent action of the Galois group π passes to a trivial action on the quotient $H^1/(H^1)^\pi$; in other words, for curves, the action can be filtered in at most two steps.

Note that since the result that an open subgroup of the inertia group acts unipotently on $T_\ell(A)$ for any given ℓ prime to the residue characteristic is equivalent to saying that A has a semi-stable model over a finite extension of its field of definition, and this latter statement is independent of ℓ, the theorem also proves that if the action is unipotent for one ℓ then it is unipotent for all ℓ. This kind of "independence from ℓ" result—in the same spirit as the original proof of the fourth Weil conjecture (on Betti numbers) saying that the dimensions of the ℓ-adic cohomology groups are given by the degrees of the factors of the rational function $Z(X,t)$, and thus implying that these dimensions are independent of ℓ—was the initial stimulation for the idea of a motive.

1964–1965: Motives

Motives made their appearance during the same exceptionally active period (in terms of letter-writing), the fall of 1964. The first mention of motives in the letters—the first ever written occurrence of the word in this context—occurs in Grothendieck's letter from August 16: "I will say that something is a 'motive' over k if it looks like the ℓ-adic cohomology group of an algebraic scheme over k, but is considered as being independent of ℓ, with its 'integral structure', or let us say for the moment its '$\mathbb{Q}$' structure, coming from the theory of algebraic cycles. The sad truth is that for the moment I do not know how to define the abelian category of motives, even though I am beginning to have a rather precise yoga for this category, let us call it $\mathbf{M}(k)$." He is quite hopeful about doing this shortly: "I simply hope to arrive at an actual construction of the category of motives via this kind of heuristic consideration, and this seems to me to be an essential part of my 'long run program' [sic: the words 'long run program' are in (a sort of) English in the original]. On the other hand, I have not refrained from making a mass of other conjectures in order to help the yoga take shape."

Serre's answer is unenthusiastic: "I have received your long letter. Unfortunately, I have few (or no) comments to make on the idea of a 'motive' and the underlying metaphysics; roughly speaking, I think as you do that zeta functions (or cohomology with Galois action) reflect the scheme one is studying very faithfully. From there to precise conjectures..." This way of expressing Grothendieck's idea underlines the similarity with the *anabelian theory* that he developed in the 1980's, several years after his grand departure from the established world of mathematics. This theory investigates algebraic varieties which are completely determined by their arithmetic fundamental group, i.e. their geometric fundamental group equipped with its canonical Galois action. The anabelian theory, however, represents a break with the linear aspects (cohomology groups=vector spaces) studied in the 1960's.

Grothendieck was not deterred from thinking directly in terms of motives in order to motivate and formulate his statements. On September 5, just days after first mentioning the idea, he calmly writes "Let M be a motive, identified if you like with the ℓ-adic cohomology space of a smooth projective scheme over the base field K." The expression 'if you like' shows how in his own mind he is moving away from considering the actual ℓ-adic cohomology, towards considering only some of its fundamental independent-from-ℓ properties—while respecting the fact that Serre may not want to follow him into so vague a terrain.

The letters of early September constitute the first technical discussion as to whether something is or is not a motive, here taken in the simple sense to mean that a family of ℓ-adic objects forms (or comes from) a motive if each member of the family is obtained by tensoring a fixed object defined over $\mathbb{Q}$ with $\mathbb{Q}_\ell$. The family they consider is the family of ℓ-adic Lie proalgebras associated to a field K. It is in these letters—precisely, in the one from September 5, 1964, that Grothendieck first introduces the motivic Galois group, which is constructed from the absolute Galois group G_K of K as follows: G_K acts on the ℓ-adic cohomology space, which can be viewed as a vector space over $\mathbb{Q}_\ell$, therefore there is a homomorphism $G_K \to \mathrm{GL}_{\mathbb{Q}_\ell}$; let G_0 denote the image, and let H denote the algebraic envelope of G_0. This H is the motivic Galois group associated to K, and Grothendieck's assertion is that H is described by 'motivic' equations, i.e. equations with coefficients in algebraic cycles which are independent of ℓ. From this, he notes that one should be able to recover the semi-simple part of G_0, which is thus motivic, but recovering the center of G_0 seems more difficult; still, Grothendieck seems optimistic that G_0 itself might be motivic. Serre, who prefers to think in terms of the Lie algebra $\mathfrak{g}_\ell$ associated to $G_0 \subset \mathrm{GL}_{\mathbb{Q}_\ell}$, produces a counterexample in his answer three days later, considering an elliptic curve E over a finite field k such that the ring of endomorphisms of E is an imaginary quadratic field K. Then he identifies the Lie algebras $\mathfrak{g}_\ell$ explicitly in terms of the ℓ-adic logarithm of the Frobenius of E, and explains that this family $\mathfrak{g}_\ell$ cannot be motivic. With some surprise, he notes that he is not able to construct a counterexample of this type over a number field.

The situation proves confusing as Grothendieck tries to get a feel, or what he terms a *yoga*, for what is motivic and what isn't: "Since my last letter I have also been giving some thought to the finite ground field case, which does seem to give rise to algebras which do not come from motives. You should look into this thoroughly. If this screws up, I do not see any other plausible yogic reason why the rank of the center of your $\mathfrak{g}_\ell$ should be independent of ℓ...As for your suggestion that maybe over a number field the $\mathfrak{g}_\ell$ do come from a motive, I have no feeling for it (except of course that it would be nice!) I am actually very annoyed not to have managed to produce any kind of yoga for number fields..."

In November, Grothendieck is continuing to develop the notion of motivic Galois group: "For the last three weeks I have been getting very excited about the interpretation of Galois and fundamental groups of all kinds in terms of algebraic groups over number fields, especially over $\mathbb{Q}$, and even in terms of group schemes over the integers. I have convinced myself that these groups, together with the general motivic yoga, are the key to a good understanding of a transcendence conjecture linked to various cohomology lattices (integral cohomology, Hodge-de Rham cohomology), the relations between the latter and the Hodge and Tate conjectures, and a better understanding of 'non-commutative' class field theory, which would realize Kronecker's 'Jugendtraum', and of a functional equation for L-functions over $\mathbb{Q}$. I will probably send you another infinite letter one of these days, but I will need some time to arrive at a coherent set of conjectures."

Several more letters are exchanged, with all of the previous subjects still being touched upon: functional equations, good reduction of abelian varieties, especially elliptic curves. Then there is a silence of several months, interrupted only by a short letter from Serre in May 1965, responding to a phone call and giving an elegant two-page exposition of the theory of the Brauer group and factor systems. Silence again until August 1965, when Grothendieck addressed to Serre one of the key letters in the history of motives: the one containing the *standard conjectures*. This letter—written four months after the birth of Mathieu, his third child from Mireille—exudes an atmosphere of intense creativity in a totally new direction. This is the period in which Mireille described him as working at mathematics all night, by the light of a desk lamp, while she slept on the sofa in order to be near him, and woke occasionally to see him slapping his head with his hand, trying to get the ideas out faster.

In order to understand the origin and motivation of the standard conjectures, it is necessary to understand two basic things: how Grothendieck envisioned his conjectural category of motives, and what Serre had done in his 1960 note *Analogues kählériens des conjectures de Weil.*

Serre noticed that in the case of Kähler varieties, if one considers correspondences which are cohomology classes (rather than cycles) respecting the Hodge structure on both sides, then Weil's proof of the Weil conjecture for curves and abelian varieties, using algebraic correspondences and positivity of a certain form, goes through directly. In 1960, ℓ-adic cohomology was not sufficiently developed to apply this idea to it. But by 1965, Grothendieck was able to see that under certain conditions of algebricity of certain ℓ-adic cohomology classes, the same proof could be transported to any smooth projective variety over any algebraically closed field. This idea fitted into the broader framework of motives.

In 1965, Grothendieck's envisioned category of motives $\mathbf{M}(k)$ concerned smooth projective varieties over a field k. He wanted it to be a semi-simple abelian category, whose objects should correspond to 'motivic cohomology groups' $h(X)$ for the varieties X, equipped with decompositions $h(X) = \oplus_i h^i(X)$, which behave like absolute cohomology groups 'covering' or 'explaining' all specific types of cohomologies of X.

The functor $\mathrm{Var}_k \to \mathbf{M}(k)$ given by $X \mapsto h(X)$ should be contravariant, as for any cohomology theory. The $h^i(X)$ should behave more or less like finite-dimensional vector spaces; for instance, for any two objects H, K of $\mathbf{M}(k)$, $\mathrm{Hom}(H, K)$ should be a finite-dimensional $\mathbb{Q}$-vector space. Most importantly— the real meaning of the idea that objects of $\mathbf{M}(k)$ are 'motivic cohomology groups'—is that there should be natural functors from them to the actual cohomology groups associated to X:

- when char. $k = 0$, a functor T_{dR} such that $T_{dR}(h^i(X)) = H^i_{dR}(X)$;
- when $k \subset \mathbb{C}$, a functor T_{Betti} such that $T_{\mathrm{Betti}}(h^i(X)) = H^i_{\mathrm{Betti}}(X)$;
- for all $\ell \neq$ char. k, a functor T_ℓ such that $T_\ell(h^i(X)) = H^i_{et}(X, \mathbb{Q}_\ell)$.

Grothendieck actually sketched out the way to define the category $\mathbf{M}(k)$ of pure motives (corresponding to smooth projective algebraic varieties). Very (very, very) roughly speaking, he used the varieties X themselves as objects and $\mathbb{Q} \otimes C(X, Y)$ for the $\mathbb{Q}$-vector space of morphisms, where $C(X, Y)$ denotes the group of algebraic cycles of $X \times Y$ up to some equivalence relation. Various other objects need to be added to make sure the category possesses some of the desired properties.

In order to prove that $\mathbf{M}(k)$ is a semi-simple abelian category, in his letter of August 27, 1965, Grothendieck formulated the two famous *standard conjectures* inspired by Serre's *Analogues kählériens.*

"Let X denote a smooth connected projective n-dimensional variety over an algebraically closed field k, and Y a smooth hyperplane section. Let me denote by $C^i(X)$ the group of cycle classes of codimension i modulo algebraic equivalence, tensored with $\mathbb{Q}$, and by $\xi \in C^1(X)$ the class of Y.

Conjecture A: for every integer i such that $2i \leq n$, the product with ξ^{n-2i} induces an isomorphism $C^i(X) \simeq C^{n-i}(X)$."[10]

[10] As pointed out by S. Kleiman, Grothendieck's conjecture is actually false when $C^i(X)$ is defined by classes modulo algebraic equivalence. P. Griffiths was the first to find counterexamples showing that algebraic equivalence does not always coincide with numerical equivalence of cycles: numerical (or homological) equivalence works better for the formulation of Conjecture A (see below).

Then after some discussion of this conjecture, he goes on to state the following conjecture, where ϵ denotes the form given by $\epsilon(xx') = L^{n-2r}(x)\cdot y$, L denoting cup product with ξ.

"**Conjecture B**: Assume that $n = 2m$, and let $P^m(X)$ be the kernel of the homomorphism from $C^m(X)$ to $C^{m+1}(X)$ given by multiplication by ξ. Then the form $(-1)^m\epsilon(xx')$ on $P^m(X)$ is positive definite."

In this letter, Grothendieck explains that conjecture A, which can be reinterpreted as a certain algebricity condition on cohomology classes of ℓ-adic cohomology giving correspondences respecting Hodge structure, makes it possible to apply Weil's original techniques for abelian varieties to prove the Weil conjectures in general (except for the absolute values of the eigenvalues, which are covered by conjecture B), just as Serre had noted in the Kähler case in 1960. He calls conjecture A "the 'minimum minimorum' to be able to give a usable rigorous definition of the concept of motive over a field." As it happens, he was not quite correct. In 1991, U. Jannsen showed that in order for $\mathbf{M}(k)$ to be abelian semi-simple, it is necessary and sufficient to use the equivalence relation of numerical equivalence. Homological equivalence would also work if, as is conjectured, it is actually the same as numerical equivalence. But algebraic equivalence does not work, and using it in the definition would yield a category which is not abelian semi-simple. Jannsen was able to show these results without proving either of the standard conjectures. The generalization of the whole theory to arbitrary varieties, which would yield the yet undefined category of mixed motives, is much more difficult; even the standard conjectures would not suffice to prove that it has the desired properties.

In his letter, Grothendieck makes some initial attempts at sketching out proofs or directions of proofs of the conjectures, which however resisted his attempts and all other attempts to prove them. The letter ends with what constitutes the major obstacle: "For the moment, what is needed is to invent a process for deforming a cycle whose dimension is not too large, in order to push it to infinity. Perhaps you would like to think about this yourself? I have only just started on it today, and am writing to you because I have no ideas."

Although not the last letter, this letter represents the end of the Grothendieck-Serre correspondence in a rather significant way, expressing as it does the mathematical obstacle which prevented Grothendieck from developing the theory of motives further; the standard conjectures are still open today. Of course he remained incredibly intense and hardworking for several more years, continuing the SGA seminar until 1969, the writing of the EGAs and ever more research. Yet this letter has a final feel to it. Only two more letters date from before the great rupture of 1970: a short one from Serre from December 1966, answering a rather precise question about representations of GL_n, probably in response to a phone conversation, then a letter from Grothendieck to Serre from January 1969, also referring to a previous conversation, about Steinberg's theorem. Then fourteen years of silence.

1984–1987: The last chapter

The six letters from these years included in the Correspondence— a selection from a much larger collection of existing letters—are intriguing and revealing, yet at the same time somewhat misleading. From the tone of some of Grothendieck's comments ("As you probably know, I no longer leave my home for any mathematical meeting, whatever it may be," or "I realize from your letter that beautiful work is being done in math, but also and especially that such letters and the work they discuss deserve readers and commentators who are more available than I am,") it may seem as though by the 1980's, he had completely abandoned mathematics. But this was in fact far from the case. Quite the contrary, although he did stop working in mathematics for months at a time, there were other months during which he succumbed to a mathematical fever in the course of which he filled thousands of manuscript pages with "grand sketches" for future directions, finally letting his imagination roam, no longer reining himself in with the necessity of advancing slowly and steadily, proving and writing up every detail. A famous text ("Sketch of a Program"), three enormous informally written but more or less complete manuscripts and thousands of unread handwritten pages from his hand date from the 80's and 90's, describing more or less visionary ways of renewing various subjects as concrete as the study of the absolute Galois group over the rationals, or as abstract as the theory of n-categories. And this does not count the many thousands of non-mathematical pages he wrote and still writes.

At the time of the exchange of letters included in the published Correspondence, Grothendieck had just completed his monumental mathematico-autobiographical work *Récoltes et Semailles* (Reaping and Sowing), retracing his life and his work as a mathematician and, over many hundreds of pages, his feelings about the destiny of the mathematical ideas that he had created and then left to others for completion. He sent the successive volumes of this work to his former colleagues and students. The exchanges between Serre and Grothendieck on the topic of this text underscore all of the differences in their personalities already so clearly visible in their different approaches to mathematics.

Serre, a lover as always of all that is pretty ("les jolies choses" is one of his favorite expressions), clean-cut, attractive and economical, viscerally repelled by the darker, messier underside of things, reacts negatively to the negative ("I am sad that you should be so bitter about Deligne..."), positively to the positive ("My way of thinking is...quite distant from yours, which explains why we complemented each other so well for ten or fifteen years, as you say very nicely in your first chapter...On the topic of nice things,

I very much liked what you say about the Bourbaki of your beginnings, about Cartan, Weil and myself, and particularly about Dieudonné...") and uncomprehendingly to the ironic ("There must be about a hundred pages on this subject, containing the curious expression 'the Good Lord's theorem' which I had great difficulty understanding; I finally realized that 'Good Lord's' meant it was a beautiful theorem."[11])

Grothendieck picks up on this at once, and having known Serre for twenty years, is not in the least surprised: "As I might have expected, you rejected everything in the testimony which could be unpleasant for you, but that did not prevent you from reading it (partially, at least) or from 'taking' the parts you find pleasant (those that are 'nice', as you write!)" After all, "One thing that had already struck me about you in the sixties was that the very idea of examining oneself gave you the creeps."

It is true enough that self-analysis in any form strikes Serre as a pursuit fraught with the danger of involuntarily expressing a self-love which to him appears in the poorest of taste. Grothendieck, trying in all honesty to take a closer look at his acts and feelings during the time of his most intense mathematical involvement, speaks of his "absence of complacency with respect to myself." Serre, disbelieving in the very possibility of self-analysis without complacency, and already struggling with the embarrassment of hundreds and hundreds of pages of self-observation, writhes at this phrase which—worse than ever—analyses the analysis, and wonders how Grothendieck could have typed it at all without laughing: "How can you?"

But where, exactly, does he perceive complacency, Grothendieck asks in some surprise. There is no need for Serre to answer. It is obvious that for him, the act of looking at oneself implies self-absorption, which as a corollary necessarily implies a secret self-satisfaction, something which perhaps exists in everyone, but should remain hidden at all costs.

And then, if one is going to do the thing at all, should one not do it completely? Hundreds of pages of self-examination, hundreds of pages of railing because the beautiful mathematical work accomplished in the fifties and sixties met a fate of neglect after the departure of its creator—mainly because basically no one, apart from perhaps Deligne, was able to grasp Grothendieck's vision in its entirety, and therefore perceive how to advance it in the direction it was meant to go—but Serre reproaches him the fact that the major question, "the one every reader expects you to answer," is neither posed nor answered: *"Why did you yourself abandon the work in question?"* Clearly annoyed by this, he goes on to formulate his own guesses as to the answer: "despite your well-known energy, you were quite simply tired of the enormous job you had taken on..." or "one might ask oneself if there is not a deeper explanation than simply being tired of having to bear the burden of so many thousands of pages. Somewhere, you describe your

[11] A misunderstanding! Grothendieck's sarcastic references to the 'Good Lord's theorem' meant that this theorem was not attributed by name to its author, whom he felt to have been neglected and mistreated by the mathematical establishment.

approach to mathematics, in which one does not attack a problem head-on, but one envelopes and dissolves it in a rising tide of general theories. Very good: this is your way of working, and what you have done proves that it does indeed work, for topological vector spaces or algebraic geometry, at least...It is not so clear for number theory...whence this question: did you not come, in fact, around 1968-1970, to realize that the 'rising tide' method was powerless against this type of question, and that a different style would be necessary—which you did not like?"

Grothendieck's answer to this letter and the subsequent exchanges are not included in the present publication, but he did answer in fact, referring to a passage in *Récoltes et Semailles* in which he powerfully expresses the feeling of spiritual stagnation he underwent while devoting twenty years of his life exclusively to mathematics, the growing feeling of suffocation, and the desperate need for complete renewal which drove him to leave everything and strike out in new directions. Reading *Récoltes et Semailles*, it is impossible to believe that Grothendieck felt that his mathematical methods were running into a dead end, whatever their efficacity on certain types of number theoretical problems might or might not have been. His visions both for the continuation of his former program and for new and vast programs are as exuberant as ever; what changed was his desire to devote himself to them entirely. *Récoltes et Semailles* explains much more clearly than his letters how he came to feel that doing mathematics, while in itself a pursuit of extraordinary richness and creativity, was less important than turning towards aspects of the world which he had neglected all his life: the outer world, with all of what he perceived as the dangers of modern life, subject as it is to society's exploitation and violence, and the inner world, with all its layers of infinite complexity to be explored and discovered. And, apart from the sporadic bursts of mathematics of the 1980's and early 90's, he chose to devote the rest of his life to these matters, while Serre continued to work on mathematics, always sensitive to the excitement of new ideas, new areas, and new results. In some sense, the difference between them might be expressed by saying that Serre devoted his life to the pursuit of beauty, Grothendieck to the pursuit of the absolute.

Did earlier thoughts inspire Grothendieck?

Frans Oort

"... mon attention systématiquement était
... dirigée vers les objets de généralité maximale ..."
Grothendieck on page 3 of [**11**]; see [**68**], page 8

Introduction

When I first met Alexandre Grothendieck more than fifty years ago I was not only deeply impressed by his creativity, his knowledge and many other aspects of his mathematics, but I also wondered where all his amazing ideas and structures originated from. It seemed to me then as if new abstract theories just emerged in his mind, and then he started to ponder them and simply build them up in their most pure and general form without any recourse to examples or earlier ideas in that particular field. Upon reading his work, I saw my impression confirmed by the direct and awe-inspiring precision in which his revolutionary structures evolved.

Where does inspiration come from ? We can ask this question in general. The question has fascinated me for many years, and it is particularly intriguing in connection with the mathematics of Alexandre Grothendieck.

Forty years ago the picture was even more puzzling for me. At that time, we had been confronted with thousands of pages of abstract mathematics from his hand. It was not easy at all to understand this vast amount of material. Hence it was a relief for me to read, much later, what Mumford wrote to Grothendieck about this: " ... I should say that I find the style of the finished works, esp. EGA, to be difficult and sometimes unreadable, because of its attempt to reach a superhuman level of completeness." See: Letter Mumford to Grothendieck, 26 December 1985, [**44**], page 750.

Those who had the privilege to follow closely these developments could see the grand new views. Here is what Mumford wrote about Grothendieck 's visit to Harvard about fifty years ago in connection with a new proof of Zariski's "Main Theorem": Then Grothendieck came along and he reproved

Mathematisch Instituut, Pincetonplein 5, 3584 CC Utrecht, The Netherlands. f.oort@uu.nl.

this result now by a *descending* induction on an assertion on the higher cohomology groups with Zariski's theorem resulting from the H^0 case: this seemed like black magic." See the paper [**45**] by Mumford, this volume.

The fact that there should exist a cohomological proof of this theorem by Zariski was conjectured by Serre; see [**1**], page 112 (here we see already where the inspiration came from). See [**73**], bottom of page 21.

The magic described by Mumford can also be found in a description by Deligne. "Je me rappelle mon effarement, en 1965-66 après l'exposé de Grothendieck [SGA5] prouvant le théorème de changement de base pour $Rf_!$: dévissages, dévissages, rien ne semble se passer et pourtant à la fin de l'exposé un théorème clairement non trivial est là." See [**23**], page 12.

About this passage Luc Illusie communicated to me: " .... base change for $Rf_!$ is a trivial consequence of proper base change, and proper base change was proved by Artin in his exposés in SGA 4, not SGA 5. ... January 2005, was the beginning of the first part of SGA 5, and as far as I remember (I wrote preliminary notes for them) Grothendieck recalled the global duality formalism, and then embarked in the local duality formalism (construction of dualizing complexes). Also, the proof of the proper base change theorem is not just a long sequence of trivial 'dévissages' leading to a trivial statement: the dévissages are not trivial, and proper base change for H^1 is a deep ingredient."

It was clear to many of us that the tools which Grothendieck developed in this branch of mathematics revolutionized algebraic geometry and a part of number theory and offered us a clear and direct approach to many questions which were unclear to us before.

But it was also frustrating for us that the maestro himself left the scene too early, with EGA unfinished and many developments that he had initiated left hanging in the air, leaving us with the feeling that now we had to find our own way.

The question of whether Grothendieck 's brilliant ideas had simply occurred to him out of the blue or whether they had some connection to earlier thought continued to puzzle me, and over the years I started to approach each of his theories or results with this particular question in mind. The results were illuminating. Every time I started out expecting to find that a certain method was originally Grothendieck's idea in full, but then, on closer examination, I discovered each time that there could be found in earlier mathematics some preliminary example, specific detail, part of a proof, or anything of that kind that preceded a general theory developed by Grothendieck. However, seeing an inspiration, a starting point, it also showed what sort of amazing quantum leap Grothendieck did take in order to describe his more general results or structures he found.

In this short note I will discuss, describe and propose the following.

§ 1. **Some questions Grothendieck asked**
In a very characteristic way Grothendieck asked many questions. Some of these are deep and difficult. Some other questions could be answered easily, in many cases with a simple example. We describe some of these questions.

§ 2. **How to crack a nut?**
Are we theory-builders or problem-solvers? We discuss Grothendieck's very characteristic way of doing mathematics in this respect.

§ 3 **Some details of the influence of Grothendieck on mathematics.**
We make some remarks on the style of Grothendieck in approaching mathematics. His approach had a great influence especially in the way of doing algebraic geometry and number theory.

§ 4. **We should write a scientific biography.**
Here we come to the question asked in the title of this paper. We propose that a scientific biography should be written about the work of Grothendieck, in which we indicate the "flow" of mathematics, and the way results by Grothendieck are embedded in this on the one hand and the way Grothendieck created new directions and approaches on the other hand. Another terminology could be: we should give a genetic approach to his work.

This would imply each time discussing a certain aspect of Grothendieck's work, indicating possible roots, then describing the leap Grothendieck made from those roots to general ideas, and finally setting forth the impact of those ideas. This might present future generations a welcome description of topics in 20th century mathematics. It would show the flow of ideas, and it could offer a description of ideas and theories currently well-known to specialists in these fields now; that knowledge and insight should not get lost. Many ideas by Grothendieck have already been described in a more pedestrian way. But the job is not yet finished. In order to make a start, I intend to give some examples in this short note which indicate possible earlier roots of theories developed by Grothendieck. We give some examples supporting our (preliminary) Conclusion (4.1), that all theory developed by Grothendieck in the following areas has earlier roots:

§ 5. **The fundamental group.**
§ 6. **Grothendieck topologies.**
§ 7. **Anabelian geometry.**
§ 8. **If the general approach does not work.**
It may happen that a general approach to a given problem fails. What was the reaction of Grothendieck, and how did other mathematicians carry on?

In this note we have not documented extensively publications of Grothendieck, because in this volume and in other papers a careful and precise list of publications is to be found. For more details see e.g. [**6**], [**31**].

In this note we only discuss research by Grothendieck in the field of algebraic geometry.

An earlier draft of this note was read by L. Illusie, L. Schneps and J-P. Serre. They communicated to me valuable corrections and suggestions. I thank them heartily for their contributions.

1. Some questions Grothendieck asked

During his active mathematical life, Grothendieck asked many questions. Every time, it was clear that he had a general picture in mind, and he tried to see whether his initial idea would hold against the intuition of colleagues, would be supported or be erased by examples. Many times we see a remarkable insight, a deep view on general structures, and sometimes a lack of producing easy examples, not doing simple computations himself.

We may ask ourselves how it was possible that Grothendieck could possibly work without examples. As to this question: now that we have the wonderful [**10**] and letters contained in [**44**] it is possible to see that there is more to the creative process of Grothendieck than I originally knew.

Also in this line of thought we should discuss what happened in case Grothendieck constructed a general machinery, which for certain applications however did not give an answer to questions one would like to see answered. Some examples will be given in Section 8.

(1.1). Local and global topological groups. In [**32**], on page 1039 of the first part we find the story of how Grothendieck in 1949, then 21 years old, came to C. Ehresmann and A. Borel during a break between lectures in the Bourbaki seminar asking: "Is every local topological group the germ of a global topological group ?" I find this typical of his approach to mathematics. Seeing mathematical structures, Grothendieck was interested in knowing their interrelations. And one of the best ways of finding out is going to the true expert, asking a question and obtaining an answer which would show him the way to proceed. See the beautiful paper of Jackson describing this episode, also characterizing Grothendieck's "social niceties" and much more. The question which was asked has a counterexample, as Borel knew. Many times we see this pattern: Grothendieck would test the beauty and coherence of mathematics by asking a question to a "real expert" and obtain an answer which either would show him the way to proceed, or save him from going on in a wrong direction.

(1.2). Correspondence with Serre. The volume [**10**] is a rich source of information. We obtain a glimpse of the exchange of ideas between these mathematicians. It is fascinating reading, it gives insight into the way they feel about mathematics, and it gives food for further thought. We highlight just a few of the many questions Grothendieck asked in these letters. Also see (3.9).

(1.2).1. See [**10**], p. 7. Grothendieck wrote on 18.2.1955: "...Sait-on si le quotient d'une variété de Stein par un groupe discret 'sans point fixe' est de Stein?"

To which Serre responds on 26.02.1955: " ...ça peut même être une variété compacte! Cf. courbes elliptiques, et autres,..."

(1.2).2. See [**10**], p. 42. Grothendieck wrote on 23.7.1956: "Quant à plonger une variété algébrique complète dans un espace projectif, j'avoue que je ne vois pas de méthode encore."

Did Grothendieck expect this to be true? In 1957 Nagata constructed an example of a *complete normal surface* which cannot be embedded into a projective space, and in his Harvard PhD-thesis in 1960 Hironaka constructed complete, *non-singular threefolds* which cannot be embedded into any projective space. See [**29**], 3.4.1.

(1.2).3. See [**10**], p. 67. Grothendieck wrote on 5.11.1958: "...me font penser qu'il est possible de remonter canoniquement toute variété X_0 définie sur un corps parfait de caractéristique $p \neq 0$..." For a further discussion see (8.3).

It is not clear what Grothendieck had in mind here. We know he was much too optimistic, see [**75**]. But we see his theory of formal liftings (not canonical, sometimes obstructed) and his "existence theorem in formal geometry" foreshadowed here.

(1.2).4. See [**10**], p. 145. Grothendieck had the hope (in 1964, or earlier) of proving the Weil conjectures by first showing that any variety could be dominated by a product of curves, see [**10**], p. 271. We can understand his insight that indeed that would solve problems. But Serre gave an example of an algebraic surface which does not satisfy this condition, see [**10**], page 145. We see the mechanism of Grothendieck asking a question before embarking on this general idea, and Serre finishing off the attempt by an example. As far as I know this example was never published. And it seems it was not known to C. Schoen in 1995, see [**70**]. It would be nice to understand Serre's example in the light of this new approach by Schoen.

(1.2).5. See [**10**], p. 169. Grothendieck wrote on 13.08.1964: "...si V est un schéma algébrique projectif et lisse sur le corps local K, et si $G(\overline{K}, K)$ opère de façon non ramifiée sur tous les $\mathrm{H}^i_\ell(\overline{V})$, on peut se demander si V n'a pas forcément une *bonne réduction.* C'est probablement un peu trop optimiste, mais tout de même, je ne vois pas de contre-example immédiat."

For every curve of genus at least two degenerating into a tree of regular curves of lower genus, its Jacobian has good reduction; hence the condition of trivial monodromy is satisfied (the local Galois group operates in a non-ramified way). However the curve does not have good reduction.

(1.2).6. See [**10**], p. 203. Grothendieck wrote on 3-5.10.1964: "...est-il connu si la fonction ζ de Riemann a une infinité de zéros?"

On which Serre later made the comment: "... Grothendieck ne s'est jamais intéressé à la théories analytiques des nombres." See [**10**], p. 277.

Already this small selection shows that some questions asked by Grothendieck have an easy answer that can be provided by anyone knowing simple examples on the one hand, and deep thoughts and attempts on the other hand.

(1.3). Correspondence between Grothendieck and Mumford. We will discuss in (8.6) a question Grothendieck asked in 1970 to Mumford. See [**44**], page 745. Mumford gave an easy example which showed that this idea by Grothendieck did not match mathematical reality. This exchange shows that Grothendieck's thoughts, without simple computations or examples for support, were geared towards new insight in the objects he was studying at that time.

Perhaps these two sentences from their correspondence characterize their interaction particularly well.

Grothendieck to Mumford 25.04.1961: "It seems to me that, because of your lack of some technical background on schemata, some proofs are rather awkward and unnatural, and the statements you give not as simple and strong as they should be." See [**44**], page 636/637.

Mumford to Grothendieck on 11.02.1986: "I hope you know how vivid and influential a figure you were in my life and my development at one time." See [**44**], page 758.

(1.4). We may ask ourselves how it was possible that Grothendieck could possibly work without examples. As to this question: now that we have the wonderful [**10**] and letters contained in [**44**] it is possible to see that there is more to the creative process of Grothendieck than I originally knew. His contacts with colleagues, such as Serre and Mumford, and the information he obtained saved him from spending time on trying to develop structures which do not exist (as follows by counterexamples). We can admire Grothendieck for asking the right questions to the right colleagues.

Here is another explanation. Serre remarked to me (private correspondence): "Grothendieck could prove such nice theorems ... the strong consistency of mathematics".

And perhaps Grothendieck knew examples better than can be concluded from his correspondence and from his style of writing. L. Illusie communicated to me: "In his filing cabinets, located behind his desk, Grothendieck kept many handwritten notes, where he had studied specific examples: he sometimes told me that he was weak on surfaces, but as everybody knows, he was not so weak in local algebra, and he knew enough of curves, abelian varieties and algebraic groups to be able to test his ideas. Also, his familiarity (and constant interest) in analysis and topology was a strong asset. All these examples appeared when you discussed with him."

But perhaps we had best cite Grothendieckhimself, where "harmony" could be the inspiring source:
"Et toute science, quand nous l'entendons non comme un instrument de pouvoir et de domination, mais comme aventure de connaissance de notre espèce à travers les âges, n'est autre chose que cette harmonie, plus ou moins vaste et plus ou moins riche d'une époque à l'autre, qui se déploie au cours des générations et des siècles, par le délicat contrepoint de tous les thèmes apparus tour à tour, comme appelés du néant.
(ReS; see [**32**], Part 1, page 1038, also for a translation).

The construction of very general ideas was a strong point of the mathematics of Grothendieck. In this line of thought we discuss what happened in case Grothendieck constructed a general machinery, which for certain applications however did not give an answer to questions one would have liked to see answered. If a counterexample showed that a general approach could not work, or that a general idea did not describe the true structure, if mathematics was not as simple and beautiful as Grothendieck would have liked to see, then what was Grothendieck's reaction? We will see some examples of this in Section 8, and describe how progress could still be made by others.

2. How to crack a nut?

(2.1). Here we study the way mathematicians try to solve a problem, or develop further mathematical insight.

In ReS, see [**13**], Grothendieck described two (extreme) ways of cracking a big nut ("...une grosse noix..."). The first way he described is basically by brute force. The second way is to immerse the nut in a softening fluid: "on plonge la noix dans un liquide émollient", until the nut opens just by itself. And Grothendieck leaves the reader to guess which is his method. See ReS, and see [**23**], pp. 11/12.

However, I think, mathematical reality is not as simple as described in this metaphor. FLT, Fermat's Last Theorem, or the Weil conjectures were not solved in not just one of these two ways.

I would like to give a description of the creative aspect of mathematical activity which has been on my mind for the last 50 years; a concept slightly different from the nut-story. To put it in an extreme form:

Method (1) One method is to construct a "machine", a general concept, find a universal truth. Then "simply" feed the problem studied into it, and wait, see what happens.

Method (2) Or, one can study special cases, make an inventory of known examples, and try to connect the problem to a general principle. Or one can at first try to find a proof, see where it gets stuck, then use the obstructions in an attempt to construct a counterexample, and by this zig-zag method discover more about the structure of the objects studied, and hope that these attempts eventually converge to a conclusion.

Does a mathematician discover or create a result? This is an interesting question on which many ideas already exist. However, this question and related lines of idea will not be further discussed in this note. The first method is very appealing. It is the one we should start with: "finding a preexisting pattern".

Yuri Manin wrote: "I see the process of mathematical creation as a kind of recognizing a preexisting pattern"; see [**38**]. In my opinion Grothendieck followed this line of research consistently. He discovered many mathematical structures, and he created important tools for us to proceed in our search for mathematical truth.

In a sense this is very reassuring: if Grothendieck studied a certain question or structure, and there is the possibility of a smooth, direct, general solution, he will have found it.

Grothendieck taught us how successful mathematical research along the lines of Method (1) can be. Also, this seems to be the heart of our profession: creating the evolution of our understanding of mathematical structures. – However, clinging only to this method has its drawbacks. If you are not successful, what can you do? – You can try to generalize the problem, and find a structure which solves the more general question. But we have learned that mathematical reality sometimes (or often? according to your taste and experience) does not fit into the approach (1). I have the impression that in many cases when this first method did not work out well, Grothendieck would let the problem rest, waiting "until the nut opens just by itself"; and he sometimes left the question completely untouched afterwards.

The second method has been applied quite often. Many results have been achieved this way.

Here is another description of this activity of mathematicians, given by Andrew Wiles. "Perhaps I could best describe my experience of doing mathematics in terms of entering a dark mansion. One goes into the first room, and it's dark, completely dark. One stumbles around bumping into the furniture, and gradually, you learn where each piece of furniture is, and finally, after six months or so, you find the light switch. You turn it on, and suddenly, it's all illuminated. You can see exactly where you were. At the beginning of September, I was sitting here at this desk, when suddenly, totally unexpectedly, I had this incredible revelation. It was the most important moment of my working life..." (BBC-documentary by S. Singh and John Lynch: Fermat's Last Theorem. Horizon, BBC 1996.)

We have seen that FLT was not proved, and as far as we know, cannot be proved by just constructing a general theory and "feeding the problem into the machine". Not only did Andrew Wiles try to "learn where each piece of furniture is", but all those attempts during more than three centuries before can be seen as "stumbling around bumping into the furniture". This *evolutionary process* is fascinating to watch and to describe. We can mention Fermat, Euler, Legendre, Dirichlet, Sophie Germain, Kummer, Serre, Shimura-Taniyama-Weil, Frey, Ribet and many others (and they paved the road for Wiles). The final achievement is a combination of growing insight, knowing which roads should not be taken, and then coming up with a combination of general concepts and deep insight on the one hand, and "tricks" and precise knowledge of all the pieces of the "furniture" on the other hand. How different from either brute force or expecting that the nut will open just by itself.

(2.2). Conclusion. Grothendieck created new tools and gave us deep insight, and we can be grateful for that. However, reality in mathematical research shows that there are problems which need more than only general insight. If Method (1) fails, it seems wise to apply Method (2) (and many mathematicians, tenaciously, have done so); we describe some examples of this in Section 8.

3. Some details of the influence of Grothendieck on mathematics

"...le jour où une démonstration nous apprend au-delà de tout doute que telle chose que nous imaginions était bel et bien l'expression fidèle et véritable de la réalité elle-même..." ReS, page 211.

In this section we describe some characteristics of the way Grothendieck was working and thinking while doing algebraic geometry in his fruitful years, and we speculate about the ways in which this formed and changed our views on these topics.

(3.1). Representable functors. We describe a general approach known in algebraic topology, algebra, and many other fields, that started already more than 70 years ago, but was adapted and used to its full consequences by Grothendieck.

There were many occasions in mathematics where a "solution satisfying a universal property" was constructed. Topologists knew that vector bundles come from a general one. Bourbaki made use of the solution of a "universal problem" (such as a tensor product).

Samuel wrote in 1948: "It has been observed" (with a footnote to unpublished work by Bourbaki) "that constructions so apparently different enter in the same frame"; see the first lines of [**64**].

To French mathematicians in the 1960s, but especially to Grothendieck , we owe the mantra that defining a functor and proving it is representable should be the heart, or at least the beginning of any construction. In algebraic geometry before Grothendieck , there were many constructions where no a priori "universal property" was formulated, or where defining conditions and corollaries of such properties appear in the same lines. For us, nowadays, it is hard to assess the influence of even this "small" aspect of the French lucidity of view, and the systematic use of it made by Grothendieck.

We taste this atmosphere in the description by Samuel of Igusa's construction of what we now would call the coarse moduli scheme $\mathcal{M}_2 \to \mathrm{Spec}(\mathbb{Z})$:

"Signalons aussitôt que le travail d'IGUSA ne résoud pas, pour les courbes de genre 2, le 'problème des modules' tel qu'il a été posé par GROTHENDIECK à diverses reprises dans ce Séminaire." (See the first lines of [**65**].) This aspect of abstract methods was to have a direct influence on our profession for years to come.

Illusie wrote to me: "...there are two aspects in the technique of representable functors:
(1) of course, defining the functor makes clear the object we are searching for,
(2) but independently of whether that functor is representable or not, what Grothendieck taught us is that we can do geometry on the functor itself: e.g. (formal) smoothness, étaleness, etc. This was the 'quantum leap' as you said before."

(3.2). Non-representable moduli functors. Grothendieck's views helped us to understand essential features much better then we knew them before. This portrays a phenomenon that we will encounter many times when observing how abstract methods of Grothendieck 's were digested, adapted and used. But several times we also see that the abstract and clean approach does not completely cover mathematical reality. E.g., sometimes we want to construct an object which does not represent a functor that is easily defined beforehand.

I remember once I met Grothendieck in a Paris street; both of us were going to the same lecture, and he was very excited by a construction made by a young American mathematician. It was Mumford, who wrote in 1961 to Grothendieck about his proof of "the key theorem in a construction of the arithmetic scheme of moduli M of curves of any genus." Grothendieck was excited about this idea, apparently completely new to him. Later Mumford pinned down the notion of a "coarse moduli scheme", necessary in case the obvious moduli functor is not representable by a variety (or by a scheme). See [**44**], pp. 635-638 where we see this excitement of Grothendieck reflected in several letters to Mumford. Grothendieck explained that for "higher levels" he could represent moduli functors, but for all levels he could not preform the necessary quotient construction, see [**44**], pp. 635/636.

Later in this note we will see instances where Grothendieck's abstract theory clarifies a lot, but sometimes "non-canonical steps" are necessary to give full access to mathematical reality; see Section 8.

(3.3). Morphisms instead of objects. "...comme Grothendieck nous l'a appris, les objets d'une catégorie ne jouent pas un grand rôle, ce sont les morphismes qui sont essentiels." See page 335 of [**76**].

One of the first theorems that Grothendieck proved in algebraic geometry, and which gave him a lot of prestige, was the Grothendieck-Hirzebruch-Riemann-Roch theorem. One of the first essential ideas is that such a theorem should not be about a variety (as all the "old" results were), but that it should describe properties of a morphism; see [**7**]; see (3.8). The idea of considering morphisms rather than objects dominated many considerations by Grothendieck in algebraic geometry, and we have seen so many results coming out of this point of view.

In many cases, it is hard now to realize how mathematicians were thinking and working some time ago, let alone long ago. For a long period of time algebraic geometry was the study of varieties. However Grothendieck has taught us to think "functorially". The way Grothendieck would start a seminar talk is well-known: "Let X vertical arrow S be a scheme over S". And since then, some of us (most of us) see the importance of this way of looking at things, although we still use the term "variety".

Illusie writes: "*Grothendieck pensait toujours en termes relatifs: un espace au-dessus d'un autre*"; see [**31**], second page. Where algebraic geometers, and certainly mathematicians working in number theory, were interested in properties of one variety, or one equation, or at best a class of varieties or equations, Grothendieck showed us the essence of changing our point of view. Certainly here we can indicate earlier roots. Just one example: a "complete variety" was defined by Chevalley, see Chap. IV of [**22**]. It opened the possibility of studying varieties which appeared naturally in constructions, which were not necessarily projective but still had the property that "no points are missing". In the hands of Grothendieck , is was no longer a variety that matters, but a morphism, and Chevalley's definition was generalized to the notion of a "proper morphism". Indeed this is a generalization: an algebraic variety V defined over a field K is complete if and only if the morphism $V \to \mathrm{Spec}(K)$ is proper.

In 1970 we had a Summer School on Algebraic Geometry. I remember Swinnerton -Dyer starting a talk by writing, in a very Grothendieckian way: X vertical arrow S, and continuing for just one minute saying very complicated things about schemes over schemes. We were amazed: even this famous number theorist had converted to the new faith? Then Swinnerton-Dyer continued his talk on "Rational points on Del Pezzo surfaces of degree 5" by saying that he wanted to compute something, that schemes for him were not very helpful, and soon equations were solved, determinants computed, and the result followed.

(3.4). The most general situation. "Alors que dans mes recherches d'avant 1970, mon attention systématiquement était dirigée vers les objets de généralité maximale, afin de dégager un language d'ensemble adéquat pour le monde de la géométrie algb́rique...pour développer des techniques et énoncés 'passe-partout' valables en toutes dimensions et en tous lieux..."; see [**11**], pp. 2/3. In many cases this has enriched our point of view. However, sometimes we feel that working on a specific problem in "maximal generality" is not always helpful.

(3.5). Commuting diagrams. Grothendieck gives us the feeling that mathematics satisfies all possible rules of simplicity and elegance. And certainly we have learned a lot from him by looking at our profession this way. However, Serre writes on 23.7.1985:

"On ne peut pas se borner à dire que les diagrammes qu'on écrit 'doivent' commuter..."; see [**10**], page 244.

Let me add to this a description of a personal episode from the time, in 1960/61, when I was a student in Paris. The goal of my research was modest: constructing the Picard scheme of X in case the Picard scheme of X_{red} is known to exist (first for curves, later for arbitrary algebraic schemes). Grothendieck had claimed in September 1960 to me that he had already proved everything I was after, which however turned out later not to be the case. After I finished my proof, Serre insisted to Grothendieck that I should give a talk on my first (small) result in the Grothendieck seminar. In my talk, I explained that in a large diagram with two quite different cohomology sequences with down arrows connecting them, the crucial square was not commutative in general, as I had checked in several examples. However, I proved that in the relevant square the two images were the same, and that was all that I needed in that situation.

The week after my performance in his seminar Grothendieck gave a talk in Cartan's seminar; there he needed my result. In [**12**], Th. 3.1 on page 16-13 we see an extra condition (not "de généralité maximale") which helps to avoid this non-commutativity. After my result appeared in print, Grothendieck used it in 1962 to prove the theorem without this extra condition, see [**4**], page 232-17.

(3.6). Schemes. Classical algebraic geometry studied varieties over a field. However, in many cases in geometry and number theory, particularly when considering varieties moving in a family, or equations together with their reduction mod p (in Grothendieck's language this amounts to just taking a special fiber of a morphism between schemes), it is necessary to use a more general machinery. Already in [**47**], and in many later publications, we find a attempt to formulate this; it was also studied by E. Kähler. When sheaf theory became available, ringed spaces, substituted for the notion of sets of solutions of polynomial equations, paved the way for a more general concept. According to Pierre Cartier, the word scheme was first used in the 1956 Chevalley Seminar, in which Chevalley was pursuing Zariski's ideas ([**17**]). Serre communicated to me: "I was well aware when I wrote FAC of the notion (but not the word) of Spec and of its use; I had read Krull's Idealtheorie, which is probably the first place where the technique of going from a ring to its local rings was systematically used (and in order to prove non-trivial theorems, such as Krull's theorems on dimension.)" In [**82**] we read on page 43: "Schemes were already in the air, though always with restrictions on the rings involved. In February 1955, Serre mentions that the theory of coherent sheaves works on the spectrum of commutative rings in which every prime ideal is an intersection of maximal ideals."

It was Grothendieck who saw the importance of the more general definition. Still, algebraic geometers in the beginning complained that the notion of a point should be related to a maximal ideal. However Grothendieck (of course) noted that a ring homomorphism $R \to R'$ in general does not give a map between the set of maximal ideals, e.g. as is the case when R is an integral domain, unequal to R', its field of fractions. General principles and thinking of morphisms instead of objects made Grothendieck replace old habits by clean new ideas.

Here we see the "earlier roots" that inspired Grothendieck, and his jump to the general concept we use now. In [**1**] on page 106 Grothendieck describes these ideas originating in work by Nagata-Chevalley-Serre and many others. See Cartier's description of the development of these ideas ([**18**], page 398).

In [**10**], page 26, Grothendieck writes on 16.1.1956 "...le contexte général des spectres d'anneau à la Cartier-Serre." And Serre writes as comment in this edition: " cela s'appellera plus tard des schémas affines." In [**10**], page 53, Grothendieck writes on 22.11.1956 "...Cartier a fait le raccord des schémas avec les variétés..." We see the inspiring atmosphere of the Paris mathematical community at that time for Grothendieck.

Did everyone adopt the theory of schemes? For some algebraic geometers it was hard to adjust to this modern terminology. And there were several reasons for that. Partly because the machinery was too general: in some cases an easy and direct approach would give a better and easier framework for understanding, for describing easy structures, and for writing things down in a plain language. Also, it was not so easy to change from old habits into the new discipline.

In 1960 I made an appointment with Néron, and I asked him to explain to me his theory of "minimal models". I had the feeling it was important, but I must confess that I understood very little of his explanation at that time. Then, reading his [**48**] I could better understand the result, but it was hard to digest the proof. I know that during that time, his colleagues tried to convince him to publish his results in the language of schemes, but in fact we can see that Néron's publication used terminology that closely followed the language of Weil and Shimura. In 1966 M. Artin wrote in his review of this result: " It would be very useful to have a clear exposition of his theory in the language of schemes." It was by reading [**77**] (see p. 494) that I obtained a clearer view of this notion. In SGA 7, Vol. I Exp. IX by Grothendieck (see IX.1.1), and in fact already in [**63**], we can see the formulation of the result in the language of schemes. But it was only in later work by Raynaud, and in 1986 (see [**15**]) that a discussion completely in modern terminology became available.

(3.7). Going on with general theory, leaving applications to others. We know that Grothendieck had a grand plan for completing the foundations of algebraic geometry in EGA; e.g. see [**10**], page 83, where Grothendieck writes in 1959 that he expects to have EGA finished in 3 or at most 4 years. I have the impression that laying these foundations became more important than having this work actually "aboutir à la démonstration des conjectures de Weil" (as in the footnote on page 9 of EGA I). The plan for the 13 chapters of EGA can be found on page 6 of EGA I. We know that he did not finish writing EGA – alas ! – only 4 chapters ever appeared.

In ReS more than once we find a sentence like "Au moment de quitter la scène mathématique en 1970 l'ensemble de mes publications (dont bon nombre en collaboration) sur le thème des schémas devait se monter à quelques deux mille pages" (ReS page 44, footnote 21). However, some of the material which should have appeared in later volumes of EGA, but was in fact never written down in that setting, was luckily already divulged in SGA and in FGA. These are rich sources of information.

(3.8). Certain applications he did not publish himself.

We can mention *the Riemann-Roch theorem*, discussed and published by Borel and Serre ([**7**]). Also see SGA 6, and [**26**], 15.2 and 18.3.

Part of the monodromy theorem: *every eigenvalue of a monodromy matrix is a root of unity*, a wonderful application of the theory of the fundamental group, which intertwines Galois theory and classical monodromy, see the appendix of [**77**]; see SGA 7 I, Exp. I, Section 1; see (5.1) for the fundamental group; see (5.2) for comments on the monodromy theorem.

CM *abelian varieties are, up to isogeny, defined over a finite extension of the prime field*; see [**55**], also published with his permission.

Dieudonné wrote: "Il ne publia pas lui-même sa démonstration..." (of the Riemann-Roch-Grothendieck theorem) "...premier example de ce qui allait devenir chez lui une coutume: poussé par les idées qui se pressaient en foule dans son esprit, il laissait souvent à ses collègues ou élèves le travail de leur mise au point dans tous les détails" (es [**27**], Vol. I, pp. 6).

We see that Grothendieck in those years 1958 - 1970 spent all his energy on the main lines of his plans, and we can be grateful for that. For other things "he was never in a hurry to publish", see [**69**], p. 22.

(3.9). "Toujours lui!" Grothendieck had contact with Serre on many occasions, mainly by phone it seems, but also by correspondence. Serre's insight, his results, and certainly his incredible ability to see through a question or a problem, and come up either with a counterexample or a critical remark, was often crucial for Grothendieck. In [**10**] we see just a small part of this interaction. Here is one of the Serre's results which had a deep influence on the work of Grothendieck(see [**74**]):

"Cétait là une réflexion qui a dû se faire vers le moment de ma réflexion sur une formulation des "conjectures standard", inspirées l'une et l'autre par l'idée de Serre (toujours lui!) d'un analogue 'kählérien' des conjectures de Weil." See ReS, pp 209/210.

(3.10). "On pourra commencer à faire de la géométrie algébrique!" In his letter of 18.8.1959 (see [**10**], page 83), Grothendieck tells Serre his schedule for the next 4 years: in those years he expects to write down the planned volumes of EGA, and also things which were later partly published in [**4**] and in volumes of SGA. And the letter concludes:

"Sans difficultés imprévues ou enlisement, le multiplodoque devrait être fini d'ici 3 ans, ou 4 ans maximum. On pourra commencer à faire de la géométrie algébrique!"

This plan for material to be published in the 12 chapters (many volumes) of EGA appeared in 1960, on page 6 of EGA 1. Now, though, we know that the first four chapters of EGA already took 7 years to be published, and contained more than 1800 pages in 8 volumes. The remaining eight chapters were never written or published.

In January 1984, Grothendieck wrote: "Mais aujourd'hui je ne suis plus, comme naguère, le prisonnier volontaire de tâches interminables, qui si souvent m'avaient interdit de m'élancer dans l'inconnu, mathématique ou non" (see [**11**], page 51).

This shows that Grothendieck did find it a heavy task to lay the foundations of algebraic geometry in his style. Indeed, as Serre writes:

" J'ai l'impression que, malgré ton énergie bien connue, tu étais tout simplement fatigué de l'énorme travail que tu avais entrepris" (see [**10**], 8.2.1986, page 250).

Although the original plan for EGA was far from finished, I think that Grothendieck did hand down enough of his ideas of these foundations to us in a way for which we can use them and proceed. Also we see that basically everything he produced in those twelve fruitful years did belong to "known territory" to him. Did he consider his activity before 1970 as "faire de la géométrie algébrique"?

Cartier remarks that Grothendieck , after leaving the field of "nuclear spaces" and everything connected with that, "in rather characteristic fashion, never paid attention to the descendant of his ideas, and showed nothing but indifference and even hostility towards theoretical physics, a subject guilty of the destruction of Hiroshima!" Was Grothendieck's behavior after 1970 with respect to the "descendance" of his ideas in algebraic geometry very different?

(3.11). Let me mention at least three very different aspects of Grothendieck's work in algebraic geometry.

Foundational work. The way Grothendieck revolutionized this field is amazing. And, how is it possible that someone writes, within say 10 years, thousands of pages of non-trivial mathematics with no flaws, theory just flowing on and on?

Imagination. His published work, say between 1960 and 1970, was based on his deep insight, which enabled Grothendieck to see clearly the structure of this material. But Grothendieck also conveyed his ideas in manuscripts of many pages. We will see how just one idea (the anabelian conjecture) gave rise to a flow of activities and results. So many more deep ideas are still not fully understood. Grothendieck supplied many starting points which will keep us busy for many years; e.g. see § 7. I think that large parts of [**14**] are still not understood.

Questions. Grothendieck was very open in asking questions spurred by his curiosity. And here we see a strange mixture of deep insight (into structures and in theory) on the one hand and some innocent ignorance (in easy examples, in very concrete matters in mathematics) on the other. For me, it has always been a puzzling mystery how someone with such deep insight can proceed in mathematics without basic contact with elementary examples, and how it is possible that someone with such deep insight could miss easy aspects which are obvious to mathematicians who are used to living with examples and finding motivation in simple and easy structures. Putting things together, one can conclude that Grothendieck was not hampered by details which could obstruct his incredible insight in abstract matters. And perhaps we can be grateful that he did not know such easy examples, so that they did not obstruct him when finding his way through the mazes of abstract thoughts. See Section 1.

(3.12). Sometimes too abstract? When examples and direct applications are not there to form an obstruction to developing abstract mathematics, sometimes theory can go too far. For just ordinary people this point comes quite soon; many times I have seen a student doing much better after being asked to produced at least one example of the theory developed. It quite often happens that I ask a former student something, and the answer is just a beautiful, complicated example illustrating what I am asking. I call it "Feynman's method": while following a talk, or reading a paper, you test every statement against a non-trivial example that you know very well.

Many attempts by Grothendieck put the right perspective on the matter at hand. But sometimes I have the feeling that he went too far. Many years ago, I asked Monique Hakim to explain to me what she worked on for her Ph.D. She explained to me some material which much later appeared in her book [**28**]. During that explanation I saw the connection with deformation theory as explained by Kodaira and Spencer, see [**37**]. Before Schlessinger's paper and the Grothendieck-Mumford deformation theory was available, the Kodaira-Spencer paper was a valuable source of information and inspiration. You can see how the authors find the right concept. However, they have to struggle with a mixture of methods: we see families where the base is a differentiable manifold, the fibers are algebraic varieties, and on the total space these structures are mixed in an obvious but not so easy way. This was all at a time when a "scheme over $\mathrm{Spec}(k[\epsilon])$" did not yet exist; David Mumford writes: "But now Grothendieck was saying these first order deformations *were actually families*, families whose parameter space was the embodied tangent vector $\mathrm{Spec}(k[\epsilon]/(\epsilon^2))$ (see [**45**]).

Quite understandably, Grothendieck tried to find a unifying framework in which such families naturally find their place. The idea is to replace every geometric object by the category of, say, coherent sheaves on it. The category of varieties then becomes a category of categories. And we see fundamental problems arise: one doesn't want to talk about "isomorphisms of categories", but rather of equivalences. The idea is nice, but I doubt whether any geometer can truly work, do computations, or consider structures in such an abstract universe. History has shown us that while we have gratefully accepted many structures handed down to us by Grothendieck, common sense and practical necessity sometimes forces us to back up our abstract theory by more concrete methods, examples and computations.

Several of the considerations above can be summarized by the following words of Leila Schneps: "...Grothendieck's style...his view of the most general situations, explaining the many 'special cases' others have worked on, his independence from (and sometimes ignorance of) other people's written work, and above all, his visionary aptitude for rephrasing classical problems on varieties or other objects in terms of morphisms between them, thus obtaining incredible generalizations and simplifications of various theories." (See [**69**], page 5.)

(3.13). Grothendieck inspired many of us. Not only did earlier results form a basis for ideas by Grothendieck, but even more, Grothendieck's new theories gave rise to many new developments. One could draw a diagram of this:

earlier ideas – structures invented/discovered by Grothendieck – later developments.

This gives a clear picture of the flow of mathematical ideas.

An answer to Grothendieck as to whether his "pupils" did continue his work could be that indeed, a lot of us did build upon the work he did, although not precisely in his style; in many cases with a different approach, in some cases with less insight, but certainly with great respect. Also see [**10**], page 244, where Serre writes: "Non continuation de ton œuvre par tes anciens élèves. Tu as raison: ils n'ont pas continué. Cela n'est guère surprenant: c'était toi qui avais une vision d'ensemble du programme, pas eux (sauf Deligne, bien sûr)."

4. We should write a scientific biography

(4.1). *We should start writing a* **scientific biography** *of Grothendieck.* It would be worthwhile to write a mathematical biography of Grothendieck in terms of his scientific ideas. This would imply each time discussing a certain aspect of Grothendieck's work, indicating possible roots, then describing the leap Grothendieck made from those roots to general ideas, and finally setting forth the impact of those ideas. This might present future generations with a welcome description of topics in 20th century mathematics. It would show the flow of ideas, and it could offer a description of ideas and theories currently well-known to specialists in these fields now; that knowledge and insight should not get lost. The present volume already is a first step in this direction.

Many ideas by Grothendieck have already been described in a more pedestrian way. But the job is not yet finished. In order to make a start, I intend to give some (well-known) examples in §§ 5, 6, 7, which indicate possible earlier roots of theories developed by Grothendieck. This is just a small and superficial selection: many more examples should be described and worked out in greater detail.

Or, should we speak of "a genetic approach to algebraic geometry"? In [**83**] we see: "Otto Toeplitz did not teach calculus as a static system of techniques and facts to be memorized. Instead, he drew on his knowledge of the history of mathematics, and presented calculus as an organic evolution of ideas beginning with the discoveries of Greek scholars such as Archimedes, Pythagoras, and Euclid, and developing through the centuries in the work of Kepler, Galileo, Fermat, Newton, and Leibniz. Through this unique approach, Toeplitz summarized and elucidated the major mathematical advances that contributed to modern calculus." I thank Viktor Blåsjö for indicating this reference to me. Instead of what I phrase as "Grothendieck and the flow of mathematics", I could also choose to say "a genetic approach to Grothendieck's results".

5. The fundamental group

" .. une définition algébrique du groupe fondamental...."
Grothendieck 22.11.56, see [**10**], page 55

For a description of this topic, see [**3**], Vol. 1, and see the paper by Murre in this volume [**46**].

We are familiar with classical ideas like Galois theory and the theory of the fundamental group of a pointed topological space. In Grothendieck's theory of the fundamental group, these two theories are combined in one framework. It is due to Grothendieck that we have this beautiful and important tool at our disposal, combining pillars of algebra and topology into a new concept, with many more applications and much more insight than were possible before.

(5.1). The arithmetic and the geometric part. In the unified fundamental group defined by Grothendieck for (say) a variety X over a ground field K, the Galois group of that field appears as a quotient:

$$1 \to \pi_1(\overline{X}, a) \longrightarrow \pi_1(X, a) \xrightarrow{p_X} \mathrm{Gal}(K^{\mathrm{sep}}/K) = \pi_1(\mathrm{Spec}(K)) \to 1$$

(see [**3**], Vol. 1, Th. 6.1 for an even more general situation): Grothendieck defined $\pi_1(X, a)$ for an arbitrary scheme X with a geometric point a. Here we see that starting with classical ideas and placing them in a new frame work, a powerful tool becomes available.

(5.2). An application: the monodromy theorem. In this theorem, we study a family over a punctured disk (or over the field of fractions of a discrete valuation ring) and we consider in which way the fundamental group of the base (or the Galois group of that field) acts on, say, the homology of the fibers. This situation was studied in many separate cases (Landman, Steenbrink, Brieskorn and many others). One version of the monodromy theorem says that

(1) *the eigenvalues of a monodromy matrix are roots of unity.*

Proofs were not easy. However as soon as Grothendieck's theory of the fundamental group combined the fundamental group of the base (or the Galois group of the field of definition) and the geometric fundamental group of a fiber into one concept, a proof was just an elementary exercise in linear algebra. See [**77**], page 515 for this idea by Grothendieck published by Serre and Tate; see [**68**], pp. 79-83 for an elementary proof of a simplified version, and for some references to earlier work. – This is a beautiful example of what Grothendieck means by: "the nut opens just by itself". Or one could say that it seems "like black magic". This theorem is proved by an easy exercise in linear algebra.

The result was proved and used in a more general setting. Usually what we call the "Grothendieck monodromy theorem" is the fact that a variety (or an ℓ-adic representation coming from algebraic geometry) over a local field is potentially semi-stable. For more explanation and references, see [**31**]. As a comment to my use of the term "monodromy theorem", Luc Illusie communicated to me:

"The monodromy theorem: 'a wonderful application of the theory of the fundamental group': here you are mixing and confusing two things:

(1) the 'exercise in linear algebra' saying that the action of inertia on ℓ-adic

representations over a local field with finite residue field (or such that the local field is small enough in the sense that it does not contain all roots of unity of order a power of ℓ) is quasi-unipotent (appendix of [**77**]);
(2) the theorem that the same statement holds for representations arising from ℓ-adic cohomology with proper supports or no supports of schemes separated and of finite type over the local field (whether or not the residue field satisfies the 'smallness' assumption).

Grothendieck gave two proofs of (2), both using much more than 'the theory of the fundamental group'. One (the 'arithmetic' one, as Grothendieck called it) consisted in a delicate reduction to (1), using the main theorems of SGA 4 and Néron's smoothification method, the second one (the 'geometric' one) was conditional, based on resolution of singularities, and only worked unconditionally in characteristic zero. This second proof was inspired to Grothendieck by Milnor's conjecture on the monodromy of an isolated singularity (Grothendieck told me he had greatly enjoyed Milnor's book), and used the full force of Grothendieck's theory of $R\Psi$ and $R\Phi$, together with the calculation of nearby cycles in the general semistable reduction case (nowadays we can make Grothendieck's proof work unconditionally, using de Jong – getting uniform bounds for the index of the open subgroup of the inertia group which acts unipotently)."

(5.3). The fundamental group under specialization. An application. (Computation of the prime-to-p part of the geometric fundamental group of a curve in characteristic p.) One of Grothendieck's results that he seemed very satisfied with was his computation of the prime-to-p part of the geometric fundamental group of a curve in positive characteristic. *Let X_0 be an irreducible, complete, non-singular algebraic curve over an algebraically closed field of characteristic p, and let Y be an irreducible, complete, non-singular algebraic curve over $\mathbb{C}$ of the same genus. Then the group $\pi_1(X_0)^{(p)}$ is isomorphic to $\pi_1(Y)^{(p)}$* (see [**3**], Vol. 1, Cor. 3.10). The structure of this group is well-known, as follows by classical, topological considerations. Note, however, that there seems to be no known proof giving this structure only using algebraic and geometric methods of algebraic geometry; this is the key to the result quoted above.

Here we see that that a question can lead naturally the discovery of new methods, new insight. Grothendieck developed "specialization of the fundamental group" (see [**3**], Vol. 1, Th. 3.8). In this theorem, for a scheme that is proper and smooth over a discrete valuation ring with residue characteristic p, the prime-to-p part of the fundamental group of the geometric generic fiber maps isomorphically onto the prime-to-p part of the fundamental group of the geometric special fiber.

This example shows in what way Grothendieck revolutionized this part of algebraic geometry "just" by describing the right concepts. Such ideas (unramified maps, coverings in topology, Galois groups) certainly were known in special cases, but the "quantum leap" from those previous ideas to the concept of the algebraic fundamental group is startling. For us, nowadays, it is hard to imagine how to proceed in algebraic geometry without such a tool at hand.

It is clear that Galois theory, the theory of the topological fundamental group, and existing monodromy-singularities considerations were a source of inspiration for Grothendieck.

(5.4). The result mentioned in (5.3) studies the geometric fundamental group of an algebraic curve, of a Riemann surface, as an abstract group. The wonderful paper [**41**] convinced me that it is even better to consider the *geometric fundamental group*, in characteristic zero, *as a subgroup* of $\mathrm{PSL}_2(\mathbb{R})^0$.

6. Grothendieck topologies

(6.1). When working in the algebraic context, the classical topology is replaced by the Zariski topology. But then cases arise that demand yet other adaptations. For example, consider a quotient by an algebraic group, such as an isogeny $\varphi : E \to E'$ of elliptic curves (a quotient by a finite group scheme). When working over $\mathbb{C}$ *in the classical topology*, this map is locally trivial. However if φ is not an isomorphism, this is not locally trivial in the Zariski topology. And this applies to many quotient maps in algebraic geometry. However, we would like to work with the notion of a fiber space, as was done earlier in so many cases in classical topology. This problem was recognized immediately after introducing the Zariski topology. Already in [**72**], we see how to circumvent this by proposing "une définition plus large, celle des espaces *localement isotriviaux*, qui échappe à ces inconvénients." The general theory was then extended by Serre to this new notion of "isotrivial", "trivial in the étale topology" in modern language. Already in that article Serre answered many questions, e.g. when is a quotient map locally trivial in the Zariski topology? See "groupes spéciaux", and the fact that *every special group is connected and linear* ([**72**], Section 4). He also observed the limitations of this new notion; e.g. see [**72**], 2.6: quotient maps which are are purely inseparable do not fall under the considerations of locally isotrivial coverings just discussed (in modern terminology, e.g. a quotient map under the action of a non-étale local group scheme). Serre also constructed a first cohomology group in this article, and asked whether one can define higher cohomology groups and whether they give the desired "vraie cohomologie" necessary for a proof of the Weil conjectures.

Note that what "localement isotriviaux" really means is "locally trivial in some Grothendieck toplogy". It was M. Artin who found the correct notion of "étale localization" (see [**31**] for a description).

The way Grothendieck approached this new concept is characteristic of his way of developing new ideas: a rather simple remark, and a need for further technique in order to solve problems becomes clear. Grothendieck sets out to develop a new method in the most general situation possible, and many pages of abstract mathematics are created (it is clear that he had a grand view of possibilities), and a new tool is created that can be applied and used in many situations.

(6.2). Here we clearly see the roots of further developments constructed and described by Grothendieck. The simple remark that a quotient map need not be locally trivial in the Zariski topology, and the remedy by Serre leads to a new concept: "Grothendieck topologies". Hundreds of pages on this topic can be found in SGA 4. It is one of the most important tools in fields like logic and algebraic geometry. Also, we can see by this example how we become accustomed to a new concept. I remember the first time I saw a topology as a set of maps which do not give necessarily subsets; it was new to me. After some time you get accustomed to it, and it seems as if it must always have been that way.

"... j'admettais de confiance que pour le plongement usuel du groupe projectif dans le groupe linéaire, il y a une section rationnelle, puisque tout le monde semblait convaincu que ça devait toujours se passer comme ça pour une fibration par un groupe linéaire..." Letter of Grothendieck to Serre of 30.1.1956, see [**10**], page 29.

7. Anabelian geometry

(7.1). After 1970 Grothendieck wrote down many new ideas: "On pourra commencer à faire de la géométrie algébrique!" Many of these ideas have not yet been unravelled and certainly many of them not at all understood. Let me describe one of these, where we can clearly indicate the "roots" and where we now have a fairly good understanding of some of the implications and general structures involved.

In order to state the idea, Grothendieck introduced the notion "anabelian". In particular this applies to the (the fundamental group of) a curve of genus at least two. Of course Grothendieck also mentions that we should prove such results more generally for arbitrary "hyperbolic" varieties. Grothendieck baptizes these curves, these situations, these groups "anabelian" because such "groupes fondamentaux...sont très éloignés des groupes abéliens..." (see [**11**], p. 14, or [**68**], page 17). Later on, a more technical definition of an "anabelian group" became available:

Definition. A group G is called *anabelian* if every finite index subgroup $H \subset G$ has trivial center.
Definition. A topological group G is called *anabelian* if every finite index, closed subgroup $H \subset G$ has trivial center.

Examples.
(1) For a number field K, i.e. $[K : \mathbb{Q}] < \infty$, its absolute Galois group $G = G_K = \mathrm{Gal}(\overline{K}/K)$ is anabelian. This follows from results known to F. K. Schmidt, see [**61**] to which Neukirch refers, see [**50**].

(2) On page 77 of [**40**] we find the definition of a *sub-p-adic field.* In particular any number field (a finite extension of $\mathbb{Q}$), or a finite extension of $\mathbb{Q}_p$ is a sub-p-adic field. Following Mochizui and Tamagawa we have:

For every sub-p-adic field K, its absolute Galois group is anabelian; see [**40**], Lemma 15.8 on page 80.

(3) For a hyperbolic curve X over an algebraically closed field, the fundamental group is anabelian. E.g. for complete curves of genus at least 2 over an algebraically closed field (of arbitrary characteristic), see [**25**], Lemma 1 on page 133.

In the terminology of S. Mochizuki – H. Nakamura – A. Tamagawa such groups are called "slim groups".

It might be that a more refined definition of an "anabelian group" is necessary in order to be able to prove the full analogue of the anabelian Grothendieck conjecture in higher dimensions.

(7.2). Let K be a field and let X be a geometrically irreducible algebraic *curve*, smooth over K. Let k be an algebraic closure of K. The following statements are equivalent:
(1) The fundamental group of X_k is non-commutative.
(2) The fundamental group of X_k is anabelian.
(3) The genus of X is either 2, or the genus is 1 and X is not proper over K, or its genus is zero and at least three geometric points have to be added to obtain a complete model.
(4) (In case $K \subset \mathbb{C}$.) The Euler characteristic is negative: $\chi(X(\mathbb{C})) < 0$.
(5) (Definition.) The curve is called *hyperbolic.*

Over an arbitrary field, (3) is usually used as the definition of a hyperbolic curve.

In [**11**], and in the letter June 27, 1983 of Grothendieck to Faltings (see [**68**], pp. 49-58) we see the following "anabelian" conjecture. For a scheme X (with base point, which will be omitted in the notation) over a field K we write

$$p_X : \pi_1(X) \to G_K := \mathrm{Gal}(K)$$

for the natural map of fundamental groups as in (5.1). For schemes X and Y over K we write

$$\mathrm{Isom}_{G_K}(\pi_1(X), \pi_1(Y))$$

for continuous isomorphisms which commute with p_X, respectively p_Y. We write $\mathrm{Inn}(\pi_1(X))$ for the group of inner automorphisms.

(7.3). Anabelian conjecture (Grothendieck). *Let K be a number field,* i.e. $[K : \mathbb{Q}] < \infty$ *and let X and Y be hyperbolic algebraic curves over K. Then the natural map*

$$\mathrm{Isom}_K(X, Y) \longrightarrow \mathrm{Isom}_{G_K}(\pi_1(X), \pi_1(Y))/\mathrm{Inn}(\pi_1(Y))$$

is bijective.

I will not describe here the rich history and the flow of ideas, proofs and results on this topic due to F. Bogomolov, Y. Ihara, S. Mochizuki, H. Nakamura, Takayuki Oda, F. Pop, Michel Raynaud, M. Saïdi, A. Shiho, A. Tamagawa, Y. Tschinkel, V. Voevodskii and many others, starting from the moment Grothendieck made his conjecture on this topic, and made public his ideas on this and other related topics. Basically this conjecture, as well as several generalizations and considerations in analogous situations, have now been proved or settled.

(7.4). Neukirch and Uchida. In trying to determine the "roots" of the anabelian conjecture, we can find at least two different sources. For the arithmetic of number fields, as far as this is encoded in the absolute Galois groups, there is a theorem of Artin and Schreier (from 1927). Then, in 1969-1977 Neukirch and Uchida proved that two number fields are isomorphic if and only if their absolute Galois groups are isomorphic as profinite groups; see [**49**], [**50**], [**84**]. This is called the Neukirch-Ikeda-Iwawasa-Uchida result. For a survey of the history of these, see [**62**].

Note, however, that the corresponding statement *does not hold for local fields*: two finite extensions of $\mathbb{Q}_p$ can have isomorphic absolute Galois groups without being isomorphic; see [**51**], XII.2, "closing remark". I thank Jakob Stix for helpful discussions and for providing references on this subject.

(7.5). Tate and Faltings. In 1966, Tate formulated a conjecture, that he proved for abelian varieties over finite fields; see [**81**]. In 1983, the conjecture was proved by Faltings over number fields (see [**24**]).

(7.6). Theorem (The Tate conjecture; Tate, Zarhin, Mori, Serre, Faltings). *Let K be a field of finite type over its prime field. Let ℓ be a prime number not equal to the characteristic of K. Let X and Y be abelian varieties over K. Then the natural map*

$$\mathrm{Hom}(X, Y) \otimes \mathbb{Z}_\ell \xrightarrow{\sim} \mathrm{Hom}(T_\ell(X), T_\ell(Y))$$

is an isomorphism.

This conjecture was generalized by Tate to the situation of algebraic cycles, but that generalization will not be discussed here.

We note that the analog of the above result *does not hold over local fields,* as was remarked by Lubin and Tate: *there exists a finite extension*

$L \supset \mathbb{Q}_p$ and an abelian variety A over L such that the natural inclusion

$$\mathrm{End}(A) \otimes \mathbb{Z}_\ell \subsetneqq \mathrm{End}(T_\ell(A))$$

is not an equality. In fact, we can choose A to be an elliptic curve with $\mathrm{End}(A \otimes \overline{L}) = \mathbb{Z}$ and $\mathrm{End}_L(T_\ell(A))$ of rank two over $\mathbb{Z}_\ell$. For details and references see [**21**], 3.17.

(7.7). Theorem (Tate). *Let K be a finite field, and let X and Y be abelian varieties over K. Then the natural map*

$$\mathrm{Hom}(X, Y) \otimes \mathbb{Z}_p \xrightarrow{\sim} \mathrm{Hom}(X[p^\infty], Y[p^\infty])$$

is an isomorphism.
See [**85**], Th. 6.

(7.8). After Neukirch and Uchida's result, which we could now call *anabelian theory for number fields [??] finite fields*, and after Faltings' proof of the Tate conjecture over number fields, we can see how Grothendieck's anabelian conjecture for hyperbolic curves *over number fields* arises naturally. Basically, this conjecture and several generalizations have been proved. However, the following result came as a big surprise (at least to me).

(7.9). Theorem (Mochizuki). *Let L be a finite extension of $\mathbb{Q}_p$, and let X and Y be hyperbolic algebraic curves over L. Then the natural map*

$$\mathrm{Isom}_L(X, Y) \longrightarrow \mathrm{Isom}_{G_L}(\pi_1(X), \pi_1(Y))/\mathrm{Inn}(\pi_1(Y))$$

is bijective.
This amazing counterpart of (7.3) can be found in [**40**].

(7.10). In the Tate conjecture/theorem (see (7.6)), it is essential to work over a number field, or over a field finitely generated over a prime field, but not over a p-adic field. Grothendieck knew this idea, and we can assume (conclude?) that his anabelian conjecture had the Tate conjecture as a stimulating source. Also see "Brief an G. Faltings", [**68**], pp. 49-58. In this letter to Faltings, Grothendieck stressed that we should work over a field of finite type over a prime field. It seems that it did not occur even to Grothendieck himself that a result like Mochizuki's theorem (see (7.9)) could be true for curves over a p-adic field; in fact the analog for abelian varieties does not hold over a p-adic field.

(7.11). The section conjecture. Consider the exact sequence in (5.1). Any K-rational point in X will give rise to a section of the map

$$p_X : \pi_1(X, a) \longrightarrow \mathrm{Gal}(K^{\mathrm{sep}}/K).$$

The *Grothendieck anabelian section conjecture* expects that the map

$$X(K) \quad \longrightarrow \quad \Gamma\left(\pi_1(X, a) \to G_K\right)/(\text{conjugation by } \pi_1(\overline{X}))$$

from $X(K)$ to the set $\Gamma(-)$ of *sections* for p_X up to conjugacy thus obtained *is a bijection in the case of hyperbolic curves.* Grothendieck already knew that this map was injective. No final result seems to be known at present about this conjecture. For a survey see [**78**].

In §§ 5, 6, 7 we have described three examples of earlier thoughts which inspired Grothendieck to create a completely new theory, or a conjecture which opened new areas of research for us.

8. If the general approach does not work

" ... obtaining even good results 'the wrong way'
– using clever tricks to get around deep theoretical obstacles –
could infuriate Grothendieck."
See [**69**], p.18.

(8.1). What happens if general patterns and theories do not suffice to settle a specific problem? Grothendieck gives us the impression that at such a point, one might need to develop a more general structure; to "escape" into a more general problem. However, there are mathematicians who, especially in specific situations, like to proceed by studying examples or making non-canonical choices, and sometimes a proof or construction comes out of all this even though it is neither expected nor obtained by general principles. A challenging difficulty in a problem, something which for years and years obstructs any solution, has always seemed to me to be more a stimulating and beautiful aspect of mathematics than a negative one. In past decades we have seen many examples of proofs that diverge from Grothendieck's "general approach" philosophy. Just to indicate the flavor, I will discuss a few of these.

All of the problems and questions in this section were studied by Grothendieck , but he did not solve them. These results show that while general theory is certainly needed, additional considerations such as a "trick" or a worked-out example were necessary in order to arrive at a solution.

Already in Section 2 we discussed various manners of finding new ways in mathematical research. Sometimes the activity of a mathematician is to create a new abstract theory, or if one prefers, to describe a structure which already exists but has not been discovered as yet. To solve a given problem, it is sometimes better to first understand the general pattern, and then just "turn the machine" in order to get the desired answer. As Grothendieck said, "Une fois cette théorie développée, j'espère bien que les conjectures de Weil viendront toutes seules": these words express what he was hoping for (see [**10**], 9.8.1960, page 104).

On 23.7.1985 Serre wrote to Grothendieck:

"Je sais bien que l'idée même de 'contourner une difficulté' t'est étrangère – et c'est peut-être cela qui te choque le plus dans les travaux de Deligne (autre exemple: dans sa démonstration des conjectures de Weil, il 'contourne' les 'conjectures standard' – cela te choque, mais cela me ravit).

(En fait, malgré ce que tu dis dans L28, mes façons de penser ne sont pas très différentes – profondeur à part – de celles de Deligne. Et elles sont assez éloignées des tiennes – ce qui explique d'ailleurs que nous nous soyons très bien complétés pendant 10 ou 15 ans, comme tu le dis très gentiment dans ton premier chapitre.)" See [**10**], pp. 244/245.

Serre communicated to me: "About theorems being proved by general methods or by tricks. The word trick is pejorative. But one should keep in mind that a 'trick' in year N often becomes a 'theory' in year $N + 20$. This is typically what has happened with Deligne's proof, and Wiles' proof."

(8.2). The Weil conjectures. As could already be seen in [**1**], the Weil conjectures were the starting point for Grothendieck to revolutionize algebraic geometry. Following this hint, we could deduce that Grothendieck was interested in "problem solving" research. We have seen that this was not true at all. Although these conjectures remained the driving force behind many of his endeavors between 1960 and 1970, we see that even when necessary methods became available, Grothendieck did not immediately sit down and try a head-on approach for a solution. As long as "the nut did not open just by itself", the time was not ripe: "j'espère bien que les conjectures de Weil viendront toutes seules". It seemed necessary to Grothendieck to develop more general theory, or more general conjectures which, once proved, would yield the Weil conjectures as an easy corollary. For the "standard conjectures" see [**5**] , [**35**], [**36**].

In 1959 Dwork proved essential parts of the Weil conjectures; for a description and for references, see [**8**]. It seems that Grothendieck was not very interested in this work; at least we have no record that he ever seriously plunged into itt; see [**69**], p. 17. Perhaps this was an essential aspect of his devotion to his own plan: examples or work by other mathematicians only partially interested him, I think, insofar as it supported his ideal view on further development, or revealed the intrinsic beauty of general structures studied, or if it could stimulate him to transform this "source" into a grand new idea.

We know how the story of the Weil conjectures did eventually proceed. Deligne proved these conjectures in the end; however, he diverged from the road proposed and wanted by Grothendieck. Instead of Grothendieck expressing admiration for Deligne for this great achievement, only a negative reaction came out (to say the least); I find this one of the most regrettable episodes in the development of modern algebraic geometry. We are grateful to Deligne for this wonderful result, for this token of insight combining abstract and deep insight on the one hand and a direct approach (sometimes called "a trick", but that is not fully adequate) on the other; it gives us confidence to try to proceed with such insight and energy. But we can also be grateful for Grothendieck formulating the 'standard conjectures', which are still a source for further inspiration. This fascinating aspect – the Weil conjectures and everything they created – of the "flow of mathematics" is a great example of the essence of our profession and the way various mathematicians work and react, in such different ways, to challenges.

(8.3). Lifting abelian varieties to characteristic zero. Suppose we have an abelian variety A_0 defined over a field κ of positive characteristic. Does there exist a lifting to an abelian variety defined over a field of characteristic zero? I.e. does there exist an abelian scheme over a mixed characteristic base having A_0 as special fiber? Grothendieck was interested in such questions as early as 1958 (see [**10**], p. 67).

There is a natural approach to this question. One studies deformations (in mixed characteristic) of A_0, like Kodaira-Spencer, and later Schlessinger, Grothendieck and Mumford taught us to do. Illusie communicated to me: "...Grothendieck studied formal deformations before Schlessinger; of course, it's Schlessinger who gave a really manageable criterion, and I remember that Grothendieck was surprised, vexed, and finally happy at that."

The result is that indeed, a *formal* abelian scheme can be constructed in mixed characteristic, and in fact, as Grothendieck showed, this problem is unobstructed (see [**54**]). However, we need to algebrize the result in order to end up with a true abelian variety in characteristic zero. As the Lefschetz-Chow-Grothendieck method is available, it suffices to make a formal deformation of (A_0, μ_0), where $\mu_0 : A_0 \to A_0^t$ is a polarization; for a beautiful description of Grothendieck 's existence theorem in formal geometry, see Part 4 written by Illusie in [**6**]. Grothendieck and Mumford proved that this problem is unobstructed in the case where A_0 admits a *principal polarization*, or at least a polarization of degree prime to p; in that case the problem is settled satisfactorily (see [**54**]).

However there are (many) cases where A_0 does not admit a principal polarization, and where the deformation problem defined by (A_0, μ_0) can be obstructed. Stepwise deformations do not give much information: if the next infinitesimal step is obstructed, how can we change the previous steps in order to be able to proceed unobstructed? It seems as though here, the machine comes to a stop. This was as far as Grothendieck could bring the state of affairs.

But at this point, ideas of David Mumford entered the scene; we see a pattern that is visible in many other cases. One uses the ideas, and the structures and tools given to us by Grothendieck, but one adds a new ingredient, which has a completely different flavor. Mumford started by describing the theory of "displays": choosing a basis for the Dieudonné module of the p-divisible group of A_0, one describes in this coordinate system an arbitrary "deformation" of the Frobenius in characteristic p (or in mixed characteristic) which still divides p; this can be done directly; from the "formula $V = p/F$" one can construct (over a perfection of the deformation ring) a p-divisible group which defines this deformation, and this can be descended to the deformation ring. This method gives access to direct computations: stepwise deformations are all encoded in one system. Later this tool was further developed by T. Zink: it gave rise to the general and very useful theory of "windows".

After developing this general theory Mumford proceeded to use it to show that any polarized abelian variety (A_0, μ_0) can be deformed in characteristic p to a polarized *ordinary* abelian variety. Note that this deformation is not canonical and not unique; it depends on choices, and is very much *not* in the style of Grothendieck.

Once this point is reached, one can use a general theory by Serre and Tate which shows that any polarized *ordinary* abelian variety admits a (canonical) lifting to characteristic zero, which concludes the proof. We see the ingenious combination of general theory, tricks, computations and general structures. This theorem was proved/expected by Mumford (see [69b] in [**44**]; also see [**52**]); this program was outlined by Mumford, and details were worked out in [**53**]:

Theorem (Mumford; Norman-Oort) **(8.3).1**. *Suppose given a polarized abelian variety* (A_0, λ) *over a field* κ *of characteristic* p. *Then there exists an integral domain* R *of mixed characteristic, with a residue class map* $R \to \kappa$ *and a polarized abelian scheme* $(A, \lambda) \to \mathrm{Spec}(R)$ *such that* $(A, \lambda) \otimes_R \kappa \cong (A_0, \lambda)$.

Remark. On several occasions, Grothendieck considered the question of the existence of (canonical) liftings. In his letter to Serre of 5 December 1958, he wrote: "...me font penser qu'il est possible de remonter canoniquement toute variété X_0 définie sur un corps parfait de caractéristique $p \neq 0$ en une sorte de 'variété holomorphe' X définie sur un anneau local complet quelconque $\mathcal{O}$ ayant le même corps résiduel. *Si* on a la chance que cette 'variété holomorphe' provient d'une variété algébrique X définie sur $\mathcal{O}$, alors cette dernière est unique, dépend fonctoriellement de X_0, etc." (see [**10**], p. 67).

It is hard to understand what Grothendieck had in mind at that moment. For an algebraic curve, it is not clear what a "canonical lift" should be. For an elliptic curve (abelian variety of dimension one) which is supersingular there is no "canonical lift" to characteristic zero. Serre gave an example of a surface which does not admit a lift to characteristic zero at all (see [**75**]). Further examples by Serre, non-singular projective varieties which could not be lifted to characteristic zero, are described by L. Illusie in [**6**], Part 4, Chapter 8: *Grothendieck's existence theorem in formal geometry with a letter of Jean-Pierre Serre.* In Coroll. 8.6.7 these examples are studied, and results are extended to varieties of dimension at least two.

The theorem alluded to in the previous paragraph, that an ordinary (polarized) abelian variety in positive characteristic admits a canonical lift to characteristic zero, was explained by Serre to Grothendieck right after the Woods Hole conference (see [**10**], pp. 161-164).

Example. In 1965, Grothendieck and Serre tried to have at least a candidate for an abelian variety in positive characteristic which could not be lifted to characteristic zero (see [**44**], page 704). Here is the idea.

Let E be a supersingular elliptic curve, say over $\mathbb{F} = \overline{\mathbb{F}}_p$ (it can be defined over $\mathbb{F}_{p^2}$). The group scheme $E[F]$, the kernel of the Frobenius map $F : E \to E^{(p)}$, is called α_p; this is a finite group scheme of rank p which is neither isomorphic with $\underline{(\mathbb{Z}/p)}$ nor with μ_p. Choose an embedding

$$i = (i_1, i_2) : \alpha_p \hookrightarrow E \times E; \quad \text{define} \quad X_i := (E \times E)/i(\alpha_p).$$

It is easy to see that X_i is a product of elliptic curves if and only if $i_1/i_2 \in \mathbb{F}_{p^2}$. Moreover, X_i is a CM abelian variety. If i cannot be defined over $\mathbb{F}$, then the CM abelian variety X_i (defined over a transcendental extension of $\mathbb{F}$) cannot be CM lifted to characteristic zero.

It might be that any X_i defined over $\mathbb{F}$ which is *not isomorphic to a product of elliptic curves* cannot be lifted to characteristic zero (and this was the example Grothendieck and Serre had in mind). This abelian surface does not admit a principal polarization, and the deformation problem might be non-smooth. However it can be shown that any such X_i can be lifted to characteristic zero. Moreover, in general (i.e. in case i generates a "large" finite field), it cannot be CM lifted to characteristic zero, as we will see in [**20**].

CM liftings. One can ask for even more. An abelian variety defined over a finite field is always a CM abelian variety, as was proved by Tate (see [**81**]). Does it admit a CM lifting to characteristic zero? Complete answers can be found in [**20**].

(8.4). A conjecture by Grothendieck. This line of thought, this partly non-canonical approach as sketched in (8.3), was also used to prove a conjecture of Grothendieck from 1970 about deformations of a p-divisible groups. He asked the following question:

Let X_0 be a p-divisible group over a field of characteristic p; let ζ be a Newton polygon under the Newton polygon of X_0; does there exist a deformation in equal characteristic where the generic fiber has Newton polygon equal to ζ? (see [**9**], page 150 of the appendix, a letter of Grothendieck to Barsotti).

This question remained unanswered for almost thirty years. The problem shows the same kind of difficulty as we saw above: one can describe a deformation space of a (quasi-polarized) p-divisible group. By a theorem of Grothendieck and Katz, a given Newton polygon describes a closed set in that space (see [**9**], pp. 149/150; see [**34**], Th. 2.3.1 on page 143). However, in general that locus is highly singular; the corresponding deformation problem is (formally) non-smooth in most of the interesting cases. A locus where we require the generic fiber to have Newton polygon equal to ζ may even be empty as can be seen on some examples (for certain non-principally quasi-polarized p-divisible groups and given ζ); for a complete description of all such examples, see **[59]**, Section 6. But also for a principally quasi-polarized (X_0, λ_0), the general approach does not give a straightforward proof for this conjecture by Grothendieck. An analog of Mumford's approach, however, proved to be successful (it was only much later that I realized this analogy between my approach to this question, and the method as described in (8.3)). A general theory was developed where for certain cases (technically speaking the case $a(X_0) = 1$) objects known as "displays" and easy linear algebra showed the Grothendieck conjecture to be true (see **[56]**). The proof was finished (the most difficult step) by showing that a deformation exists with the same Newton Polygon and with $a = 1$ in the generic fiber (a non-canonical, non-unique choice is needed). For details and references see **[60]**, especially § 8, and the discussion in § 9.

We note that the developments described in (8.3) and the methods described here not only use ideas and structures developed by Grothendieck, but also show the necessity (sometimes) of supplementing these by new insight and non-canonical constructions. For the tricky step (deformation to $a = 1$) in this approach to this conjecture by Grothendieck , we still do not have an "easy" proof; we do not have a structure or a general method which avoids any computation and study of special cases. This aspect of mathematics, considered as not very elegant by some people, has an appealing beauty to me, "cela me ravit" (I find this exciting).

Theorem (8.4).1. *Let X_0 be a p-divisible group over a field κ of characteristic p. Let $\gamma := \mathcal{N}(X_0)$ be its Newton polygon. Assume that β is a Newton polygon such that all points of β lie on or below γ. Then there exists an integral domain of characteristic p, a residue class map $R \to \kappa$ and a p-divisible group $X \to \mathrm{Spec}(R)$ with $X \otimes_R \kappa \cong X_0$ such that the Newton polygon of its generic fiber equals $\mathcal{N}(X_\eta) = \beta$.*

An analogous theorem holds for principally quasi-polarized p-divisible groups, and for principally polarized abelian varieties. An analogous statement for quasi-polarized p-divisible groups and for polarized abelian varieties admits many counterexamples. For references see **[60]** or **[59]**.

We remark that in (8.3), reduction to the case $a = 1$ (the case of monogenic Dieudonné modules for the local-local component of the p-divisible group) was done by an appropriate Hecke correspondence (see **[43]**, page 141, see **[53]**, Lemma 3.4). However, this is of little help for a proof of this conjecture by Grothendieck : a Hecke correspondence might drastically change the local deformation space. Moreover, the non-principally polarized analogue of this conjecture by Grothendieckdoes not hold in general. Hence a new method, deformation to $a = 1$ keeping the Newton polygon fixed, had to be developed for this case.

(8.5). Extending homomorphisms between p-divisible groups. *Let X and Y be p-divisible groups over a discrete valuation ring R with field of fractions K. Suppose a homomorphism $\beta_K : X_K \to Y_K$ is given. Does this extend to a homomorphism $\beta : X \to Y$?*

In case the characteristic of K equals zero, this question was answered in a positive way by Tate in 1966 (see [**79**], Theorem 4). For a long time, any answer to this question in the remaining cases was unknown. On page V of the introduction of Exp. IX by Grothendieck in [**3**] 7I (page 317 in that volume) we find this question in the general setting.

Once someone said to me that Grothendieck tried to prove that indeed such an extension should exist in general, that he did not succeed, and that this was his reason for leaving algebraic geometry; this seems unlikely to me, but I do not know. Johan de Jong solved this affirmatively for all cases in 1998 in [**33**]. Also here, we see that at least as far as we know at present, no general theory, no "general machinery" can decide for us what the answer should be (also see [**39**]).

Theorem (Tate, A. J. de Jong) **(8.5).1.** *Suppose we are given a discrete valuation ring R with field of fractions K, and p-divisible groups $X, Y \to \mathrm{Spec}(R)$. Then any homomorphism $\beta_K : X_K \to Y_K$ extends to a homomorphism $\beta : X \to Y$.* (See [**79**], Th. 4 on page 180; [**33**], Coroll. 1.2.)

(8.6). Truncations of p-divisible groups. On several occasions Grothendieck considered Barsotti-Tate groups, also called p-divisible groups. Such a group, or rather ind-p-group scheme, is an inductive limit, a union, $\{G_i\}$ of group schemes:

$$G = \cup_i G_i = \text{limind} \;\; G_i, \quad \text{with} \;\; G[p^i] = G_i;$$

we refer to [**30**] for definitions and certain properties; also see [**21**], 1.15. On 5.1.1970 (see [**44**], p. 745) Grothendieck wrote to Mumford:

"I wonder if the following might be true: assume k algebraically closed, let G and H be BT groups, and assume $G(1)$ and $H(1)$ are isomorphic. Are G and H isomorphic? This is true, according to Lazard, if G is a formal group of dimension 1."

Here Grothendieck writes $G(i)$, which we can also denote by G_i or by $G[p^i]$. Note that $G_{i+1}/G_i \cong G_1$, i.e. G is a "tower of which all building blocks all isomorphic to the same G_1". Mumford answers right away that the answer to this question is negative, as already is shown by 2-parameter formal BT groups. In [**60**], Section 12, we find an explicit infinite set of mutually different isomorphism classes of BT groups over $\overline{\mathbb{F}}_p$ which all have, up to an isomorphism, the same p-kernel.

This exchange of ideas shows that Grothendieck could ask a question that could be answered by giving an easy example, and reveals that Grothendieck had an expectation that mathematical reality would show a simple and beautiful structure (understanding BT groups would be elegant if this were true). But, it also shows that Grothendieck could lose interest as soon as the pattern could be more intricate (or less elegant) than he expected at first. I think this little episode is quite characteristic of his way of thinking and working: test by an easy question (e.g. to Serre or to Mumford), and only proceed when the original idea shows that mathematics indeed is simple.

As far as we know, Grothendieck dropped this idea. Was this topic not as beautiful and elegant as he wished? One could, however, proceed by asking: for which $G(1)$ can we conclude $G(1) \cong H(1) \Rightarrow G \cong H$? The answer is not obvious, not simple and elegant, but the technique developed in this way is very useful. Not knowing anything about this correspondence between Grothendieck and Mumford until 2010, I myself considered this problem; a complete answer can be found in [**57**]; also see [**60**], Section 12.

Here is an elegant and simple answer to this question (although the proof I know is neither obvious nor trivial). For any Newton polygon ζ, define over $\mathbb{F}_p$ a p-divisible group $H(\zeta)$ which we call *the minimal p-divisible group* attached to ζ. Such a p-divisible group can easily be described explicitly (e.g. in terms of Dieudonné modules), but we will not do that here. A minimal p-divisible group H can be characterized over $\mathbb{F} = \overline{\mathbb{F}}_p$ by requiring that H be a direct sum of its simple factors, and that for any simple summand H_i over $\mathbb{F}$ we have that $\mathrm{End}(H_i)$ is the maximal order in $\mathrm{End}^0(H_i) := \mathrm{End}(H_i) \otimes_{\mathbb{Z}_p} \mathbb{Q}_p$.

Theorem (8.6).1. *Let $k \supset \mathbb{F}_p$ be an algebraically closed field. Let X be a p-divisible group over k. Then*

$$(\forall\, Y,\quad X[p] \cong Y[p] \Longrightarrow X \cong Y) \Longleftrightarrow (X \ \text{ is minimal}).$$

See [**57**].

Another characterization that can be found in [**58**] gives the following elegant result.

Theorem (8.6).2. *Let $k \supset \mathbb{F}_p$ be an algebraically closed field. Let X be a simple p-divisible group over k. Then X is minimal if and only if $X[p]$ is* BT1 *simple* (i.e. there is no smaller, non-zero BT1 group scheme contained in $X[p]$).

(8.7). Conclusion of this section. Grothendieck constructed an impressive theory, a foundation for a new way of doing algebraic geometry, and handed down to us new tools. In many cases, all this leads directly to results and proofs. However, in some cases general theory can only be applied if special choices and non-canonical constructions are also supplied. Although this seems to contradict what Grothendieck taught us, sometimes such roads have to be taken. In fact, it is very often the combination of methods constructed by Grothendieck and insight we owe to him together with the study of special cases and the use of examples and "tricks" that lead us to new results.

In 1966 Grothendieck wrote to Mumford:

" .. I found it kind of astonishing that you should be obliged
to dive so deep and so far in order
to prove a theorem whose statement looks so simple-minded."

See [**44**], p. 717. Of course, we should always look for a simple proof, a proof which uses more structure and less tricks. But the beautiful reality, and the real beauty (I think) of mathematics is that you sometimes really do have to "dive so deep and so far".

References

[1] A. Grothendieck, *The cohomology theory of abstract algebraic varieties.* Proceed. ICM 1958, Cambridge Univ. Press, 1960.

[2] A. Grothendieck & J. Dieudonné, *Éléments de géométrie algébrique.* Inst. Hautes Ét. Sci. Publ. Math. **4, 8, 11, 17, 20, 24, 28, 32**. Volumes of this work will be cited as EGA.

[3] A. Grothendieck (and many co-authors), *Séminaire de géométrie algébrique de Bois-Marie.* Cited as SGA.
SGA1 – *Revêtements étales et groupe fondamental,* 1960 – 1961. Lect. Notes Math. **224**, Springer-Verlag 1971; Documents Math. 3, Soc. Math. France, 2003.
SGA2 – *Cohomologie locale des faisceaux cohérents et théorèmes de Lefschetz locaux et globaux*, 1961-1962. North-Holland Publ. Comp., Masson et Cie, 1986; Documents Math. 4, Soc. Math. France 2005.
SGA3 – *Schémas en groupes*, 1962-1964. Lect. Notes Math. **151, 152, 153**, Springer-Verlag 1970.
SGA4 – *Théorie des topos et cohomologie étale des schémas*, 1963-1964. Lect. Notes Math. **269, 270, 305**, Springer-Verlag 1972, 1973.
SGA4$\frac{1}{2}$: P. Deligne et al., *Cohomologie étale.* Lect. Notes Math. 569, Springer-Verlag 1972, 1977.
SGA5 – *Cohomologie ℓ-adique et fonctions L*, 1965-1966. Lect. Notes Math. **589**, Springer-Verlag 1977.
SGA6 – *Théorie des intersections et théorème de Riemann-Roch*, 1966-1967. Lect. Notes Math. **225**, Springer-Verlag 1971.
SGA7 – *Groupes de monodromie en géométrie algébrique*, 1967-1969. Lect. Notes Math. **288, 340**, Springer-Verlag 1972. 1973.

[4] A. Grothendieck, *Fondéments de la géométrie algébrique* (Extraits du Séminaire Bourbaki, 1957-1962). Exp. 149, 182, 190, 195, 212, 221, 232, 236. Cited as FGA.

[5] A. Grothendieck, *Standard conjectures on algebraic cycles.* In: Algebraic Geometry (Internat. Colloq., Tata Inst. Fund. Res., Bombay, 1968), Oxford University Press, pp. 193-199.

[6] *Fundamental algebraic geometry.* Grothendieck's FGA explained. (B. Fantechi, L. Göttsche, L. Illusie, S. Kleiman, N. Nitsure, A. Vistoli). Math. Monogr. Surveys **123**, Amer. Math. Soc. 2005.

[7] A. Borel, J-P. Serre, *Le théorème de Riemann-Roch* (d'après des résultats inédits de A. Grothendieck). Bull. Soc. Math. France **86** (1958), 97-136.

[8] N. Katz, J. Tate, *Bernard Dwork (1923-1998).* Notices of the AMS **46** (1999), 338-343.

[9] A. Grothendieck, *Groupes de Barsotti-Tate et cristaux de Dieudonné.* Sém. Math. Sup. **45**, Presses de l'Univ. de Montreal, 1970.

[10] *Correspondance Grothendieck-Serre.* Edited by Pierre Colmez and Jean-Pierre Serre. Documents Mathématiques (Paris), **2**. Soc. Math. France, Paris, 2001.

[11] A. Grothendieck, *Esquisse d'un programme* (January 1984). Published in [**68**].

[12] Séminaire H. Cartan **13** 1960/61. *Familles d'espaces complexes et fondements de la géométrie analytique.*
Exp. I – VI: A. Grothendieck, *Techniques de construction en géométrie analytique.*

[13] A. Grothendieck, *Récoltes et semailles: réflexions et témoignage sur un passé de mathématicien.* Univ. Sc. Techn. Languedoc (Montpellier) et CNRS (1985). Cited as ReS.

[14] A. Grothendieck, *La Longue Marche 'a travers la théorie de Galois.* We hope it becomes available:
`http://www.math.jussieu.fr/~leila/grothendieckcircle/LM/`

[15] M. Artin, *Néron models.* In: Arithmetic geometry (Eds G. Cornell, J. Silverman), pp. 213-230, Springer-Verlag, 1986,

[16] S. Bosch, W. Lütkebohmert & M. Raynaud, *Néron models.* Ergebn. Math. Grenzgebiete (3) **21**, Springer-Verlag, Berlin, 1990.

[17] H. Cartan & C. Chevalley, *Séminaire E. N. S.* 1955/56: Géométrie algébrique.

[18] P. Cartier, *A mad day's work: from Grothendieck to Connes and Kontsevich. The evolution of concepts of space and symmetry* In: Les relations entre les mathématiques et la physique théorique, pp. 23-42. Inst. Hautes Ét. Sci., Publ. Math., Bures-sur-Yvette, 1998. Translated from the French by Roger Cooke. Bull. Amer. Math. Soc. **38** (2001), 389-408.

[19] P. Cartier, *A country of which nothing is known but the name: Grothendieck and "motives"*, in this volume.

[20] C.-L. Chai, B. Conrad & F. Oort, *CM liftings of abelian varieties.* To appear.

[21] C.-L. Chai & F. Oort, *Moduli of abelian varieties and p-divisible groups: density of Hecke orbits, and a conjecture of Grothendieck.* Arithmetic Geometry, Proceeding of Clay Mathematics Institute 2006 Summer School on Arithmetic Geometry, Clay Mathematics Proceedings **8**, eds. H. Darmon, D. Ellwood, B. Hassett, Y. Tschinkel, 2009, 441-536.

[22] C. Chevalley, *Fondements de la géométrie algébrique.* Fac. Sc. Paris 1857/1958. Sécr. Math. Paris, 1958.

[23] P. Deligne, *Quelques idées maîtresses de l'œuvre de A. Grothendieck.* Matériaux pour l'histoire des mathématiques au XXe siècle, Actes du colloque à la mémoire de Jean Dieudonné (Nice 1996), Soc. Math. France, 1998, pp. 11-19.

[24] G. Faltings, *Endlichkeitssätze für abelsche Varietäten über Zahlkörpern.* Invent. Math. **73** (1983), 349-366.

[25] G. Faltings, *Curves and their fundamental groups* [following Grothendieck, Tamagawa and Mochiziuchi]. Sém. Bourbaki **50** (1997/98), Exp. 840. Astérisque 242, Soc. Math. France 1998, 131-150.

[26] W. Fulton, *Intersection theory.* Ergebn. Math. Grenzgeb. (3) **2**. Springer-Verlag, Berlin, 1984.

[27] *The Grothendieck Festschrift*, 3 Volumes (Eds P. Cartier et al). Progress Math. **86-88**, Birkhäuser, 1990.

[28] M. Hakim, *Topos annelés et schémas relatifs.* Ergebn. Math. Grenzgebiet, Band **64**. Springer-Verlag, Berlin-New York, 1972.

[29] R. Hartshorne, *Algebraic geometry.* Graduate Texts in Mathematics, **52**. Springer-Verlag, New York-Heidelberg, 1977.

[30] L. Illusie, *Déformations de groupes de Barsotti-Tate.* Exp.VI in: Séminaire sur les pinceaux arithmétiques: la conjecture de Mordell (L. Szpiro), Astérisque **127**, Soc. Math. France 1985; pp. 151-198.

[31] L. Illusie, *Grothendieck et la cohomologie étale,* in this volume.

[32] A. Jackson, *"Comme appelé du néant" – As if summoned from the void: the life of Alexandre Grothendieck.* Notices of the AMS **51** (2004), first part: pp. 1038-1056; second part: pp. 1196-1212.

[33] A. de Jong, *Homomorphisms of Barsotti-Tate groups and crystals in positive characteristic.* Invent. Math. **134** (1998), 301-333.

[34] N. Katz, *Slope filtration of F-crystals.* Journ. Géom. Alg. Rennes, Vol. I, Astérisque **63** (1979), Soc. Math. France, 113-164.

[35] S. Kleiman, *Algebraic cycles and the Weil conjectures.* In: Dix exposés sur la cohomologie des schémas, Amsterdam: North-Holland, 1986, pp. 359-386.

[36] S. Kleiman, *The standard conjectures,* in: Motives (Seattle, WA, 1991), Proceedings of Symposia in Pure Mathematics **55**, Amer. Math. Soc., 1994, pp. 3-20.

[37] K. Kodaira & D. Spencer, *On deformations of complex analytic structures.* I, II. Ann. Math. **67** (1958), 328-466.

[38] Yu. Manin, *Good proofs are proofs that make us wiser.* Interview by Martin Aigner and Vasco A. Schmidt. The Berlin Intelligencer, 1998, pp. 16-19.

[39] W. Messing & T. Zink, *De Jong's theorem on homomorphisms of p-divisible groups,* manuscript 2001.
`http://www.math.uni-bielefeld.de/~zink/z_publ.html`

[40] S. Mochizuki, *The local pro-p anabelian geometry of curves.* Invent. Math. **138** (1999), 319-423.

[41] S. Mochizuki, *Correspondences on hyperbolic curves.* J. Pure Appl. Algebra **131** (1998), 227-244.

[42] D. Mumford, *An elementary theorem in geometric invariant theory.* Bull. Amer. Math. Soc. **67** (1961), 483-487.

[43] D. Mumford, *Bi-exensions of formal groups.* In: Algebraic Geometry (Internat. Colloq. Tata Inst. Fund. research, Bombay, 1968). Oxford Univ. Press, 1969; pp. 307-322. See [**44**], 69b.

[44] *David Mumford, Selected papers.* Volume II. *On algebraic geometry, including correspondence with Grothendieck.* Edited by Ching-Li Chai, Amnon Neeman and Takahiro Shiota. Springer-Verlag, New York, 2010.

[45] D. Mumford, *My introduction to schemes and functors,* in this volume.

[46] J. Murre, *On Grothendieck's work on the fundamental group,* in this volume.

[47] M. Nagata, *A general theory of algebraic geometry over Dedekind domains, I: the notion of models.* Amer. Journ. of Math. **78** (1956), 78-116.

[48] A. Néron, *Modèles minimaux des variétés abéliennes sur les corps locaux et globaux.* Inst. Hautes Études Sci., Publ. Math. **21** (1964).

[49] J. Neukirch, *Kennzeichnung der p-adischen und der endlichen algebraischen Zahlkörper.* Invent. Math. **6** (1969), pp. 296-314.

[50] J. Neukirch, Über die absoluten Galoisgruppen algebraischer Zahlkörper. In: Journeés Arithmétiques de Caen, Astérisque **41-42** (1977), 67-79.

[51] J. Neukirch, A. Schmidt & K. Wingberg, *Cohomology of number fields.* Second edition. Grundl. Math. Wissensch. **323**. Springer-Verlag, Berlin, 2008.

[52] P. Norman, *Lifting abelian varieties.* Invent. Math. **64** (1981), 431-443.

[53] P. Norman & F. Oort, *Moduli of abelian varieties.* Ann. Math. **112** (1980), 413-439.

[54] F. Oort, *Finite group schemes, local moduli for abelian varieties and lifting problems.* Compos. Math.**23** (1971), 265-296. Also in: Algebraic geometry Oslo 1970 (F. Oort, editor). Wolters-Noordhoff 1972; pp. 223-254.

[55] F. Oort, *The isogeny class of a CM-type abelian variety is defined over a finite extension of the prime field.* Journ. Pure Appl. Algebra **3** (1973), 399-408.

[56] F. Oort, *Newton polygons and formal groups: conjectures by Manin and Grothendieck.* Ann. Math. **152** (2000), 183-206.

[57] F. Oort, *Minimal p-divisible groups.* Ann. Math. **161** (2005), 1021-1036.

[58] F. Oort, *Simple p-kernels of p-divisible groups.* Advances in Mathematics **198** (2005), Special volume in honor of Michael Artin: Part I - Edited by Aise Johan De Jong, Eric M. Friedlander, Lance W. Small, John Tate, Angelo Vistoli, James Jian Zhang; pp. 275-310.

[59] F. Oort, *Foliations in moduli spaces of abelian varieties and dimension of leaves.* Algebra, Arithmetic and Geometry: In Honor of Yu. I. Manin (Manin Festschrift; Eds: Y. Tschinkel and Yu. Zarhin), Vol. II, Progress in Mathematics Vol. **270**, Birkhäuser, (2009); pp. 465-501.

[60] F. Oort, *Moduli of abelian varieties in mixed and in positive characteristic.* Handbook of moduli (Eds Gavril Farkas & Ian Morrison), Vol. III, pp. 75-134. Advanced Lectures in Mathematics **25**, International Press, 2013.

[61] F. K. Schmidt, *Körper, über denen jede Gleichung durch Radicale auflösbar ist.* Sitzungsbericht Heidelberger Akademie der Wissensch., 1933, 37-47.

[62] F. Pop, *On Grothendieck's conjecture of birational anabelian geometry.* Ann. Math. **139** (1994), 145-182.

[63] M. Raynaud, *Modèles de Néron.* C. R. Acad. Sci. Paris Sect. A-B **262** (1966), A345-A347.

[64] P. Samuel, *On universal mappings and free topological groups.* Bull. Amer. Math. Soc. **54** (1948), 591-598.

[65] P. Samuel, *Invariants arithmétiques des courbes de genre 2, d'après Igusa.* Sém. Bourbaki **14** (1961/62), No 228.

[66] W. Scharlau, *Wer ist Alexander Grothendieck?* Teil 1: Anarchie; 2nd printing 2010; `scharla@uni-muenster`.

[67] W. Scharlau, *Wer ist Alexander Grothendieck? Anarchie. Mathematik. Spriritualität, Eisnsamkeit. Eine Biographie.* Teil 3: Spiritualität. Books on demand GmbH, 2010.

[68] L. Schneps & P. Lochak, *Geometric Galois actions I: Around Grothendieck's Esquisse d'un programme.* Cambridge University Press, 1997.

[69] L. Schneps, *A biographical reading of the Grothendieck-Serre correspondence,* in this volume. A short version of this article appeared as a book review in the Mathematical Intelligencer, **29** (2007).

[70] C. Schoen, *Varieties dominated by product varieties.* Internat. J. Math. **7** (1996), 541-571.

[71] J-P. Serre, *Faisceaux algébriques cohérents.* Ann. Math. **61** (1955), 197-278.

[72] J-P. Serre, *Espaces fibrés algébriques.* In: Sém. C. Chevally E.N.S. **2** (1958), *Anneaux de Chow et applications* pp. 1-01 – 1-37. Also in: J-P. Serre, *Exposés de séminaires (1950-1999), deuxième édition, augmentée.* Docum. Math. **1**, Soc. Math. France 2008, pp. 107-140

[73] J-P. Serre, *Géométrie algébrique et géométrie analytique.* Ann. Inst. Fourier, Grenoble **6** (1955-1956), 1-42.

[74] J-P. Serre, *Analogues kählériens de certaines conjectures de Weil.* Ann. Math. **71** (1960), 392-394.

[75] J-P. Serre, *Exemples de variétés projectives en caractéristique p non relevable en caractéristique zéro.* Proc. Nat. Acad. Sc. USA **47** (1961), 109-109.

[76] J-P. Serre, *Motifs.* In: Journ. arithm. Luminy 1989 (Ed. G. Lachaud). Astérisque **198-200**, Soc. Math. France, 1991, pp. 333-349.

[77] J-P. Serre & J. Tate, *Good reduction of abelian varieties.* Ann. Math. **88** (1968), 492-517.

[78] J. Stix, *Evidence for the section conjecture in the theory of arithmetic fundamental groups.* Preprint, January 2011.

[79] J. Tate, *p-divisible groups.* Proc. Conf. Local Fields (Driebergen, 1966) pp. 158-183 Springer-Verlag, 1967.

[80] J. Tate, *Algebraic cycles and poles of zeta functions.* Arithmetical algebraic geometry (Proc. Conf. Purdue Univ., 1963; Ed. O. Schilling), 93-110; Harper, 1965.

[81] J. Tate, *Endomorphisms of abelian varieties over finite fields.* Invent. Math. **2** (1966), 134-144.

[82] J. Tate, *Correspondance Grothendieck-Serre.* Bookreview. Nw. Archief Wiskunde **5**, 42-44.

[83] O. Toeplitz, *The calculus: a genetic approach.* The University of Chicago Press, 2007.
[84] K. Uchida, *Isomorphisms of Galois groups of algebraic function fields.* Ann. Math. **106** (1977), 589-598.
[85] W. C. Waterhouse & J. S. Milne, *Abelian varieties over finite fields.* Proc. Sympos. pure math. Vol. XX, 1969 Number Theory Institute (Stony brook), AMS 1971, pp. 53-64.
[86] O. Zariski, *Algebraic sheaf theory.* Bull. Amer. Math. Soc. **62**, (1956), 117-141.

A country of which nothing is known but the name: Grothendieck and "motives"

Pierre Cartier

To the memory of Monique Cartier (1932–2007)

Introduction

It is superfluous to introduce Alexandre Grothendieck[1] to mathematicians: he is recognized as one of the greatest scientists of the 20th century. For other audiences, however, it is important to explain that Grothendieck is much more than his rather sulphurous reputation, that of a man in a state of rupture, committing what one could call the suicide of his work, or at any rate consciously destroying the scientific school that he had created. What I want to discuss here are the interactions between his scientific work and his extraordinary personality. Grothendieck's story is not absolutely unique in the history of science; one may think of Ludwig Boltzmann for example. But there are essential differences: Boltzmann's work was rejected by the scientific community of his time and remained unrecognized until after his death, whereas Grothendieck's scientific work was immediately and enthusiastically accepted in spite of its innovative nature, and developed and continued by top-notch collaborators. The path traveled by Grothendieck appears different to me: a childhood devastated by the effect of Nazi crimes, an absent father who soon perished in the torments of the time, a mother who held her son in thrall and permanently affected his relationship with other women; all of this compensated for by an unlimited investment in mathematical abstraction, until psychosis could no longer be held off and came to drown him in the anguish of death – his own and the world's.

Institut des Hautes Etudes Scientifiques, 35 route de Chartres, 91440 Bures-sur-Yvette, France. cartier@ihes.fr.

[1]While Grothendieck was given the name "Alexander"at birth, he eventually chose to use the French spelling "Alexandre."

The case of Georg Cantor is an intermediate one, which has been beautifully analyzed by Nathalie Charraud.[2] After encountering violent opposition to his ideas, the support of great mathematicians such as Dedekind and Hilbert allowed him to reach an apotheosis at the International Congress of Mathematicians[3] in 1900 in Paris. The French school of analysis, from Poincaré to Borel, Baire and Lebesgue, was converted with enthusiasm to Cantor's ideas. Cantor's ultimate mental shipwreck may perhaps be attributed to the "Nobel syndrome", by which term I mean a type of depression which has been observed to occur in certain Nobel prize winners. Incapable of confronting their own individuality and the life that remains before them – especially when the prize has been attributed at a young age – with the world-renowned public figure they have now become, they fear that they have already given the best of themselves and will never again be able to reach the same height. There is an echo of self mockery in this feeling.

The typology of Grothendieck is incredibly complex. Like Gauss, Riemann, and so many other mathematicians, his major obsession was with the idea of space. But Grothendieck's originality was to deepen the idea of a geometric point[4]. As futile as such research might appear, it is nevertheless of considerable metaphysical importance, and the philosophical problems related to it are far from entirely solved. But what kind of intimate concerns, what secret fears are indicated by this obsession with the point? The ultimate form of this research, that of which Grothendieck was proudest, was that concerning the concept of a "motive", considered as a beam of light illuminating all the incarnations of a given object in its various guises. But this is also the point at which his work became unfinished: a dream rather than an actual mathematical creation, contrarily to everything else I will describe below in his mathematical work.

Thus, his work eventually opened onto an abyss. But Grothendieck's other originality is that of fully accepting this. Most scientists are careful to efface their footprints on the sand and to silence their fantasies and dreams, in order to construct their own inner statue, in the words of François Jacob. André Weil was typical in this: he left behind a perfectly finished product in the classical style, in two movements: his *Scientific Works*, graced by a compelling *Commentary* written by himself, and a fascinating but carefully filtered autobiography, *Memories of an apprenticeship*, in which the effects of privacy and self-censorship are veiled by the appearance of a smooth and carefree tale.

---

[2]Nathalie Charraud, Infini et Inconscient, Essai sur Georges Cantor, Economica (1994).

[3]This is the official name of the world congress in mathematics that takes place every four years. Note the shift from mathematics to mathematicians in the title.

[4]On the occasion of the 40th anniversary of the IHES, a "Festschrift" was published as a special issue of the *Publications Mathématiques* which was not widely circulated. My contribution, entitled *La folle journée*, was an analysis of the notion of a geometric point, where Grothendieck's ideas are largely present. An English translation appeared in the *Bulletin of the AMS* (October 2001).

Grothendieck played at a different game, nearer to Rousseau's *Confessions*. From the depths of his self-imposed retreat, of over two decades – which it would be indecent to attempt to force – he sent us a vast introspective work[5]: *Récoltes et Semailles*. I will make use of this confession to try to clarify some of the main features of his work. But let us not fool ourselves: Grothendieck reveals himself in all his nakedness, exactly as he appears to himself, but there are clear signs of well-developed paranoia, and only a subtle analysis could reveal all the partly unconscious blockages and silences. The existence of *Récoltes et Semailles* aroused a somewhat unhealthy curiosity in the eyes of a certain public, akin to the sectarian devotion to a guru, an imaginary White Prince. For myself, I prefer to stick to an analysis of the work and of the biography of the author, remaining as rational and honest as possible, before letting *Récoltes et Semailles* illuminate this exceptional body of work from within.

Acknowledgments. As always, I first wish to thank my wife Monique and my daughter Marion, for their help in transcribing my talk at Cerisy, for their typing and their critical re-reading. I must also thank Nathalie Charraud, my associate in this intrusion onto foreign ground, for her tenacity in publishing my notes. Finally, warm thanks to the whole team at Cerisy-la-Salle, for the quality of their welcome, which greatly contributed to the success of the conference. The translation of this text from French to English was done by Leila Schneps.

Birth of the mathematical work

To present Grothendieck's scientific work in a few pages to a non-specialist audience is something of a challenge. To do it, I will make use of the analysis given by Jean Dieudonné – for years Grothedieck's closest associate – in his introduction to the "Festschrift" produced on the occasion of Grothendieck's 60th birthday[6].

The inheritance of Cantor's Set Theory allowed the 20th century to create the domain of "Functional Analysis". This comes about as an extension of the classical Differential and Integral Calculus (created by Leibniz and Newton), in which one considers not merely a particular function (for example the exponential function or a trigonometric function), but the operations and transformations which can be performed on all functions of a certain type. The creation of a "new" theory of integration, by Émile Borel and above all Henri Lebesgue, at the beginning of the 20th century, followed by the invention of normed spaces by Maurice Fréchet, Norbert Wiener and

[5]Grothendieck was a very close friend, and we also collaborated scientifically, but I have not seen him in more than 20 years. He sent me only a part of *Récoltes et Semailles*, the part that he judged I would be able to comprehend. For the missing part, I borrowed the copy belonging to the library of the IHES.

[6]"The Grothendieck Festschrift", 3 volumes edited by P. Cartier, L. Illusie, N.M. Katz, G. Laumon, Y. Manin and K.A. Ribet, Birkhäuser, Boston-Basel-Berlin, 1990.

especially Stefan Banach, yielded new tools for construction and proof in mathematics. The theory is seductive by its generality, its simplicity and its harmony, and it is capable of resolving difficult problems with elegance. The price to pay is that it usually makes use of non-constructive methods (the Hahn-Banach theorem, Baire's theorem and its consequences), which enable one to prove the existence of a mathematical object, but without giving an effective construction. It is not surprising that a beginner, infatuated with generality, reacted with enthusiasm at what he learned about this theory in Montpellier, during the course of his undergraduate studies under somewhat old-fashioned professors. In 1946, Lebesgue's theory of integration was nearly 50 years old, but it was still hardly taught in France, where it was considered as a high precision tool, reserved for the use of especially able artisans [7].

Upon his arrival in the mathematical world of Paris, in 1948 at the age of 20, he had already written a long manuscript in which he reconstructed a very general version of the Lebesgue integral. Once he was received into a favorable milieu, in Nancy, where Jean Dieudonné, Jean Delsarte, Roger Godement and Laurent Schwartz (all active members of Bourbaki) were attempting to go beyond Banach's work, he revolutionized the subject, and even, in a certain sense, killed it. In his thesis, defended in 1953 and published in 1955, he created from scratch a theory of tensor products for Banach spaces and their generalizations, and invented the notion of "nuclear spaces". This notion, created in order to explain an important theorem of Laurent Schwartz on functional operators (the "kernel theorem"), was subsequently used by the Russian school around Gelfand, and became one of the keys of the application of techniques from probability theory to problems from Mathematical Physics (statistical mechanics, "constructive" quantum field theory). Grothendieck left this subject, after a deep and dense article on metric inequalities, which fed the research of an entire school (G. Pisier and his collaborators) for 40 years. But, in rather characteristic fashion, he never paid attention to the descendence of his ideas, and showed nothing but indifference and even hostility towards theoretical physics, a subject guilty of the destruction of Hiroshima!

Starting in 1955, at the age of 27, he began a second mathematical career. It was the golden age of French mathematics, where, in the orbit of Bourbaki and impelled above all by Henri Cartan, Laurent Schwartz and Jean-Pierre Serre, mathematicians attacked the most difficult problems of geometry, group theory and topology. New tools appeared: sheaf theory and homological algebra (invented by Jean Leray on the one hand, Henri Cartan and Samuel Eilenberg on the other), which were admirable for their generality and flexibility. The apples of the garden of the Hesperides were

[7]In the same period, Quantum Mechanics, the other pillar of twentieth century science, whose mathematical basis makes extensive use of functional analysis, was banished from French teaching for very similar reasons.

the famous conjectures[8] stated by André Weil in 1949: these conjectures appeared as a combinatorial problem (counting the number of solutions of equations with variables in a Galois field) of a discouraging generality (even though several significant special cases were already known). The fascinating aspect of these conjectures is that they assume a sort of *fusion of two opposite poles*: "discrete" and "continuous", or "finite" and "infinite". Methods invented in topology to keep track of invariants under the continuous deformation of geometric objects must be employed to enumerate a finite number of configurations. André Weil caught sight of the Promised Land, but unlike Moses, he was unable to cross the Red Sea on dry land, nor did he have an adequate vessel. For his own work, he had already reconstructed "algebraic" geometry on a purely algebraic basis, in which the notion of a

---

[8]Here is a simple presentation of the problem. Consider a prime number p and an equation of the form $y^2 = x^3 - ax - b$, where a and b are integers modulo p. We want to count the number N_p of solutions of this equation, where x and y are also integers modulo p. According to Hasse (1934), we have the inequality $|N_p - p| \leq 2\sqrt{p}$; this inequality has recently found applications in coding theory. In a result announced in 1940 and completely proven in 1948, André Weil considered the case of a more general equation, of the form $f(x, y) = 0$, where f is a polynomial with integral coefficients modulo p. Here the inequality takes the form $|N_p - p| \leq 2g\sqrt{p}$, where the new element is the integer g, the *genus* (which is equal to 1 in the case above). The genus is an algebraic invariant of the equation $f = 0$, whose significance was discovered by Riemann: the initial equation can also be written as a congruence $F(x, y) \equiv 0 \bmod p$, where the polynomial F has integer coefficients. Now we consider the set of solutions to the equality $F(x, y) = 0$ where x and y are *complex numbers*; these solutions form a "Riemann surface" which is obtained by adding g handles to a sphere.

The final inequality was proven by Weil and Lang in 1954: if we consider a system of m equations $f_1 = \cdots = f_m = 0$ in n variables $x_1, \ldots, x_n$, the number N_p of solutions satisfies an inequality $|N_p - p^d| \leq Cp^{d-1/2}$, where the integer d is the *algebraic dimension*, usually given by $d = n - m$. The constant C is more difficult to describe explicitly. But in the case above, we have $n = 2$, $m = 1$, $d = 1$ and $C = 2g$.

The challenge proposed by Weil in 1949 was to give an *exact formula*, not just an inequality. To do this, one has to count in N_p also the points at infinity (in the sense of projective geometry) of the variety V defined by the equations $f_1 = \cdots = f_m = 0$, giving a new number $\overline{N}_p$ of solutions. By a generalization of the construction given above, in which the congruences modulo p were replaced by equalities of complex numbers, one can associate to V a space S of dimension $2d$, locally parametrized by d complex numbers (recall that for Riemann surfaces, we have $d = 1$, so $2d = 2$). The space S has geometric invariants called *Betti numbers*, denoted by $b_0, b_1, \ldots, b_{2d}$. Weil conjectured that

$$\overline{N}_p = S_0 - S_1 + S_2 - \cdots - S_{2d-1} + S_{2d}$$

$$S_i = a_{1,i} + \cdots + a_{b_i,i} \quad \text{with} \quad |a_{j,i}| = p^{i/2}, \quad \text{for } i = 0, 1, \ldots, 2d.$$

In particular, we have $b_0 = b_{2d} = 1$, and $S_0 = 1$, $S_{2d} = p^d$. In the case of dimension $d = 1$, we have $b_0 = 1$, $b_1 = 2g$, $b_2 = 1$ and $\overline{N}_p = 1 - (a_1 + \cdots + a_{2g}) + p$ with $|a_i| = \sqrt{p}$, from which we immediately deduce that $|\overline{N}_p - 1 - p| \leq 2g\sqrt{p}$ (but here $\overline{N}_p = 1 + N_p$ in the standard case). Weil then gave a complete treatment of a certain number of classical examples, in accordance with this conjecture, and Chevalley applied these counting methods to the theory of finite groups.

"field" is predominant. To create the required "arithmetic" geometry[9], it is necessary to replace the algebraic notion of a field by that of a commutative ring, and above all to invent an adaptation of homological algebra able to tame the problems of arithmetic geometry. André Weil himself was not ignorant of these techniques nor of these problems, and his contributions are numerous and important (adeles, the so-called Tamagawa number, class field theory, deformation of discrete subgroups of symmetries). But André Weil was suspicious of "big machinery" and never learned to feel familiar with sheaves, homological algebra or categories, contrarily to Grothendieck, who embraced them wholeheartedly.

Grothendieck's first foray into this new domain came as quite a thunderclap. The article is known as "Tohoku", as it appeared in the *Tohoku Mathematical Journal* in 1957, under the modest title "Sur quelques points d'algèbre homologique". Homological algebra, conceived as a general tool reaching beyond all special cases, was invented by Cartan and Eilenberg (their book "Homological Algebra" appeared in 1956). This book is a very precise exposition, but limited to the theory of modules over rings and the associated functors "Ext" and "Tor". It was already a vast synthesis of known methods and results, but sheaves do not enter into this picture. Sheaves, in Leray's work, were created together with their homology, but the homology theory is constructed in an *ad hoc* manner imitating the geometric methods of Elie Cartan (the father of Henri). In the autumn of 1950, Eilenberg, who was spending a year in Paris, undertook with Cartan to give an axiomatic characterization of sheaf homology; yet the construction itself preserves its initial *ad hoc* character. When Serre introduced sheaves into algebraic geometry, in 1953, the seemingly pathological nature of the "Zariski topology" forced him into some very indirect constructions. Grothendieck's flash of genius consisted in solving the problem from above, as he would do again and again in the years to come. By analyzing the reasons for the success of homological algebra for modules, he unearthed the notion of an abelian category (invented simultaneously by D. Buchsbaum), and above all the condition he labels as AB5*. This condition guarantees the existence of "enough injective objects". The sheaves satisfying this condition AB5*, and along with it, the method of injective resolutions which is fundamental for modules, extends to sheaves *without the need for any artifice.* Not only does it give a sound basis for the construction of sheaf homology, but it provides an absolutely

[9]The German geometer Erich Kähler published an article in 1958 (in Italian) entitled "Geometria Aritmetica" (Annali di Matematica, t. XLV, 368 pages), and the name was an immediate success. The domain is also called "Diophantine analysis", after the Greek mathematician Diophantus. It is the theory of polynomial equations such as $x^3 + y^3 + z^3 = t^3$, which possess an infinite number of solutions in real or complex numbers, but for which one restricts oneself to seeking the integer (or rational) solutions. The very existence of such solutions is then in question, and the properties of divisibility and prime numbers play a large role, giving an "arithmetic" character to the subject.

parallel development for modules and sheaves, bringing the Ext and Tor functors over to sheaves. Everything has now become entirely *natural*.

After this first "initiation" (1955-58), Grothendieck in 1958 stated his research program: to create arithmetic geometry via a (new) reformulation of algebraic geometry, seeking maximal generality, appropriating the new tools created for the use of topology and already tested by Cartan, Serre and Eilenberg. He dared attack the synthesis that none of the actors of the time (Serre, Chevalley, Nagata, Lang, myself) had dared, throwing himself into it with his own characteristic energy and enthusiasm. The time was ripe; world science was living its most intense phase of development during the 1960's, and the disenchantment of the years following the 1968 social movement had not yet begun. Grothendieck's undertaking throve thanks to unexpected synergies: the immense capacity for synthesis and for work of Dieudonné, promoted to the rank of scribe, the rigorous, rationalist and well-informed spirit of Serre, the practical know-how in geometry and algebra of Zariski's students, the juvenile freshness of the great disciple Pierre Deligne, all acted as counterweights to the adventurous, visionary and wildly ambitious spirit of Grothendieck. The new Institut des Hautes Études Scientifiques (IHES), created for him and around him, set in motion a constellation of young international talent. Organized around the key notion of a "scheme"[10], Grothendieck's theory ended up annexing every part of geometry, even the newest parts such as the study of "algebraic groups". Using a gigantic machine: Grothendieck topologies (etale, crystalline,...), descent, derived categories, the six operations, characteristic classes, monodromy and so on, Grothendieck arrived halfway down the path he had set himself, whose final goal was the proof of the Weil conjectures. In 1974, Deligne put the final touch on the proof, but in the meantime, Grothendieck had dropped everything since 1970, after 12 years of an undisputed scientific reign over the IHES.

What were the reasons for this total abandonment in the middle of everything? Put bluntly, his psychosis caught up with him, but at the time, it was stimulated by more direct reasons: the despair of being surpassed by his favorite disciple Deligne, the "Nobel syndrome", the revelation by the "1968 revolution" of the contradiction between the free spirit he believed himself to be and the university "top brass" he appeared in the eyes of others, a feeling of failure faced with some of his aborted mathematical efforts (the Hodge conjecture, the standard conjectures), weariness and exhaustion after 20 years of total devotion, day and night, to the service of his mathematical muse? A mixture of all of those.

---

[10]This word results from a typical epistemological shift from one thing to another: for Chevalley, who invented the name in 1955, it indicated the "scheme" or "skeleton" of an algebraic variety, which itself remained the central object. For Grothendieck, the "scheme" is the focal point, the source of all the projections and all the incarnations.

It remains to make some observations on Grothendieck's "posthumous" work. After his break with the mathematical world, which essentially occurred at the ICM in Nice (September 1970), and two further years of *Wanderung*, he became an ordinary professor at the very same average-level university (Montpellier) where he had studied as an undergraduate. He had a few more students, none of which attained the level of his team at the IHES, and whose fortunes were varied. Until his official retirement, in 1988 at the age of 60, he continued to work at mathematics in occasional spurts, leaving a "posthumous" body of work not without importance. There are three main texts:

– *Pursuing Stacks* (written in 1983) is a 600-page reflection on higher categories. Combinatorics, geometry and homological algebra come together in a grandiose project. After more than 15 years of the combined efforts of many, three (probably nearly equivalent) definitions have been proposed for multidimensional categories (in the widest sense[11]), using a cascade of composition laws. For general categories (called "lax"), the point is the following: when one wants to formulate an identity at a certain level, say $A = B$, one has to create a new object on the level just above, which realizes the transformation from A to B. It is a kind of dynamic theory of relations. In spirit, it is analogous to the Whitehead-Russell theory of types, but with a geometric aspect; in fact, Grothendieck conceives of his "stacks" as generalizations of homotopy theory (which studies deformations in geometry). The fusion of logic and geometry, whose beginnings are visible in the theory of stacks and toposes, is one of the most promising directions indicated by Grothendieck. Their importance is not just for "pure" mathematics, since a good theory of "assemblages" would have many potential applications in theoretical computer science, statistical physics, etc.

– The *Esquisse d'un Programme* was a text written in 1984 for inclusion in his application for a position with the CNRS. In it, Grothendieck sketches (the word is exact) the construction of a tower (or a game of Lego) describing deformations of algebraic curves.

– *The Long March through Galois Theory*, written before the previous one (in 1981), gives partial indications about some of the constructions suggested in the *Esquisse*.

These texts have circulated by being passed from hand to hand, with the exception of the *Esquisse* which was finally published, thanks to the insistence of a group of "devotees". Curiously, the true heirs of Grothendieck's work are essentially members of a Russian mathematical school (Manin, Drinfeld, Goncharov, Kontsevitch, to cite just a few), who have had little if any direct contact with Grothendieck, but who inherited and made use

[11] It is not difficult to define a *strict* multidimensional category

of methods from mathematical physics – a domain which he loathed and of which he was totally ignorant.

Biographical elements

The first thing is to describe something of Grothendieck's family origins, in order to place him in a proper perspective. There were three central characters: the *father*, the *mother* and the *son*, each remarkable in his own way, and a ghost – an older half-sister, on the mother's side, who died recently in the United States, and whom he did not know very well[12].

According to my information, the father's name was Shapiro – which indicates a Jewish origin.[13] He was apparently born in Belyje-Berega, which is today situated in Russia, near the border with Belorussia and the Ukraine, now independent countries. At the time, it was a Jewish town located in the Ukraine and inhabited by very pious Hassidic Jews. Breaking away from this background, Shapiro frequented the revolutionary Jewish circles in Russia, and took part at the very young age of 17 in the aborted revolution of 1905 against the tsars. He paid for this participation with more than 10 years in prison, and was only freed at the 1917 revolution. This was the beginning of a long period of revolutionary wandering, and the first of a long series of imprisonments. His son told me one day, with pride and exaltation, that his father had been a political prisoner under 17 different regimes. I answered that he should have been included in the *Who's Who* of the Revolution, and he didn't deny it! But a sign of the Bolshevik taboos which still exist is that in fact, most of the histories of socialism – including the ones written by Trotskyists like Pierre Broué – give virtually no information about Shapiro or his companions. There is still quite a bit of historical research to be done there.

According to what I know, in 1917, he belonged to the left-wing S.R. (Revolutionary Socialists), one of the factions which was struggling for power in Saint Petersburg. We know that in the end Lenin crushed all the factions except for the Bolsheviks, not to mention their own internal purges. One of the best descriptions of these events, although obviously partially romanticized, is the famous book by John Reed: *Ten Days that Shook the World.* Grothendieck always told me that one of the people in the book was his father.[14] After Lenin's purges, Shapiro was to be found everywhere that an extreme left-wing revolution broke out in Europe in the 1920's – and there were many! Naturally, he was with Bela Kun in Budapest, with Rosa Luxemburg in Berlin, with the Soviets in Munich. When Nazism was on the rise in Germany, he struggled with the S.A.P. (Left-wing Socialist Party)

[12] Is it merely a coincidence that there was also such a ghost in Einstein's life: a girl, born before his first marriage, whose trace has been entirely lost as neither of the two parents wanted to find their child?

[13] Shapiro later went under the name of Tanaroff as attested by a number of official documents.

[14] But there may be confusion between two different individuals of the same name.

against the Nazis, and he was compelled to leave Germany when Hitler came to power. Then, naturally, he was to be found in the Spanish Civil War, in the International Brigades (with the P.O.U.M – worker's Marxist unification party) like Simone Weil, in a surprising parallel. After Franco's victory in Spain, he joined his wife Hanka and their son Alexandre, refugees in France.

The end of his story is a manifestation of the shame of our country. When he returned to France, he was a broken man, according to his son. He drifted without energy for a while, and then, like so many other antifascist refugees, emigrants from Germany or Spain, he was interned, late in 1939, in the Camp du Vernet. This was not, of course, an extermination camp, although many of the prisoners died of malnutrition or lack of medical care (like, for example, in Gurs). But where exactly is the boundary between a refugee camp, an internment camp, and a concentration camp[15]? In any case, without ever recovering his freedom, he was handed over to the Nazis by the Vichy authorities, and finally perished in Auschwitz. The last concrete sign of his life that exists is a rather hallucinatory portrait in oils, painted by another prisoner in Vernet, which his son preserved like a talisman – the similarity is so striking that it could almost be a portrait of the son.

Hanka Grothendieck – that was the name of Alexandre's mother – came from Northern Germany. In the 1920's, she militated in various left-wing groups, and tried to be a writer. She had a daughter, mentioned above, and then met Shapiro, and Alexandre was born in Berlin[16] in March 1928. She emigrated to France when Hitler came to power, and managed to scrape a survival in the circles of German emigrants, which Simone Weil frequented around that period. In September 1939, when war was declared, the situation of these refugees – already very difficult – became worse, as they were henceforth considered "enemy citizens". In any case, Hanka and her son were interned in Mende in 1939, and knew no respite until after the catastrophe of June 1940.

Alexandre – who adopted the French spelling of his given name at about this time – was left behind by his parents when they left Germany in 1933. He remained hidden on a farm in Northern Germany until about 1938, when he was 10 years old, raised by a pastor in the style of Freinet, who believed in a "return to Nature". This "natural" ideology (inherited from Romanticism) was shared by the most diverse political groups in Germany, from the Nazis to the Socialists, and anticipated the concerns of ecological groups fifty years

[15]My colleague Szpiro has confirmed this point; his father was interned at Vernet for analogous reasons. Sixty years later, the testimonies begin to emerge!

[16]The *Götterdämmerung* of Berlin in 1945 saw the destruction of all public records. Because of this, Grothendieck had continual administrative problems. Until the beginning of the 1980's, he had to travel with a "Nansen passport" from the United Nations, documents which were parsimoniously offered to stateless people. After 1980, convinced that he could no longer be called up to serve in the French army, he consented to apply for French citizenship.

later. But he preferred to talk about the period of his life that he spent in le Chambon-sur-Lignon, from 1942 to 1944. The true nature of the resistance in the Cévenol region is much better understood nowadays. Le Chambon-sur-Lignon, an agreeable village and vacation site frequented mainly by Protestants, has a private high school called Collège Cévenol, which until 1939 was nothing more than a prep school for wealthy Protestant youth. During the war, however, the decisive hand of Pastor Trocmé transformed the Collège Cévenol into the center of a movement of spiritual resistance to Nazism, consonant with the historical military resistance rooted deep within the Huguenot tradition, which performed splendid rescue work for the Jewish children who came under its care. Grothendieck was a boarder at the Foyer Suisse and a student at the Collège, and made an impression so strong that even at the end of the 1950's I was able to obtain some personal memories from people who remembered him.

* * * * * * * *

His childhood ended there. Thanks to the Collège Cévenol, he obtained his baccalaureate and became a student in Montpellier in 1945. Then began the period of his *scientific training*. With the help of *Récoltes et Semailles*, I will now examine from the inside the gestation of the mathematical work that I described from the outside in the paragraphs above.

It was in Montpellier, during his undergraduate days, that he underwent his first real mathematical experience. He was very unsatisfied with the teaching he was receiving. He had been told how to compute the volume of a sphere or a pyramid, but no one had explained the definition of volume. It is an unmistakeable sign of a mathematical spirit to want to replace the "how" with a "why". A professor of Grothendieck assured him that a certain Lebesgue had resolved the last outstanding problems in mathematics, but that his work would be too difficult to teach. Alone, with almost no hints, Grothendieck rediscovered a very general version of the Lebesgue integral. The genesis of this first mathematical piece of work, accomplished in total isolation, is beautifully described in *Récoltes et Semailles*: he discovered that he was a mathematician without knowing that there was such a thing as a mathematician. Of course, he was surrounded by mathematics students and professors who taught mathematics properly enough, but who could not be taken for mathematicians: in all simplicity, he thought he was the only one in the world.[17]

[17] I experienced some of the same feelings during my provincial youth (in Sedan): I had a taste for mathematics, but I wasn't aware that they could actually constitute a profession. With a grandfather who had graduated from the engineering school Arts-et-Métiers and an uncle from the engineering school École Centrale, the ambition of my family was to see me enter the École Polytechnique! I would have been continuing the dynasty of engineers in the family, and that, it seemed, was the purpose of mathematics!

Grothendieck's "public period" (as we speak of the "public period" of that other rabbi, Jesus) began on his arrival in Paris in 1948, with a bachelor's degree in his pocket. His professor from Montpellier, who had much earlier completed a master's degree with Élie Cartan, had given him a letter of recommendation to his former teacher. He was unaware that Élie Cartan, three years before his death, was much diminished, and that his son Henri, a mathematician as famous as his father had been, was now the dominating figure on the Parisian – and French – mathematical scene.

But there was not too much chemistry between the eminent Protestant university professor and the young self-taught rebel. André Weil suggested sending Grothendieck to Nancy, where Jean Delsarte, one of the founding fathers of Bourbaki and a skilled organizer, had pushed the department of which he was Dean into becoming the first stage in Bourbaki's march towards the conquest of the universities. Jean Dieudonné and Laurent Schwartz were able to discipline Grothendieck just enough to prevent him from running off in all directions, and to restrain his excessive attraction to extreme generality. They gave him problems which led him in the direction of his first work on the Lebesgue integral. It would be an understatement to say that the disciple surpassed his masters: he pulverized the domain of Functional Analysis via a *solitary work* during the course of which he had no companions, and which subsequently found no continuers.

It was in Nancy, also, that he became an adult in the popular sense of the word. From a relationship with his landlady, a son named Serge was born. Serge had several older half brothers and sisters, and later on, when Grothendieck conceived the desire to take care of Serge himself, he was quite ready to adopt the entire family. He flung himself into a lawsuit to obtain paternal custody which was very unlikely to succeed – and which he sabotaged even further by insisting on taking advantage of the legal right to act as his own counsel. This was only the beginning of his chaotic family life: in all, he had five children from three mothers, and was as absent from their lives as his own father had been from his.

His mathematical work in Nancy made him famous, and he could have continued along the path he started on there. But he described himself very well as a builder of houses which it was not his vocation to inhabit. He embarked on the customary career of a researcher, recruited and supported by the CNRS, then spending a few years abroad after his thesis. But when he returned from São Paulo, he had closed the chapter on Functional Analysis.

* * * * * * * * *

That was the beginning of his *great period*, from 1958 to 1970, which coincides with the heyday of Bourbaki. What precipitated this period was the daring idea of Léon Motchane, who flung himself heart and soul into the adventure of creating the IHES (Institut des Hautes Études Scientifiques) in Bures-sur-Yvette. Léon Motchane, who had dreamed of being a mathematician, already had a successful career in business behind him, but he wanted to create something that would survive him. Dieudonné, who had left Nancy to spend some years in the United States, wanted to return to France. Motchane offered him the first chair in mathematics in the future institute, and Dieudonné accepted on condition that he hire Grothendieck as well. He himself was at a turning point in his career: he had reached the key age of 50 at which members of Bourbaki were required to quit the group, and he had already produced his most original piece of research, on "formal groups". Dieudonné, who was at heart a man of order and tradition[18], placed himself for the second time at the service of a revolutionary enterprise: after Bourbaki, the dual adventure of Motchane and Grothendieck.[19]

In an extraordinary organization of division of labor, the young Grothendieck created one of the most prestigious mathematical seminars that has ever existed. He attracted all the talented students[20] and threw himself with passion into mathematical discovery, in sessions that lasted ten or twelve hours (!) He formulated a grandiose program destined to fuse arithmetic, algebraic geometry and topology. A builder of cathedrals according to his own allegory, he distributed the work amongst his teammates. Every day, he sent an interminable pile of illegible notes to the older Dieudonné who, sitting at his worktable from 5 to 8 o'clock each morning, transformed the scribbles into an imposing collection of volumes signed by both Dieudonné and Grothendieck, which came out in the "Publications Mathématiques" of the IHES. Dieudonné had no personal ambitions and placed himself entirely at the service of this work as selflessly as he had done for the work of Bourbaki. Dieudonné did not remain at the IHES for many years; upon the creation of the University of Nice, he accepted the position of Dean of Sciences. But that did not stop his collaboration with Grothendieck. Only in 1970, when Dieudonné, no longer young, still managed to find the energy to organize the International Congress of Mathematicians in Nice, did the two mathematicians finally rupture their relationship.[21]

The legendary duo was, in fact, a trio. Jean-Pierre Serre, with his sharp sense of mathematics, his deep and broad mathematical culture, his quickness of thought and his technical prowess, was always there, protectively.

---

[18]We used to have epic political discussions during which he attacked what he called my "communism", which was really nothing more than an adherence to "progressive Christian" tendencies: the nuance escaped him, but maybe he was right after all.

[19]Though with very different destinies, both Motchane and Grothendieck were heirs of the revolutionary Jews of Saint Petersburg under the tsar. And in fact, both of Léon Motchane's sons became left-wing militants.

[20]In *Récoltes et Semailles*, Grothendieck counts his *twelve* disciples. The central character is Pierre Deligne, who combines in this tale the features of John, "the disciple whom Jesus loved", and Judas the betrayer. The weight of symbols!

[21]At that point there was nothing but total incomprehension between the believer in science for the sake of science and the libertarian militant who wanted to use the Congress as a tribunal for his social ideas.

He acted as an intermediary between Weil and Grothendieck when they no longer wished to communicate directly, and contributed greatly to the clarification of the above-mentioned Weil conjectures. At a time when the rate for suburban telephone calls was the same as the local rate, Serre and Grothendieck talked between Bures and Paris for hours each day. Serre was the perfect beater (I was going to say matchmaker), scaring the mathematical prey straight into Grothendieck's nets – and in nets as solid as those, the prey did not struggle for long.

Their success was immediate and smashing. As early as 1962, Serre was declaring that algebraic geometry was one and the same with scheme theory[22]. Publications on the subject, direct or indirect, grew into the thousands of pages; every newcomer to the domain needed to have read everything, and forty years later, a simple and concise yet complete exposition of the entire subject still does not exist. As Grothendieck described in his allegories, a certain know-how risks disappearing altogether from lack of fresh blood. After Grothendieck's departure from mathematics, Deligne and Illusie did a masterly archival job in completing the publication of the "Séminaire de Géométrie Algébrique", but Grothendieck was not grateful. It is true that what remains of Grothendieck's school has become a closed circle; a certain generosity has been lost, a certain breeze has died away – but then, the same is true for Bourbaki.[23]

* * * * * * * *

But the Tarpeian Rock lies not far from the Capitol! Grothendieck's scientific fame reached its peak in 1966. At the International Congress of Mathematicians, he was awarded the crowning honor: the Fields Medal[24]. The Soviet authorities were not very eager to give him a visa (his father had become an "enemy of the people" after the 1917 revolution). This was the time of the Vietnam War, and many mathematicians were against the war; not only Grothendieck but also, for example, Steve Smale, another winner of the Fields Medal in 1966. In the context of the Cold War between the USSR and the USA, which was particularly virulent at that time, certain Soviets may have hoped to make use of these mathematicians. But the press conference organized by Steve Smale in Moscow[25] (during which he not only denounced the Vietnam War but also compared it to the Soviet invasion of Hungary) must have shown them that mathematicians are not always easy to manipulate.

---

[22]Grothendieck's creation!

[23]Common destiny of institutions and civilizations!

[24]That is often compared to the Nobel Prize (which doesn't exist in mathematics), but is limited to 3 or 4 laureates every 4 years.

[25]Of which Grothendieck would have very probably approved – however he elected not to set foot in Moscow, and to have his Medal formally collected by Léon Motchane in his stead.

If I am allowed, as an absolute novice in the domain of psychoanalysis, to formulate a hypothesis, *it was in Moscow that the abyss opened for Grothendieck*, or rather, his fundamental wound reopened. This wound was that of the *absent father*, victim of Stalinists and Nazis, the Russian Jewish father recalled by the connection with a country in which anti-Semitism underwent a significant revival in the 1960's (if it had ever actually disappeared). Of course, there is also what I termed the "Nobel syndrome" above, and Grothendieck must certainly have said to himself that the Medal crowned an unfinished achievement, and suspected that he would never arrive at the end of his scientific ambitions.

This was the time of the great social fracture in France, which – following the feverish atmosphere in Berkeley in 1965 – led to the famous events of May 1968[26]. Grothendieck had not become politically involved in the Algerian war[27], and perhaps he wished to make up for that, and felt the pressure of the revolutionary past of the father he so admired. In any case, the social fracture revealed to him his own inner contradictions. Certainly Grothendieck had known what it was to be "undesirable", to be interned in camps, and he always lived very modestly, even when he was one of the gods of international mathematics. He was always very attentive to homeless people and to those who are tossed to the side of the road by the march of society; his house was always a sort of court of miracles (which was not always easy for his family), and he had not forgotten the difficult years of his childhood.

Yet, in 1968, he – whose mental self-image was anchored in the identity of the outlaw, the anarchist – suddenly discovered that he was a revered pontiff of international science, invested with great authority over both ideas and people. During this time, in which all authority was challenged, even intellectual authority, he became aware of the coexistence of two personalities within himself, and that was the beginning of a period of wavering which lasted for four or five years. His temporary response was to found a tiny group, which put forth a newsletter called *Survivre* and later *Survivre et Vivre*. This movement resembled one of those ecolo-catastrophe-oriented sects which sprang up everywhere in the 1970's: the danger (quite real at the time) of a nuclear war fitted together with obsessions about pollution and overpopulation. The absolute pacifism inherited from his father expressed itself within "Survive", and he put all of his scientific fame to use in the furtherance of his ecological aims. He seemed to believe that social issues can be settled with the same kind of proofs as mathematical ones, and in fact he often ended up actually irritating people even when they

[26]Now that the "events in Algeria" have been officially rebaptized "the Algerian War", which they were, perhaps we can look for an adequate name for the "events of 1968"?

[27]His insistence on not becoming a French citizen had enabled him to avoid being drafted during the Algerian war, but he paid a price. I only remember that he asked me once, in the early 60's, why I had not deserted. I myself took part in that cursed war, even though it was only for a short time.

were aware of his importance as a mathematician and perfectly receptive to the ideas he was expressing. I recall two quite painful incidents, one in Nice in 1970 and the other in Antwerp in 1973, during which his deliberately provocative attitude wrecked the patient efforts of others who had been working in the same direction as he was, but with more of a political vision.

This period of Grothendieck's life was followed by a few years of wandering: he resigned from the IHES in September 1970, on a rather minor pretext[28]; then he travelled abroad, took a temporary position at the Collège de France[29], and finally accepted a position as professor at the University of Montpellier, the university of his youth, for which he felt but moderate esteem.

From his years in Montpellier, one particular event stands out: that of his trial. As I already said, Grothendieck was very welcoming to the rejects of society. In the 1970's, the regions of Lozère and Larzac became a kind of Promised Land for numerous hippie groups, and seen from the outside, Grothendieck's house must have resembled a phalanstery with himself as the guru. Following some real or exaggerated incidents, the local police was becoming nervous, and one day they raided Grothendieck's house. The only "offence" they could pin on him was the presence of a Japanese Buddhist monk – a former mathematics student at the Tata Institute in Bombay and a most inoffensive person – whose residence permit in France had expired three weeks earlier. This was the kind of problem that a university professor can usually settle quite easily with a few contacts in the right places, but Grothendieck's philosophy prevented him from adopting this approach. The unexpected result was a summons to the Magistrate's Court of Montpellier six months later, the Japanese monk having in the meantime disappeared to the antipodes. Was it a preliminary test of Pasqua's laws?[30] Or did the local authorities think that Grothendieck was a suspicious hippie? What ought to have been a quickly expedited ten-minute procedure blew up into a major event. Grothendieck appeared at the Bourbaki Seminar in Paris in order to alert some of his colleagues to the situation; in particular Laurent

[28]The discovery of modest financial support given to the IHES, on the recommendation of the Prime Minister Michel Debré, by the D.R.E.T. (an organization financing military research). The financial support of the IHES was quite opaque for a long time, but military funding never played anything more than a modest role. It isn't totally absurd, however, to imagine that there might have been a world plan for the drafting of scientists into a new world war (this time against the USSR), and the IHES might have been part of that network. Léon Motchane was the only person who could have answered that question.

[29]He was an "associated professor" (a position reserved for foreigners) there from 1970 to 1972. Just when he could have received tenure, he explained clearly that he would use his chair as a vehicle for his ecological ideas. This resulted in a curious three-way competition between Grothendieck, Tits and myself, very unusual for the Collège de France, which ended by Tits being nominated to a chair in Group Theory.

[30]Interior Minister Charles Pasqua later instituted some very severe anti-immigration laws.

Schwartz, Alain Lascoux and myself. We set in motion some activity: string-pulling in some intellectual groups, mobilization of a few networks, calling upon the League of Human Rights. On the day of the trial, the judge had received 200 letters in favor of the accused, and a specially chartered airplane disgorged a medley of supporters wearing Dean's robes (with Dieudonné at their head), or the robes of prominent lawyers. Grothendieck, who was appearing in court for the second time, had once again decided to act as his own lawyer. He gave a magnificent speech for the defence, which I still have somewhere. Naturally quoting Socrates, he concluded with the following exhortation: *"I am being prosecuted in the name of a law passed in 1942 against foreigners. I was interned during the war in the name of this law, and my father died in Auschwitz because of it. I am not afraid of prison. If you apply the law, I may go to prison for two years. I am legally guilty and therefore will accept the punishment. But on a deeper level, I plead innocent. It is up to the judge to choose: the letter of the law and prison, or universal values and freedom."* This was followed by a "setting the argument in legal form" by the lawyer Henri Leclerc, who later became the president of the League of Human Rights. This was the result of a laboriously negotiated compromise with Grothendieck, who would have preferred condemnation to compromise. Alas, as Grothendieck had predicted, the judge yielded to pressure, and compromised by giving him a suspended sentence of six months' imprisonment. The sentence was confirmed on appeal, but by then the public emotion had died away.

As I already said, he retired in 1988, and has lived since then in self-imposed exile. At first he lived near the Fontaine de Vaucluse, in the middle of a little vineyard that he cultivated, and near to his daughter Johanna and his grandchildren. But later he broke off every family relation. He didn't seem to mind that the place where he lived was located so near to the infamous Camp du Vernet which played a sad role in his childhood. He lived for years without any contact with the outside world and only a few people even knew where he was. He chose to live alone, considered by his neighbors as a "retired mathematics professor who's a bit mad". He expressed his spirituality through a series of experiments, in the Buddhist tradition and others; perhaps his orthodox Jewish ancestry played some role in his adherence to strict dietary rules. He was at one time an extreme vegetarian, to the detriment of his health. The parallel with the destiny of Simone Weil is a double one: the desire to be at the level of the poorest of the poor, and a kind of mental anorexy. It is conceivable that his end, like hers, could result from a complete refusal to eat.

Autopsy of his work

Grothendieck's mathematical work in algebraic geometry totals more than 10,000 pages, published in two series. The first one, entitled *Elements of Algebraic Geometry* (EGA) with reference to Euclid's *Elements,* was written entirely by Dieudonné, and has remained incomplete since only 4 parts have been completely written, out of an initially projected 13. The second series is called *Seminar of Algebraic Geometry* (SGA) and consists in 7 volumes. The composition of SGA is less regular. At the start, there were the Seminars in the Bois-Marie (from the name of the domain where the IHES eventually found its home), which he led from 1960 to 1969. The first two volumes were written by Grothendieck or under his control, and he directed their publication: the third seminar was essentially written by Pierre Gabriel and Michel Demazure (whose thesis was part of the work). Afterward, things became more complicated. When Grothendieck abandoned mathematics in 1970, he left an incomplete worksite behind, and it was, in fact, a worksite in a pitiful state. There were manuscripts (literally) by Grothendieck which were difficult to decipher, mimeographed lectures from the seminar, and notes ready for publication. It would have been necessary to make a synthesis, plug the (sizeable) holes, and furthermore undertake an enormous writing job; all rather ungrateful tasks which would not bring any particular glory to their author. All was done, in the end, with faithfulness and filial piety, by Luc Illusie and Pierre Deligne. The central piece, in view of the Weil conjectures, is SGA 4, devoted to the most innovative ideas (toposes in particular; I will talk more about them below). In fact, when Deligne announced his complete proof of the Weil conjectures in 1974, experts considered that the foundations were insufficient, and (at the same time as the missing link of the Grothendieck seminar, SGA 5), he published an additional volume, essentially due to himself, under the curious name of SGA 4 1/2. Grothendieck took this new publication very badly, and took advantage of it to denigrate the entire enterprise: naturally, since it was not what he himself had in mind, his plans had been truncated, he had been betrayed... He describes this with a very strong image: the team of builders who, with their Master dead, disperse, each one carrying away his own sketches and tools. It is a beautiful image, but it has one problem: here, the Master abandoned his team by deliberately committing suicide. I will now undertake the autopsy of this "assassinated" work.

* * * * * * * *

As Grothendieck possesses a taste for symbolism, he claimed exactly *twelve* disciples: to arrive at this number, he cheated a little, because there is no real mathematical definition of a disciple, and he forgot the "posthumous" disciple (Z. Mebkhout), whom he welcomed and later rejected during the course of a rather inglorious polemic. In *Récoltes et Semailles*, he groups his work into *twelve* themes: I will not list them here, but I will comment on some of them.

The first theme he mentions is that of his thesis: *Functional Analysis*. He says himself that retrospectively, it seems to him rather like a school exercise, an intellectual warm-up. Certainly the perspective that Grothendieck gave to Functional Analysis is no longer modern; the major problems from within the theory have been solved, mostly by Grothendieck, and the subject has become one that serves others, its methods used to nourish the subjects of Fourier analysis (or its more recent form of wavelets) and partial differential equations. Grothendieck was pulled along by the current of "qualitative" topology of the time (which was very suited to his temperament), but today "quantitative" methods are more appreciated[31].

But of course, all the other themes concern Grothendieck's grand enterprise: *algebraic geometry.* One of the sources of mathematical development consists of the great problems, the great enigmas whose relatively simple formulation doesn't give any handle to latch onto to get started. What was improperly known as Fermat's Last Theorem was a conjecture of Biblical simplicity, expressed in symbols as: "the relation $a^n + b^n = c^n$ is impossible if a, b, c, n are non-zero integers unless $n = 2$". It was proven in 1994 (by A. Wiles and R. Taylor), via the construction of a large and complex edifice, largely based on methods due to Weil and Grothendieck. Now, the most prestigious and most perplexing open problem is the *Riemann hypothesis.* These two problems, Fermat and Riemann, are in some sense rather futile: Fermat's problem concerned a very particular equation, and the Riemann hypothesis can be interpreted in terms of very subtle regularities in the apparently random distribution of prime numbers. In itself, a counterexample to the Riemann hypothesis, given the present state of our knowledge, would have very small "practical" consequences and would certainly not be a catastrophe.

What is important to us here is a certain perception of the problems. Faced with the impossibility of proving the Riemann hypothesis, we have fled ahead. Following Artin and Schmidt, Hasse in 1930 formulated and later solved a problem analogous to the Riemann hypothesis by translating it into the form of an inequality (see footnote 8). The next step occupied Weil from 1940 to 1948. In all of these cases, by analogy with the Riemann zeta function and its relation to prime numbers[32], one associates zeta functions to the most varied geometric and arithmetic objects, and then off one goes on the way to proving the property analogous to Riemann – it has been done frequently

[31] His last text on the subject, which appeared in the Bulletin of the Mathematical Society of São Paulo in 1956, appears at first as a study of functors between Banach spaces (a premonition of his later investment in the theory of categories), even though the term "functor" does not appear. The central result is formulated as the equivalence of two of these functors. In a Bourbaki seminar, I reformulated his result as an inequality concerning matrices (with the "Grothendieck constant"), a "quantitative version" which was the starting point of later work (by G. Pisier). But Grothendieck did not accept my reformulation and considered himself betrayed.

[32] Let us recall one of its definitions:

$$\zeta(s) = \prod_p (1 - p^{-s})^{-1},$$

where the product runs over all prime numbers $p = 2, 3, 5, 7, 11, \ldots$

with great success! All of these zeta functions have contributed greatly to structuring the field of arithmetic, and Weil was guided by these ideas when he formulated his conjectures in 1949. Weil was a classical mind, attached to clarity and precision, and his conjectures have these characteristics. But for Grothendieck, the Weil conjectures are interesting rather as a test of his basic vision than in and for themselves. Grothendieck distinguished between mathematician-builders and mathematician-explorers, but saw himself as being both at the same time (André Weil was certainly less of a builder than Grothendieck, and he detested "big machinery" even if he had to construct it on occasion).

Grothendieck's favorite method is not unlike Joshua's method for conquering Jericho. The thing was to patiently encircle the solid walls without actually doing anything: at a certain point, the walls fall flat without a fight. This was also the method used by the Romans when they conquered the natural desert fortress Masada, the last stronghold of the Jewish revolt, after spending months patiently building a ramp. Grothendieck was convinced that if one has a sufficiently unifying vision of mathematics, if one can sufficiently penetrate the essence of mathematics and the strategies of its concepts, then particular problems are nothing but a test; they do not need to be solved for their own sake.

This strategy worked very well for Grothendieck, even if his dreams tended to make him go too far at times, and he needed the correcting influence of Dieudonné and Serre. But I already explained that Grothendieck only went three-quarters of the way, leaving the final conclusion to Deligne. Deligne's method was totally perpendicular to Grothendieck's: he knew every trick of his master's trade by heart, every concept, every variant. His proof, given in 1974, is a frontal attack and a marvel of precision, in which the steps follow each other in an absolutely natural order, without surprises. Those who heard his lectures had the impression, day after day, that nothing new was happening–whereas every lecture by Grothendieck introduced a whole new world of concepts, each more general than the one before–but on the last day, everything was in place and victory was assured. Deligne knocked down the obstacles one after the other, but each one of them was familiar in style. I think that this opposition of methods, or rather of temperament, is the true reason behind the personal conflict which developed between the two of them. I also think that the fact that "John, the disciple that Jesus loved" wrote the last Gospel alone partly explains the furious exile that Grothendieck has imposed upon himself.

* * * * * * * *

We now arrive at the very heart of Grothendieck's mathematical method, his unifying vision. Of the twelve grand ideas of which he was justly proud, he sets three above all the others: he gives them in the form of a progression:

$$\text{SCHEME} \rightarrow \text{TOPOS} \rightarrow \text{MOTIVE}$$

in an increasingly general direction. All of his scientific strategy was organized around a progression of increasingly general concepts. The image that occurs to me is that of a Buddhist temple that I visited in Vietnam in 1980. According to tradition, the altar was a series of rising steps, surmounted by a prone figure of Buddha – also traditional – a Buddha with a gigantic face, but whose features were actually those of a sage which local tradition described as the Vietnamese Montaigne of the 11th century. When one follows Grothendieck's work throughout its development, one has exactly this impression of rising step by step towards perfection. The face of Buddha is at the top, a human, not a symbolic face; a true portrait and not a traditional representation.

Before explaining the meaning of the trilogy displayed above, it is important to talk about Grothendieck's stylistic qualities. He was a master at naming, and he used that ability as one of his main intellectual strategies. He had a particular talent for naming things before possessing and conquering them, and many of his terminological choices are quite remarkable. But also in this, his personal experience was unusual. His mother tongue was German, and he spoke only German with his mother, during the many years that he lived with her in symbiotic closeness before her death. When I met him, around 1953, I felt when I spoke with him that he was thinking in German– and my lotharingian ear heard this correctly.

He had a remarkable sense of æsthetics. Yet I never understood his attraction to ugly women. I also cannot understand why he always lived in frightful homes: he worked at night, in general in a horrible room with the plaster falling off the walls, and turning his back to the window (seeking what secret humiliation?) And yet, when he sought mental images to explain his scientific ideas, he spoke of "the beautiful perfect manor", the "lovely inherited castle", all these allegories about beautiful homes. He even described himself as a builder. All these images are remarkably fitting and suitable. He may have continued to think for many years in German, but he certainly acquired a real sense of the French language, and his bilingualism enabled him to play on Germanic words. In French, he has a fantastically varied use of language ranging from the most familiar to the most elaborate, with an absolutely extraordinary sense of words.

His strategy, then, was to *name*. That is where I took the title of this article: "A country of which nothing is known but the name", because that was truly his way of going about things. "Motives" represented the final step for him, the one that he did not reach, although he had successfully passed the two intermediate steps of schemes and toposes.

* * * * * * * *

It is out of the question to give here a technical introduction to the notion of a *scheme*. The term itself is due to Chevalley, in a more restrictive sense than Grothendieck's (see footnote 10). In his *Foundations of Algebraic*

Geometry, André Weil had extended to abstract algebraic geometry (i.e., over an arbitrary base field, not necessarily the real or complex numbers) the method of gluing via local charts that his teacher Élie Cartan had used in differential geometry (following Gauss and Darboux). But Weil's method was not intrinsic, and Chevalley had asked himself what was invariant in a variety in the sense of Weil – a question characteristic of Chevalley's style. The answer, inspired by previous works of Zariski, was simple and elegant: the scheme of an algebraic variety is the collection of local rings of the subvarieties, inside the rational function field. No mention of explicit topology, unlike Serre, who introduced his algebraic varieties using Zariski topologies and sheaves at just about this time. Each of the two approaches had its own advantages, but also its limits:

- Serre needed an algebraically closed base field;
- Chevalley needed to work only with irreducible varieties.

In both cases, the two fundamental problems of products of varieties and base change could only be approached indirectly. Chevalley's point of view was better adapted to future extensions to arithmetic, as Nagata discovered early on.

Galois was certainly the first person to notice the polarity between equations and their solutions. One must distinguish between the *domain*, in which the coefficients of the algebraic equation are chosen (the "constants") and the *domain* in which the solutions must be sought. Weil kept this distinction between the "field of definition" of a variety and the "universal domain", but he was not very explicit about whether the field of definition had an intrinsic meaning, obsessed as he was with his ideas of specialization. For Serre, there was to be only one domain (which was necessarily algebraically closed), which is satisfying for "geometric" problems, but masks a number of interesting questions. For Chevalley (following Zariski), the central object is the rational function field, with its field of definition appearing as the field of constants, and the universal domain is practically eliminated.

Grothendieck created a synthesis of all these ideas, essentially based on the conceptual presentation of Zariski-Chevalley-Nagata. Thus, schemes are a way of encoding systems of equations and the transformations they can undergo; ideal theory, developed at the beginning of the century by Macaulay and Krull, had already had some of the same ambitions, and we owe it a large number of technical results.

The way in which Grothendieck presented the Galois problem is as follows. A scheme is an "absolute" object, say X, and the choice of a field of constants (or a field of definition) corresponds to the choice of another scheme S and a morphism[33] π_X from X to S. In the theory of

[33]From the start, this is based on the philosophy of categories: one defines the *category of schemes*, with its objects (schemes) and its transformations (morphisms); a morphism f links two schemes X and Y, which is symbolized by $X \xrightarrow{f} Y$.

schemes, a commutative ideal is identified with a scheme, its spectrum[34], but to a homomorphism from the ring A to the ring B, there corresponds a morphism in the other direction from the spectrum of B to the spectrum of A. Moreover, the spectrum of a field has only a single underlying point (but there are many "different" points of this type); consequently, giving the field of definition as being included in the universal domain corresponds to giving a scheme morphism π_T from T to S. A solution of the "system of equations" X, with the "domain of constants" S, with values in the "universal domain" T, corresponds to a morphism φ from T to X such that π_T is the composition of φ and π_X, given symbolically as:

$$\begin{array}{ccc} T & \xrightarrow{\varphi} & X \\ & {\scriptstyle \pi_T} \searrow \quad \swarrow {\scriptstyle \pi_X} & \\ & S & \end{array}$$

Admirable simplicity – and a very fruitful point of view – but a complete change of paradigm! The point of view of "modern" mathematics is based on the central role of sets. Once one has accepted the existence of sets (simple "classes" or "collections"), and the constructions one can make with them (of which the most important is to be able to consider the subsets of a set as elements of a new set), every mathematical object is a set, and coincides with the set of its points[35]. Transformations are in principle transformations of points[36]. In the various forms of geometry (differential, metric, affine, algebraic), the central object is the variety[37], considered as a set of points. Already in the 19th century, mathematicians became used to distinguishing the real points from the complex points of curves or surfaces defined by polynomial equations. Even more: in the study of Diophantine equations, one considers a system of equations $f_1 = \cdots = f_m = 0$ in unknowns $x_1, \ldots, x_n$, where the polynomials $f_1, \ldots, f_m$ have coefficients which are integers. The study of these equations led people to distinguish real and complex solutions, but also integer or rational solutions; one can also consider a less orthodox kind, such as solutions in a Galois field (for example, the integers modulo a prime number p), or even, following Kummer and Hensel, a p-adic field. It was already common usage to look for the solutions of an equation considered more or less simultaneously everywhere. For Grothendieck, *the scheme is the*

---

[34]It was Gelfand's fundamental idea to associate a normed commutative algebra to a space. Grothendieck recalled his initial approach to functional analysis, exactly at the time, following 1945, when Gelfand's theory had come to occupy a central position. The term "spectrum" comes directly from Gelfand.

[35]This set must be "structured", which is done by using a set theoretic version of Russell's theory of types.

[36]But the possibility of considering, say, lines (or circles) in space as points of a new space makes it possible to incorporate the geometry of transformations of points into lines (or circles).

[37]In the etymological sense: "domain of variation".

internal mechanism, the matrix[38], *which generates the points of the space*: the diagram above expresses this, by saying that φ is a T-point of the S-scheme X, and this for every S-scheme T.

In a recent article (see footnote 4), I studied the problem of the geometric point in a very mathematical manner, and I will not repeat that analysis here. Let us simply say that the purely mathematical analysis, by Gelfand and then by Grothendieck, of the notion of a point has recently crossed paths with a fundamental reflection in mathematical physics about the status of the point in quantum physics. The most systematic expression of this last reflection is Alain Connes' "non-commutative geometry". The synthesis is far from complete. The slowly emerging close relationship between the *Grothendieck-Teichmüller group*[39] on the one hand and the *renormalization group* from quantum field theory[40] on the other is surely only the first manifestation of a symmetry group of the fundamental constants of physics – a kind of cosmic Galois group! Grothendieck had not predicted this development, and surely would not even have wished it, because of his prejudice against physics (essentially due to his violent rejection of the military-industrial complex). It is possible that these connections could have been investigated earlier if the constraints of the Soviet system had not put the brakes on the transmission of ideas across the iron curtain.

Somewhere in *Récoltes et Semailles*, Grothendieck compares himself to Einstein for his contribution to the problem of space. He is right, and his contribution is of the same depth as Einstein's[41]. Einstein and Grothendieck both deepened our vision of space, so that this is no longer an empty receptacle for phenomena, a neutral stage, but the main actor in the life of the world and the history of the Universe. This distant descendant of Descartes' theory of vortices is the principal motor of our comprehension of the physical world at the dawn of a new century.

* * * * * * * *

Let us now examine *toposes*[42]. We saw that the geometry of schemes is a geometry with a plethora of points, at least with the very generalized notion of point shown in the diagram above. Toposes, on the contrary, realize a *geometry without points.* The idea of a geometry without points is not new: in fact, it is the oldest one. From the Euclidean point of view, one considered geometric figures of which some were points, but there were also

---

[38] I am using the word "matrix" here in its usual sense, not in the mathematical sense of a table of numbers.

[39] Thus baptized by Drinfeld, who is one of the mathematicians who penetrated the deepest into the above-mentioned *Esquisse d'un Programme* by Grothendieck

[40] Above all in the recent reformulation due to Connes and Kreimer.

[41] We should not forget Einstein's personal investment in the struggle against the military, from a political viewpoint quite close to Grothendieck's!

[42] Certain purists would like the plural to read "topoi" as in classical Greek. I will follow Grothendieck, writing "topos" and "toposes".

lines, planes, and circles; it was only in modern times, after the successes of set theory, that we adopted the habit of considering every component of a geometric figure as a set of points. Nowadays, a line is the set of its points: it is not a primitive object, but a composed one. However, nothing prevents one from proposing an axiomatic framework for geometry in which points, lines, planes and so forth would all be equal players, such as Birkhoff's axiomatic system for projective geometry, in which the primitive notion is that of a "plate" (a generalization of lines, planes etc.) and the fundamental relation is that of containment: the point is in the line, the line is in the plane, etc. Mathematically, one considers a family of partially ordered sets called lattices[43], and a geometry corresponds to one of these lattices.

In the geometry of a topological space, and particularly in the use of sheaves, the lattice of the open sets plays a major role, and points are relatively secondary. Thus, one can replace a topological space by the lattice of its open sets without losing much, and this idea was considered at various times. But Grothendieck's originality was to pick up Riemann's idea that multivalued functions actually live, not on open sets of the complex plane, but on spread-out Riemann surfaces. The spread-out Riemann surfaces project down to each other and thus form the objects of a category. Now, a lattice is just a special case of a category; one in which there is at most one transformation between two given objects. Grothendieck proposed to replace the lattice of open sets by the category of spread-out open sets. Adapted to algebraic geometry, this idea solves a fundamental difficulty linked to the absence of an implicit function theorem for algebraic functions. This is how he introduced the "etale site" associated to a scheme. Sheaves can be considered as particular functors on the lattice of open sets (viewed as a category), and can thus be generalized to etale sheaves, which are particular functors of the etale site.

Grothendieck produced a number of variations on this theme, with remarkable success, in various problems of geometric construction (for example, the "moduli" problem for algebraic curves). His greatest success was the possibility of defining the cohomological theory he needed to attack the Weil conjectures: it is called the ℓ-adic etale cohomology of schemes.

But there is one more level in this movement towards abstraction. Consider the progression:

$$\text{SCHEME} \rightarrow \text{ETALE SITE} \rightarrow \text{ETALE SHEAVES}$$

Grothendieck realized that one can place oneself directly at the last and highest level, and that all the geometric properties of a scheme are encoded in the *category of etale sheaves*. This category belongs to a particular type of categories that he called "toposes".

---

[43]The hypotheses needed here are the existence of a largest and smallest element (the empty set and the universal set), and of intersections and joins of two plates. In the last thirty years, this point of view was redeveloped under the name of "matroid" or "combinatorial geometry" (mainly by Rota and Crapo).

This, then, is the final act of the play. It was typical of the wild generosity of Grothendieck's ideas, and also of the lightheartedness with which he abandoned his (mathematical) children. Our hero had noticed that the sheaves on a given space formed a category which appeared to have all the same properties as "the" category of sets. But after the undecidability results of Gödel and Cohen in set theory, we know that there is not just *one* category of sets, but *many non-equivalent models* of set theory (in the logical sense of "model"). It was thus natural to explore the relations between toposes and models of set theory. Grothendieck was as ignorant – and perhaps as contemptuous – of Logic as his Master, Bourbaki, and his attitude towards mathematical physics was no different. It was for others to solve the puzzle (above all Bénabou, Lawvere, and Tierney): toposes are exactly models of set theory, but in a very particular logic, called 'intuitionist', in which the principle of excluded middle is not valid. It is remarkable that this logic was invented by a famous topologist, Brouwer, and that with a little perspective, it arises very naturally by virtue of the fact that the interior of the closure of an open set is not equal to the set itself.[44]

But the invention of toposes gives unheard-of freedom to the mathematical game, and makes it possible to break the yoke of the "only" set theory. To play a familiar mathematical piece in the new decor of a somewhat exotic topos can bring new surprises, and reveal new accents in well-known verses, and sometimes this new representation actually brings forth a mathematical treasure. From a more general point of view, a topos carries its own logic within it[45], and thus defines a kind of modal logic, or rather a *hic* and *nunc* logic, a spatio-temporal logic in which the truth value of an assertion can depend on the place and time.[46]

From a more technical point of view, Peter Freyd successfully applied methods from toposes to simplify Cohen's "forcing" method and his proof of the undecidability of the continuum hypothesis. It would be just as desirable to use methods from toposes together with recent results from model theory concerning the Mordell-Lang conjecture.[47]

We understand better now why Grothendieck considered the notion of toposes as central, while the more general concept of categories was nothing more for him than a tool.

---

[44]A topological version of the fact that the double negation of a property is not necessarily equivalent to it (in intuitionist logic), in violation of the hypothesis of excluded-middle.

[45]In technical terms, in every topos, the set of subobjects of the final object is a Heyting lattice, an intuitionist version of an algebra of propositions (Boolean lattices being the "classical" logical version).

[46]On my suggestion, the lawyer Mireille Delmas-Marty and the mathematician Jean Bénabou met to examine the possibility of establishing a theoretical basis for federal law (of the European type) on the theory of toposes. I do not believe these efforts were actually successful, but the idea would be worth revisiting.

[47]On this subject, see a recent lecture by Elisabeth Bouscaren at the Bourbaki Seminar (March 2000, exposé 870).

* * * * * * * *

It remains for us to make some remarks on the subject of *motives.* The image that Grothendieck himself described was that of a rocky coastline at night, illuminated by a lighthouse. The beam from the lighthouse turns, illuminating first one part of the coast and then another. In a similar manner, the various known cohomological theories, of which several were invented by him, are what we see, and it is necessary to go back to the source and build the lighthouse which will unify the representation of the entire coastline. In a certain sense, the scientific strategy is the inverse of the one he used in scheme theory. In the diagrammatic representation given above, the S-scheme X was given, and from there, one could realize its diverse incarnations: for every S-scheme T, one can construct the set of T-points of X. Here, the starting place is unknown, and only the various incarnations are in our possession: is this a theological image?

Grothendieck published nothing on this subject, contenting himself with a few remarks. I believe that Manin was the first to provide a real contribution, and then there was a long silence. Over the last few years, there has been an increase of activity and the program has become much more precise. The most ambitious contribution has been Voevodsky's: he contructs a category of objects, called motives, which is the locus of geometric invariants, and each scheme defines a particular motive. But in such a category, "pieces of objects" can migrate; the image of a genetic inheritance migrating through different beings is a good one. That this is possible follows from the definition of "weight" given by Deligne, which was the main ingredient in his proof of the Weil conjectures.

The tool created by Voevodsky undoubtedly corresponds to what Grothendieck expected, but it is going to be difficult to use. Good tools should be easy to use. Thus, the progress that has been made has been accomplished by restricting ambition to less general notions, called "mixed Hodge structures" or "mixed Tate motives", each of which is the expression of a fundamental group of symmetries, like the Grothendieck-Teichmüller group mentioned above. In fact, even within this limited scope, there is already an enormous amount of work to be done, and inestimable treasures to be unearthed. Grothendieck complained that all this was too economical, too reasonable, and from the heights of his visionary attitude, he heaped reproaches on the workers. But it seems to me that in the presence of mathematical visionaries like Grothendieck – or Langlands – who formulated wildly ambitious but sometimes imprecise programs of research, the right scientific strategy consists in isolating one piece which is sufficiently precise and restricted that one can actually work with it, and sufficiently vast to yield interesting results. The worker's philosophy?

Anatomy of an author

I will not venture upon a diagnosis of our patient, not being really competent to do so; but I will make some remarks, guided by sympathy. What is striking about Grothendieck is the expression of *suffering*: suffering because of having left an unfinished work, and the feeling of having been betrayed by his collaborators and followers. In a moment of true lucidity, he said something like: "I was the only person to have the inspiration; and what I transmitted to those around me was the task. I had workers around me, but none of them really had the inspiration!" The comment is deep and true, but it doesn't answer the question of why he deliberately closed the source of that inspiration! From what we know of his life today, he is subject to periodic crises of depression. It seems to me that his capacity for scientific creation was the best antidote to depression, and that the immersion in a living scientific milieu (Bourbaki and the IHES) helped this creation to take place by giving it a collective dimension; contrarily, in the relative scientific desert of Montpellier, and even more in his fiercely defended retreat, the isolation and the lack of minds at his own level with whom to discuss and compare himself no longer protect him from these eruptions of suffering.

To remain on a more secure terrain, I would like to say something about the religious aspect of his life. That it is permanent and deep emerges from what he says. He has had experiences of visual and auditive hallucinations; he has described divine apparitions, and speaks of canticles sung by two simultaneous voices, his own and that of God. It was following a series of these hallucinations – or apparitions – that he sent out a public eschatological message, which received no answer!

What were his antecedents? I have already noted that his father was born in a Hassidic community in the Ukraine, there where for the glory of God, hermits used to have themselves walled into towers with nothing but a tiny opening to the outside through which food might come in the form of alms from the faithful. But Grothendieck was never attached to Judaism, in any of its established forms. He felt closer to the Buddhist tradition; I don't know who first introduced him to that way of thinking, but I have already mentioned that visitor who unintentionally provoked his trial. At the end of the 60s, Grothendieck visited Vietnam which was at that time the butt of American bombardments, and he had a long-term affair in France with a Vietnamese student (officially a good Communist, but...) One of his main obsessions was about food, and at times he practiced an extreme form of vegetarianism. That is certainly an area where the Judaic and Buddhist traditions meet.

His personal Trinity was composed of God the Father, the goddess-mother, and the Devil. He calls the first "le bon Dieu": I don't know why he uses that term which corresponds to a somewhat outmoded popular usage, but it seems clear that he is not referring to Buddha, but rather to the image of the absent father (and in any case, in Buddhist orthodoxy, Buddha is not God!) The main character is the goddess-mother, whom he describes somewhere as a seductive female figure named Flora. The goddess-mother is present in many religions (including officially monotheistic Christianity), but a fairly recent phenomenon is the development in Japan and in Vietnam of the cult of Kannon (or Kan-Eum, or Lady of Mercy[48]). The relation between this worship of Grothendieck and his own mother is obvious. He lived in close symbiosis with her as she became progressively weaker and more disabled after her experiences in French detention camps, during the nearly two decades from his arrival in Paris in 1939 to her death in 1957. His name is her name[49], he dedicated his thesis to her and together they shared her native language of German. According to his own testimony, his wild passion for women had to wait until his mother's death to fully emerge, from the end of the 1950's to the end of the 1980's.

The most worrisome symptom is his obsession with the Devil. According to his most recent visitors, he, who never theologized his religion, has plunged into the writing of a sermon on the Devil's action in our world (he always was obsessional about writing!) His "catastrophism" is by no means new, and his concerns about the terrors of global nuclear war and pollution came at the right time in the 1970's. More recently, there was the incident mentioned earlier where, like Paco Rabanne, he received a revelation of the date of the end of the world, and he made it quite clear that it was going to be the end of "everything", not just our little Earth. Out of charity, he communicated this date to 200 or 300 people drawn from his list of scientific correspondents, and exhorted them to repentance before the final explosion, for afterward there were to remain but a few chosen. That letter, which received no response, was followed by a gloomy retraction after the fatidic date.

I will only describe two episodes to illustrate to what point he has become distanced from a rational and scientific point of view. About 25 years ago, during a visit to Montpellier, he showed me the work of two of his students: a long enumeration, using colored pencils, of configurations of lines (the problem was serious). When I pointed out that a computer calculation would have been quicker and surer, he responded sadly that that suggestion showed me up as an envoy of the Devil (in his military-industrial version!) More recently, he plunged into a series of long reflections in order to understand how the 300,000 km/sec that divine harmony would require for the speed of

[48]During a recent trip to Vietnam, I took note of certain curious phenomena of imitation between the Virgin Mary of the (still very numerous) Catholics and the Kannon of the Buddhists, not counting when they were even adopted by the Communist regime (for example, in a votive stela installed at the place of the last tsunami in Hué). After all, sculptors can work for various masters. The number of new sanctuaries dedicated to Kannon is amazing!

[49]It was not without difficulty that I finally discovered the name of his father, which he never mentioned. I thank some Russian friends for having helped me accomplish this research, after *perestroika*.

light had managed to become 298,779 km/sec by the corrupting influence of the Devil. He, whose mathematical work was so anchored around the notion of invariance and the naturalness of concepts refused to perceive the conventional nature of the metric system[50]. It isn't a question of error or of scientific ignorance: it is just an example of the other side of his own personal logic, that very same unstable logic which gave us his prodigious work.

In lieu of a conclusion

Mathematics sees itself as the most *objective* of all the sciences. At the very least, its *intersubjectivity* requires that the mathematical experience be as detached as possible from the affect of the mathematician, in order to be communicated without distortion, respecting its collective nature. The mathematical subject, conceived to be the mathematician subject present behind the creation, is required to disappear, and in practice, this disappearance is quite effective.

In this situation, Grothendieck represents an extremely special case. He, whose father was at the heart of every social combat for half a century, lived outside of the world, even much more so than the traditionally absent-minded professor. Even in his mathematical milieu, he wasn't quite a member of the family, and essentially pursued a kind of monolog, or rather, a dialog with mathematics. . . and God, as he did not separate the two things. His work is unique in that his fantasies and obsessions are not erased from it, but live within it, and it takes its life from them: at the same time as he gave us a strictly mathematical work, he also delivered to us, in a Freudian sense, what he believed to be its meaning.

His life was burned by the fire of the spirit, like that of Simone Weil, with whom he shares a strange kinship. He searched for a country, and for a name. Was it the Promised Land, that Judaea flowing with milk and honey, barely perceived across the Dead Sea, the accomplishment of a promise, the revelation of the whole and entire truth? I believe that the mythical land was the land of his father, that Jewish Ukraine that the sorrows of Eastern Europe rendered more inaccessible even than the Soviet Empire, already one of the most closed-off places in the world. The name – is that of the father.

[50]The creators of the metric system insisted on the *rational and natural* nature of their system. It took some time to really understand the degree to which convention played a role within it, and this is what gave rise to the permanent efforts to base our international system of units on a truly more natural foundation. But then, even the base 10 used in the decimal system is conventional!

Forgotten motives: The varieties of scientific experience

Yuri I. Manin

Le gros public:
A poêle, Descartes! à poêle!

(*R. Queneau, Les Oeuvres complètes de Sally Mara)*

When I arrived in Bures–sur–Yvette in May 1967, the famous seminar SGA 1966–67, dedicated to the Riemann–Roch theorem, was already drawing to an end. Mlle Rolland, then Léon Motchane's secretary at the IHÉS, found a nice small apartment for me in Orsay. Each early morning, awoken to the loud chorus of singing birds, I walked to Bures, anticipating the new session of private tutoring on the then brand–new project of motives, by the Grand Maître himself, Alexandre Grothendieck. Several pages, written by his hand then, survive in my archive; in particular, the one dedicated to the "Standard Conjectures". These conjectures remain unproven after half a century of vain efforts. Grothendieck himself saw them as the cornerstone of the whole project. In his letter to me dated March 20, 1969, he wrote:

Je dois avouer à ma honte que je ne sais plus distinguer à première vue ce qui est démontrable (voire plus ou moins trivial) sans les conjectures standard, et ce qui ne l'est pas. C'est évidemment honteux qu'on n'ait pas démontrées les conjectures standard!

Still, during the decades that have passed since then, the vast realm of motives kept rewarding the humility of many researchers prepared to be happy with what they could do using the tools they could elaborate.

Several times Grothendieck invited me to his house at rue de Moulon. He allowed me to browse through his bookshelves; I borrowed a few books to read at home. When I last visited him a day or two before my departure,

Max Planck Institute for Mathematics, Vivatsgasse 7, 53111 Bonn, Germany. manin@mpim-bonn.mpg.de.

I asked him to sign a book or paper for me. To my amazement, he opened "Les Œuvres complètes de Sally Mara" by Raymond Queneau and scribbled on the first page:

Hommage affectueux R. Queneau

Early history of motives

Having returned to Moscow in June 1967, after five or six weeks of intense training with Grothendieck, I spent several months writing down his main definitions related to motives and studying necessary background material in the literature. I was very pleased when it turned out that I could answer one of his questions and calculate the motive of a blow–up without using standard conjectures. My paper [Ma68] containing this exercise was submitted next summer and published in Russian. It became the first ever publication on motives, and Grothendieck recommended it to David Mumford (in his letter of April 14, 1969) as *"a nice foundational paper"* on motives.

Grothendieck wrote a letter in Russian to me about this paper (05/02/1969). This seems to be the only document showing that he had some Russian, probably learned from his father.

The first step in the definition of a category of (pure) motives is this. We keep objects of a given algebraic–geometric category, say of smooth projective varieties over a fixed field Var_k, but replace its morphisms by *correspondences.* This passage implies that morphisms $X \to Y$ now form an *additive group,* or even a K–module, rather than simply a set, where K is an appropriate coefficient ring. Moreover, correspondences themselves are not just cycles on $X \times Y$ but *classes* of such cycles modulo an "adequate" equivalence relation. The coarsest such relation is that of numerical equivalence, when two equidimensional cycles are equivalent if their intersection indices with each cycle of complementary dimension coincide. The finest one is the rational (Chow) equivalence, when equivalent cycles are deformations over a base which is a chain of rational curves. Direct product of varieties induces tensor product structure on the category.

The second step in the definition of the relevant category of pure motives consists in a formal construction of new objects (and relevant morphisms) that are "pieces" of varieties: kernels and images of projectors, i.e. correspondences $p : X \to X$ with $p^2 = p$. This produces a *pseudo–abelian,* or *Karoubian* completion of the category. In this new category, the projective line $\mathbf{P}^1$ becomes the direct sum of the (motive of) a point and the Lefschetz motive $\mathbf{L}$ (intuitively corresponding to the affine line).

The third and last step of the construction is one more formal enhancement of the class of objects: they now include *all* integer tensor powers $\mathbf{L}^{\otimes n}$, not just non–negative ones, and tensor products of these with other motives.

Various strands of intuition are interwoven in this fundamental pattern discovered by Grothendieck, and I will now try to make them (more) explicit.

The basic intuition that guided Grothendieck himself was the image of the category of pure Chow motives Mot_k as the receptacle of the *"universal cohomology theory"* $V_k \to Mot_k$: $V \mapsto h(V)$. The universal theory was needed in order to unite various cohomological constructions, such as Betti, de Rham–Hodge, and étale cohomology.

What looked paradoxical in this image was the following observation about transcendental cycles on an algebraic variety X. One could get hold of these cycles for $k = \mathbf{C}$ by appealing to algebraic topology, or else to complicated constructions of homological algebra involving all finite covers of X.

But in the category of pure motives, from the start one dealt *only* with algebraic cycles, represented by correspondences, and it was intuitively not at all clear how on earth they could convey information about transcendental cycles. Indeed, the main function of the "Standard Conjectures" was to serve as a convenient bridge from algebraic to transcendental. Everything that one could prove without them was indeed *"plus ou moins trivial"* – until people started treating correspondences themselves using sophisticated homological algebra (partly generated by the development of étale cohomology and Grothendieck–Verdier's introduction of derived and triangulated categories).

However, the passage from the set of morphisms to the K–module of correspondences involves one more intuitive idea, and it can be most succinctly invoked by referring to *physics*, namely the great leap from the classical mode of description of nature to the quantum one. This leap defined the science of the 20th century. Its basic and universal step consists in the introduction of a *linear span* of everything that in classical physics was only a set: points of a phase space, field configurations over a domain of space–time etc. Such *quantum superpositions* then form linear spaces on which Hilbert–like scalar products are defined, that in turn allow one to speak about probability amplitudes, quantum observations etc.

I have no evidence that Grothendieck himself thought then about quantum physics in relation to his algebraic geometry project. We do know that concerns about weapons of mass destruction and collaborationist behavior of scientists towards their governments and military–industrial complexes inspired in him deep disturbance and aversion. The most direct source of his inspiration might have been algebraic topology which, after the 1940s, laid more stress on chains and cochains than on simplices and the ways they are glued together.

However, in my personal development as a mathematician in the 1970's–80's and later, the study of quantum field theory played a great role, and feedback from theoretical physics – that was ahead – to algebraic geometry became a great source of inspiration for me. I was and remain possessed by a Cartesian dream, poetic rationalism, whatever history has yet to say about *Der Untergang des Abendlandes.*

Below I will sketch a map of two branches of the development of Grothendieck's ideas about motives that approximately followed two intuitions invoked above: from homological algebra and from physics respectively. The references at the end of this essay constitute the bare minimum of the relevant research, but the reader will be able to find much additional bibliographical material in the survey collection [Mo91] and in [A04], [VoSuFr00], [Ta11].

Motives and homological algebra

The most common linear objects are modules over rings in algebra and sheaves of modules in algebraic geometry. *Free modules/locally free sheaves* are the closest to classical linear spaces.

A general algebraic variety X, or a scheme, is a highly non–linear object.

In classical algebraic geometry over, say, the complex numbers, the variety X used to be identified with the topological space $X(\mathbf{C})$ of its $\mathbf{C}$–points, and could be studied by topological methods involving triangulations or cell decompositions. In the geometry over, say, finite fields, this did not work, and when in 1949 André Weil stated his famous suggestion that point counting over finite fields should be done using the trace of the Frobenius endomorphism acting upon appropriately defined (co)homology groups of X, it generated a flow of research.

The first product of this research was the creation of the cohomology theory of coherent sheaves of modules $\mathcal{F}$ on varieties X or more general schemes. Now, in a constructive definition of $H^*(X,\mathcal{F})$, one could either stress combinatorics of covers of X by open sets in the Zariski topology (Čech cohomology) or, alternatively, "projective/injective resolutions" of $\mathcal{F}$, that is special exact complexes of sheaves $\dots \to F_2 \to F_1 \to F_0 := \mathcal{F} \to 0$ or similarly with arrows inverted. This passage from the dependence of $H^*(X,\mathcal{F})$ on the non–linear argument X to the dependence on the linear argument $\mathcal{F}$ was very characteristic of the early algebraic geometry of 1950's and 1960's. "Homological Algebra" by H. Cartan and S. Eilenberg, and the famous FAC, "Faisceaux Algébriques Cohérents" by J.-P. Serre, became the standard handbooks for every aspiring young algebraic geometer.

David Mumford and I started our training as algebraic geometers at the same time, about 1956, he at Harvard, I at Moscow University. David reminisces that his teacher Zariski "was motivated by the need to make the work of the Italian school rigorous by using the new methods of commutative algebra". My teacher Shafarevich also suggested to us to study glorious Italian algebraic geometry, approaching it armed with modern insights and techniques developed by Serre, Grothendieck and their school.

I had no time nor use for a course in "Instant Italian", so I tried to read two books simultaneously, "Le Superficie Algebriche" by Federigo Enriques (Zanichelli, 1949) and "La Divina Commedia", and each time that I opened

Enriques (or for that matter, SGA), I recited mournfully: *...lasciate ogni speranza voi ch'entrate...*

Nevertheless, it worked. When I brought xeroxed papers by Gino Fano back from Bures in 1967, Vassya Iskovskikh and I were able to read them without bothering much in which language they were written, and then produce the first examples of birationally rigid varieties, and unirational but not rational threefolds using Fano methods.

Homological algebra proved more resistant, and here I learned most of what I understand now from the next generation of eager young Moscow students, who by now have been mature researchers themselves for a long time.

We first learned, of course, about the basic Grothendieck–Verdier presentation of homological algebra as the theory of derived, and more generally triangulated categories. Passage from the Bourbaki language of structures to the now dominating language of categories (and then polycategories) involved several radical changes of intuition, and as is now clear, led into the garden of forking paths. The passage from one crossroad to another one always involved a decision about what should be disregarded, and later it could happen – and always did happen – that one was obliged to turn back again and recollect some forgotten ideas.

The story of *derived categories* started with categories, whose *objects* were complexes (of abelian groups/sheaves/objects of an abelian category) considered modulo homotopy.

In the framework of Grothendieck–Verdier triangulated categories, one forgot about initial objects–complexes and focused on an abstract additive category, endowed with a translation functor and a class of diagrams called distinguished triangles. But the problem of non-functoriality of cones led back to the complexes of abelian groups, this time upgraded to the level of *morphisms* rather than objects. This was, of course, a special case of *enriched categories*, which in the simplest incarnation postulate Bourbaki–structured morphism sets Hom (X, Y), but with an upgrading: this time one clearly had to deal with the case of *categorified* morphism sets. However, when one allows morphisms to be objects of a category, then morphisms of this second floor category might form a category as well ... and we find ourselves ascending a Tower of Babel that could cause despair even in Grothendieck himself.

For the limited purposes of this note, I will disregard subtleties and various versions of the notion of triangulated/dg–categories, and will only sketch several basic discoveries of the last decades relating such categories with motives.

Roughly speaking, starting with a category of varieties (or schemes) X, one may consider either the replacement of each X by a triangulated category $D(X)$ of complexes of (quasi)–coherent sheaves on X, or else return to the initial insight of Grothendieck, but replace correspondences by *complexes of correspondences.*

The latter approach led to Voevodsky's motives ([VoSuFr00]). I will focus on some achievements of the first one.

One of the first great surprises was Alexander Beilinson's discovery ([Be83]) that a derived category of a projective space can be described as a triangulated category made out of modules over a Grassmann algebra. In particular, a projective space became "affine" in some kind of non–commutative geometry! The development of Beilinson's technique led to a general machinery describing triangulated categories in terms of exceptional systems and extending the realm of candidates to the role of non–commutative motives.

D. Orlov ([Or05]) proved a general theorem to the effect that if X, Y are smooth projective k–varieties and if there is a fully faithful functor $F : D^b(X) \to D^b(Y)$, then the Chow motive $h(X)$ is a direct summand of $h(Y)$ "up to translations and twists by Lefschetz/Tate motives".

M. Kontsevich formalized the properties of dg–categories, expressing properness and smoothness in case of the derived categories of varieties, and defined the respective class of categories (modulo homotopy) as "spaces" in non–commutative algebraic geometry. He then defined the respective class of Chow motives and has shown that there exists a natural fully faithful functor embedding Grothendieck's Chow motives (modulo twists) into non–commutative motives. These ideas were further developed by Tabuada, Marcolli, Cisinski et al., cf. the recent survey [Ta11] and references therein.

Motives and physics

In the mid–1970s and later, algebraic geometry interacted with physics more intensely that ever before: self–dual gauge fields (instantons), completely integrable systems (Korteweg–de Vries equations), emergence of supergeometry (based upon formal rules of Fermi statistics), the Mumford form and the Polyakov measure on moduli spaces of curves (quantum strings) were all discussed at joint seminars and local and international conferences of physicists and mathematicians.

Motives did not yet appear in this picture. However, in 1991 something new and unexpected happened.

B. Greene in his book "The Elegant Universe. Superstrings, Hidden Dimensions and the Quest for the Ultimate Theory" tells the following story:

"At a meeting of physicists and mathematicians in Berkeley in 1991, Candelas announced the result reached by his group using string theory and mirror symmetry: 317 206 375. Ellingsrood and Strømme announced the result of their very difficult mathematical computation: 2 682 549 425. For days, mathematicians and physicists debated: Who was right? [...]

About a month later, an e–mail message was widely circulated among participants in the Berkeley meeting with the subject heading: Physics Wins! Ellingsrood and Strømme had found an error in their computer code that, when corrected, confirmed Candelas's result. "

The problem about which Greene speaks is this. Consider a smooth hypersurface V of degree 5 in $\mathbf{P}^4$. Denote by $n(d)$ the (appropriately defined) number of rational curves of degree d on V. Calculating $n(d)$ looks like a perfectly classical problem of enumerative algebraic geometry, and in fact the numbers $n(1) = 2875$ and $n(2) = 609250$ were long known. Using the machinery and heuristics of quantum string theory, the physicists Ph. Candelas, X. C. de la Ossa, P. S. Green, and L. Parkes did not just calculate $n(3)$, but gave an analytic expression for a total generating function for these numbers, using the so called Mirror Conjecture. The mathematicians G. Ellingsrood and S. A. Strømme produced a computer code calculating $n(3)$.

Omitting a lot of exciting developments of this rich story, I will briefly explain only the part that refers to the new and highly universal motivic structure that emerged in algebraic geometry. I will speak about varieties, although in fact Deligne–Mumford stacks form the minimal habitat for this structure, and the respective extension of the construction of pure motives for them is needed; this was done by B. Toën.

Roughly speaking, we now treat the general problem, inherited from classical enumerative geometry: given a projective variety V, (define and) calculate the number of algebraic curves of genus g on V, satisfying additional incidence conditions that make this number finite, as in the Euclidean archetype: "one line passes through two different points of a plane". After considerable efforts, one can define for all stable values of (g, n) a Chow class $I_{g,n}$ on $V^n \times \overline{M}_{g,n}$ with coefficients in the completed semi–group ring, say $\mathbf{Q}[[q^\beta]]$ where β runs over integral classes in the Mori cone of V. This class expresses the virtual incidence relation described above, by reducing it to the positions of the respective points in V^n on the one hand, and to the position of the respective curve in the Deligne–Mumford stack of curves of genus g with n marked points.

When this is done, a list of universal properties of the classes $I_{g,n}$ treated as motivic morphisms essentially defines the (co)action of the modular (co)operad with components $h(\overline{M}_{g,n})$ in the category of motives upon each *total* motive $h(V)$ (I use the word total in order to stress that we are not allowed to pass to pieces here, although twisting and translations are in fact present, cf. [BehM96]).

The sophistication of both theoretical (and imaginative) physics and abstract mathematics that cooperated to discover this picture is really amazing, and I would like to draw attention to the fact that our traditional (mis)representation of mathematics as a language and technical tool needed to make physical intuition precise was reversed here: physical intuition helped discover mathematical structures that were not known before. One remarkable result of this was Deligne's generalization of the Tannakian Galois formalism ([De02]): it turned out that motivic Galois groups are actually supergroups, so that Fermi statistics, which up to then was "purely bosonian", now firmly resides in algebraic geometry as well.

Of course, such reversals have happened many times in history, but here the contemporary status of both theory of motives and quantum strings adds a strong romantic touch to the story. The beautiful two–volume cooperative project of the two communities trying to enlighten each other, [QFS99], carries two epigraphs. The epigraph to the first volume is a quotation from Grothendieck's "Récoltes et Semailles": *Passer de la mécanique quantique de Newton à celle d'Einstein doit être un peu, pour le mathématicien, comme de passer du bon vieux dialecte provençal à l'argot parisien dernier cri. Par contre, passer à la mécanique quantique, j'imagine, c'est passer du français au chinois.*

(In the pre–post–modern times one would have said: "It's all Greek to me!").

The second volume starts with an epigraph, written in Chinese logograms, from Confucius' "Analects", 17:2. Here I give its translation:

The Master said: "Men are close to one another by nature. They drift apart through behavior that is constantly repeated".

This is the collective riposte of the two communities, arguing their closeness, but in a language that is foreign to both.

***

In his letter to me from Les Aumettes dated March 8, 1988, the last letter I ever received from Grothendieck, he wrote:

... thanks for your letter of birthday congratulations, and please excuse my being late in replying to this letter, as well as the previous one and thanking for the reprint with dedication of november last year. Your letter struck me as somewhat formal and kind of ill at ease, and surely my silence has contributed to it. What I had to say about the spirits in mathematics today I said in the volumes I sent you and a number of other former friends. I am confident that before the year 2000 is reached, mathematicians (and even non–mathematicians) will read it with care and be amazed about times strange at last left behind ...

I met Grothendieck almost half a century ago. Thinking back on his imprint on me then, I realize that it was his generosity and his uncanny sense of humor that struck me most, the carnivalistic streak in his nature, which I later learned to discern in other anarchists and revolutionaries.

On the front cover of the issue no 14 of *"Survivre ... et Vivre" (Octobre–novembre 1972)* that miraculously reached me by post in Moscow, I read:

2 FRANCS
Canada 50 c
Communautés:
1 fromage de chèvre.

References

[Gr 69] A. Grothendieck. *Standard conjectures on algebraic cycles.* In: Algebraic Geometry, Bombay Colloquium, 1968, Oxford UP, 1969, 193–199.

[Ma68] Yu. Manin. *Correspondences, motives and monoidal transformations.* Math. USSR–Sb., 6 (1968), 439–470.

[Ja92] U. Janssen. *Motives, numerical equivalence, and semi-simplicity.* Invent. Math. 107 (1992), 447–452.

[Mo91] *Motives.* Proc. Summer Research Conference, 1991. Eds. U. Janssen, S. Kleiman, J.–P. Serre. Proc. Symp. in Pure Math., vols. 55.1, 55.2, AMS 1994.

[VoSuFr00] V. Voevodsky, A. Suslin, E. Friedlander. *Cycles, transfers, and motivic homology theories.* Ann. Math. studies, Princeton UP, 2000.

[A04] Y. André. *Une Introduction aux Motifs (motifs purs, motifs mixtes, périodes).* Panoramas et Synthéses, Nr. 17, Soc. Math. de France, 2004.

[Or05] D. Orlov. *Derived categories of coherent sheaves and motives.* arXiv:math/0512620

[Be83] A. Beilinson. *The derived category of coherent sheaves on* $\mathbf{P}^n$. Selecta Math. Soviet. 3, 1983, 233–237.

[Ta11] G. Tabuada. *A guided Tour through the Garden of Noncommutative Motives.* arXiv:1108.3787

[QFS99] *Quantum Fields and Strings: A Course for Mathematicians. Vols. 1, 2.* Ed. by P. Deligne, Ed Witten, et al., AMS, 1999.

[De02] P. Deligne. *Catégories tensorielles.* (Dedicated to Yuri I. Manin on the occasion of his 65th birthday.) Mosc. Math. J. 2 (2002), no. 2, 227–248.

[BehM96] K. Behrend, Yu. Manin. *Stacks of stable maps and Gromov–Witten invariants.* Duke Math. J., 85, no.1, 1996, 1–60.

Photographs

Luc Illusie met Grothendieck in 1964, and dived headlong into the SGA seminar, already in its fifth year. Learning the material was an intense experience, and frequently Illusie would arrive at two o'clock to spend the afternoon at Grothendieck's house, and not leave until 11:30 at night, just in time to catch the last train to Paris. Illusie followed SGA through to its end in 1969, and completed his doctoral dissertation under Grothendieck in 1971. By then, Grothendieck had left the IHES and was already deeply involved in ecological activities; at Illusie's thesis defense in May 1971, he asked Illusie a question about the consequences of his work on the future of humanity.

This photograph of Max Karoubi was taken in 1970, at the time when Grothendieck had just left the IHES and started his ecological movement. At that time, Karoubi was a 32-year-old assistant professor at the University of Strasbourg. He saw Grothendieck regularly in Paris, and was one of the relatively small number of mathematicians who took his involvement in ecological issues seriously (if not always agreeing fully with his views) and spent time discussing them with him.

In this picture from 1966, a 31-year-old Frans Oort is sporting with his three sons during an academic year spent visiting Harvard. At that time he was still digesting all that he had learned from his earlier stays in Paris, reading and studying Grothendieck's writings while developing his own research directions and bringing up his family.

This photo of Yuri Manin at 30 was taken just a few months before Manin first visited the IHES and met Grothendieck in May 1967. At that time, Grothendieck's work was already very well-known and the focus of intense study in Moscow, particularly in the seminars run by Shararevich and by Manin himself at Moscow University.

This picture was taken in January 1968, at a garden party celebrating the opening of a conference on Algebraic Geometry at the Tata Institute in Bombay. David Mumford, aged 31 at the time, is shown sitting next to André Weil, who had just come from staying with the President of India, a friend of his from the 1930s. Grothendieck was also at the meeting, which turned out to be a good venue for the assessment of his work at that time.

This picture shows a 31-year-old Robin Hartshorne at his wedding on December 20, 1969. Next to him is his wife, Edie, and just behind Edie is Judi Moore, who was at that time secretary of the mathematics department at Harvard, and later married mathematician Henry Pinkham, a student of Mumford. Behind them, from left to right, are mathematicians Bruce Bennett, David Mumford, Mike Artin, John Tate, and Raoul Bott. At the time of his wedding, Hartshorne was on leave from Harvard, visiting Japan and then the Tata Institute in Bombay. After having worked quite intensively with Grothendieck between 1960 and 1967, their correspondence had faded somewhat, but in 1970 he did receive Grothendieck's wishes for "much happiness" in his marriage.

This photo shows Steve Kleiman, aged 27, sitting in a café during one of his visits to France in 1969–1970. At that time, Kleiman had already completed two projects for Grothendieck, but returned to visit the IHES, where what would turn out to be the very last set of lectures for the legendary seminar SGA were in the process of being written up by Nick Katz. Kleiman recalls having dinner at Grothendieck's home, just months before Grothendieck stormed away from the IHES for good.

This picture of Joe Diestel dates from 1972, when Diestel, together with J. Jerry Uhl, was working on the manuscript that later appeared as the joint book Vector Measures, *largely devoted to presenting results of Grothendieck from a vector measure point of view. Diestel and his colleagues met Grothendieck at SUNY in Buffalo, where Grothendieck spent the summer of 1973 giving a 100-hour course on Functorial Algebraic Geometry (a set of course notes due to the efforts of F. Gaeta exists and can be found online). Diestel recalls being particularly struck by a remark made by Grothendieck in conversation, to the effect that during his years of working in functional analysis he had about ten real theorems, but when he was working in algebraic geometry he had several years in which he had almost a theorem a day.*

This picture of Carlos Simpson aged 25 was taken in Taos, New Mexico in 1987, the year that he was awarded his Ph.D. from Harvard University (and not long after he and L. Schneps first became acquainted as undergraduates at Harvard, in a course on algebraic geometry and schemes based, naturally, on R. Hartshorne's famous textbook Algebraic Geometry*). Simpson later became one of the pioneers in exploring and furthering ideas from Grothendieck's post-disappearance manuscript* Pursuing Stacks *(1983).*

This photo of Leila Schneps was taken in December 1993 at the home of Jean Malgoire, Grothendieck's last student, in a village in the south of France. Jean was and still is in possession of many large boxes of handwritten mathematical manuscripts by Grothendieck, some of which can be seen on the table. At that time Leila had recently read Esquisse d'un Programme and Récoltes et Semailles, and was eager to meet Grothendieck, but as it turned out, he had disappeared in 1991, leaving only a pile of manuscripts behind.

Participants in the conference "Alexandre Grothendieck: Biography, Mathematics, Philosophy", convened in August 2008 at Peyresq, France. The conference—organized by Leila Schneps, Pierre Lochak, and Winfried Scharlau—led ultimately to this book. Top row, left to right: William Messing, Jacob Murre, Aleksander Pełczyński, François Nicolas, Winfried Scharlau, Luc Illusie, Yves André, and Steven Kleiman. Second row: Michel Demazure, Pierre Cartier, Pierre Lochak, Guerino Mazzola, Christian Houzel, Matthias Künzer, and Jean Malgoire. Bottom row: Robin Hartshorne, Jean Bénabou, Leila Schneps, and Max Karoubi. Not pictured: Alexis Pinchard, Bruno Pinchard, and Carlos Simpson.